CHEMICAL REACTION NETWORKS

A Graph-Theoretical Approach

CHEMICAL REACTION NETWORKS

A Graph-Theoretical Approach

Oleg N. Temkin
Lomonosov Academy of Fine Chemical Technology, Moscow

Andrew V. Zeigarnik
Lomonosov Academy of Fine Chemical Technology, Moscow

Danail Bonchev
Higher Institute of Chemical Technology in Burgas
and *Texas A & M University, Galveston, Texas*

CRC Press
Taylor & Francis Group
Boca Raton London New York

CRC Press is an imprint of the
Taylor & Francis Group, an **informa** business

CRC Press
Taylor & Francis Group
6000 Broken Sound Parkway NW, Suite 300
Boca Raton, FL 33487-2742

© 1996 by Taylor & Francis Group, LLC
CRC Press is an imprint of Taylor & Francis Group, an Informa business

First issued in paperback 2019

No claim to original U.S. Government works

ISBN-13: 978-0-367-44847-9 (pbk)
ISBN-13: 978-0-8493-2867-1 (hbk)

Visit the Taylor & Francis Web site at
http://www.taylorandfrancis.com

and the CRC Press Web site at
http://www.crcpress.com

Preface

By virtue of its status as the classical language of constitutional chemistry, graph theory is regarded nowadays as the best formalism for the description of chemical reactions and their intrinsic reaction mechanisms. The growing interest in topological structure and features of chemical reaction mechanisms is due to (i) the intensive application of computers in chemical studies, mathematical modeling of chemical processes, and mechanistic studies and (ii) the increasing attention to reaction dynamics: chemical oscillations, deterministic chaos, and so on.

Many studies on the classification and coding of chemical reaction mechanisms have appeared. In this book, we attempt to highlight some selected aspects of these studies, while emphasizing the application of graph theory to mechanistic theory and chemical kinetics. The work describes various graph-theoretical approaches to the canonical representation, numbering, classification, coding and complexity estimation of chemical reaction mechanisms, as well as the analysis of their topological structure. We discuss topologically distinctive features of multiroute catalytic, noncatalytic, and chain reactions.

The Introduction describes the area of research and basic ideas, which led chemists to apply combinatorics and graph theory in mechanistic and kinetic studies. The application of these branches of mathematics will further the development of a rational methodology of mechanistic studies.

Chapter 1 is devoted to a combinatorial and graph-theoretical approach to elementary steps, which are the building blocks of reaction mechanisms. The focus on elementary reactions is due to their special role in advancing mechanistic hypotheses. We review some enumeration and classification methods applied to chemical reactions and some recent results obtained by comparing the *a priori* combinatorial variety of chemical reactions with elementary reactions extracted from the empirical databases. This chapter also reviews topological and other heuristic principles for determining whether or not a reaction is elementary.

Chapter 2 outlines current developments in application of graphs to reaction networks. Section 2.1 contains a review of different graph-theoretical models of multistage reaction mechanisms and networks. Section 2.2 deals with linear reaction networks, their classification, coding, and use in deriving steady-state rate laws and hypothesis generation. In Section 2.3, these general ideas are extended to nonlinear networks.

Chapter 3 summarizes our recent results in classifying nonlinear networks. We show how to define various types of reaction mechanisms in terms of reaction graphical counterparts, the networks. We anticipate that the nonmainstream ideas of this chapter may look debatable. Our goal was to clarify the chemical terminology of reaction mechanisms by means of a formalization that takes into account topological features of reaction mechanisms, which are not usually considered when giving a mechanism (or a reaction) a name like *catalytic, chain, noncatalytic conjugated,* etc. We beg for the reader's tolerance, which we hope to reward with the methods that will allow chemists to take a fresh look at the common mechanisms and paradigms of mechanistic chemistry.

Chapter 4 reviews different approaches to the assessment of graph complexity based on Shannon's entropy measures and topological indices. Two indices are discussed in detail. The first one is a kinetic complexity index applicable to linear mechanisms. The second is an information-theoretic index based on the stoichiometric matrix of intermediates of a mechanism, which can be applied to any mechanisms.

In Chapter 5, we discuss kinetic methods for identifying the topological structure of a mechanism. The method of dimensionless rate equations and the analysis of conjugation nodes are shown to be very useful in revealing the topological structures of mechanisms.

This book is written for experimental and theoretical chemists involved in kinetic and mechanistic studies and for mathematical chemists who are interested in application of graph theory and combinatorics in chemistry. It should also find an appropriate audience among graduate students of theoretical physical chemistry or of the theory of reaction mechanisms. The exposition makes use of the terminology of graph theory and assumes an acquaintance with its elementary concepts.

Some original material that appears here is the fruit of our joint efforts. However, particular chapters are written by different authors: Chapter 1 is written by A. V. Zeigarnik and O. N. Temkin; Chapter 2 is written by all three of us; Chapter 3 is written by A. V. Zeigarnik; Chapter 4 is written by D. Bonchev; Chapter 5 is written by O. N. Temkin, who also served as the chief editor of the entire contents.

We wish to thank Dr. Lev Bruk for his invaluable criticism. We also thank Professor Dimitar Kamenski, Dr. Ekaterina Gordeeva, and Dr. Sergei Blagov who contributed to our algorithmic studies and programming. We are grateful to Dr. Sergei Tratch, Professor Gregory Yablonskii, Dr. Evgenii Babaev, and Dr. Raul Valdés-Pérez for fruitful discussions. We wish to thank Svetlana Landau for useful advice and assistance. O. N. Temkin and A. V. Zeigarnik acknowledge financial support from the Russian Foundation for Basic Research for their graph-theoretical studies of reaction mechanisms and support from the Open Society Institute (International Soros Science Education Program). D. Bonchev acknowledges financial support from the Robert A. Welch Foundation of Houston, Texas.

Oleg N. Temkin and Andrew V. Zeigarnik
Moscow, Russia

Danail Bonchev
Galveston, Texas

Contents

Introduction

During the last 25 years, some of the most dramatic results in the areas of physical and theoretical chemistry were achieved in the understanding of reaction mechanisms. Of special note are enzymatic catalysis and catalysis with metal complexes.[1-7]
In these fields, researchers have studied the mechanisms of basic types of reactions and have identified the structures of organometallic intermediates. Currently, major types of elementary steps in catalysis are well understood. It has became apparent that, in solutions of metal complexes, reactions occur via multiple steps and routes. To say nothing of homogeneous catalysis, researchers in the field of heterogeneous catalysis often propose mechanisms having 2 to 4 reaction routes, 6 to 8 intermediates and 10 to 12 elementary steps, whereas a mere 10 to 15 years ago, mechanisms having as few as 2 to 3 steps were quite common. The presence of several routes for the same reaction is a characteristic feature of catalytic and noncatalytic reaction mechanisms. Contemporary chemical kinetics and the theory of chemical reaction networks have led not only to increased complexity, but also to a multitude of hypothetical mechanisms of any particular reaction.

The ensuing complexity of reaction mechanisms and a large number of mechanistic hypotheses for a specific reaction motivate revisiting the methodology of mechanistic studies and kinetic modeling[8] using, as a powerful new tools, formal methods for generating and analyzing hypothetical mechanisms.[9]

A reaction mechanism is "*a detailed description of the pathway leading from the reactants to the products, including as complete a characterization as possible of the composition, structure and other properties of reaction intermediates and transition states. An acceptable mechanism of a specified reaction (and there may be a number of such alternative mechanisms not excluded by the evidence) must be consistent with the reaction stoichiometry, the rate law, and all other available experimental data...*"[10] Mechanistic studies are important for several reasons:

- Improved understanding of chemical reactivity of different compounds can be achieved only with the knowledge of the mechanisms of specific reactions and the nature of key intermediates.
- The choice of the best catalyst for a particular process is practical if some hypotheses for the reaction mechanism are available.[8]
- Kinetic models, which form the basis for the models of commercial reactors, should be based on comprehensive information about the mechanism.[8]

The goal of a mechanistic study is to clarify the nature of intermediates and their interrelations (how they react with, or transform into, each other) and to determine the rates of these transformations. Thus, a mechanistic study has two aspects. First, a mechanism has a topological structure (or a structure of the reaction network). The term *topological* is used to emphasize that this structure is an invariant of the chemical content of a mechanism. The topological structure reflects the relations between intermediates, the stoichiometry of steps and that of the overall mechanism, and the presence of several routes and their conjugation. Second, a mechanism is characterized by the structure and composition of intermediates and transition states, as well as by the rate constants of formation and transformation of intermediates.

For a long time, the prevailing idea in mechanistic chemistry and chemical kinetics was that these questions can be answered in stages.[8] First, the methods of formal kinetics are used to derive the mechanistic skeleton or scheme (the reaction network) and, then, the skeleton is fattened with specific chemical content, thus arriving at the reaction mechanism.[11,12] Currently, there is no doubt that this method was somewhat arbitrary from the standpoint of philosophy of inference. In addition, the new methodology changes the status of the chemical component of the reaction mechanism concept, which was regarded as subordinate to the topological component. Within the framework of hypothetico-deductive method,[13] topological and chemical components are logically interrelated and can be studied only together. The essence of the new approach is to proceed from hypotheses to experiments able to discriminate among them. Our philosophy of hypothesis generation claims that the two types of mechanistic information (the structural and the chemical) should not be separated.

The knowledge accumulated about mechanisms of reactions suffice to formulate rather complex hypotheses. It is reasonable to advance hypotheses with the assistance of computer programs or at least with formal methods.[9] These programs resemble those designed for computer-assisted synthesis planning.[14] However, several characteristics of the latter prevent their advantageous use for hypotheses generation in mechanistic studies.[9]

The large number of mechanistic hypotheses have led to advances in the methods of mechanistic classification and coding,[15-21] as well as in the formulation of complexity indices. These problems have been studied extensively during the last 15–20 years. Several methods were proposed for the description and representation of linear and nonlinear mechanisms.[22] Recently, bipartite graphs have been used to classify nonlinear mechanisms and systematize some common terms used by mechanistic chemists.[23-25] Various complexity indices have proven useful in the analysis of mechanisms. Several such indices were proposed for linear[19,26,27] and nonlinear[28] mechanisms.

Recently it was shown that a fruitful method to study chemical reactions is to correlate the results of combinatorial enumeration of reaction types with those from the search in databases of elementary reactions.[20,29] This method enables finding new (topological) heuristics relating to elementary steps of complex reactions. These heuristics can be further tested using the sophisticated quantum-chemical calculations.

The graph-theoretical view of reaction mechanisms serves well to define exactly many commonly used concepts in mechanistic chemistry and to apply them in a systematic manner.[23,24] These include concepts such as autocatalysis, chain reactions, noncatalytic conjugated reactions, and others.

In the case of linear mechanisms, several methods have been proposed for the derivation of steady-state and non-steady-state rate laws and to predict the dynamic behavior of chemical reactions.[30]

The goal of this book is to survey the graph-theoretical methods applied to elementary and complex reactions and to chemical kinetics.

Graph theory is a common language of chemists. It is applied to molecular structures, quantum chemistry, stereochemistry, and many other fields. This book summarizes the results originating from a view of reaction mechanisms as graph-like structures. The chapters dealing with nonlinear mechanisms discuss mainly the basic ideas of mechanistic classification. The studies on coding and complexity of nonlinear mechanisms are in progress and their results will be reported in the near future.

References

1. Poltorak, O. M.; Chukhrai, E. S. *Physical and Chemical Fundamentals of Enzymatic Catalysis*; Vysshaya Shkola: Moscow, 1971 (in Russian).

2. Klesov, A. A.; Berezin, I. V. *Enzymatic Catalysis*; Moscow State Univ.: Moscow, 1980 (in Russian).

3. Moiseev, I. I. *π-Complexes in the Liquid-Phase Alkene Oxidation*; Nauka: Moscow, 1970 (in Russian).

4. Henry, P. M. Palladium Catalyzed Oxidation of Hydrocarbons. In *Catalysis by Metal Complexes*; James, B. R., Ed.; Reidel: Dordrecht, 1980, Vol. 2.

5. Temkin, O. N.; Shestakov, G. K.; Treger, Yu. A. *Acetylene: Chemistry, Reaction Mechanisms, and Technology*; Khimiya: Moscow, 1991 (in Russian).

6. *Comprehensive Organometallic Chemistry*; Wilkinson, G.; Stone, F. G. A.; Abel, E. W., Eds.; Pergamon: Oxford, 1982, Vol. 8.

7. A more extensive list of publications devoted to organometallic catalysis is given in Chapter 1.

8. Temkin, O. N.; Bruk, L. G.; Zeigarnik, A. V. Some Aspects of the Methodology of Mechanistic Studies and Kinetic Modeling of Catalytic Reactions. *Kinet. Katal.* **1993**, *34*, 445–462; *Kinet. Catal. Engl. Transl.* **1993**, *34*, 389–405.

9. Zeigarnik, A. V.; Bruk, L. G.; Temkin, O. N.; Likholobov, V. A.; Maier, L. I. Computer-Assisted Mechanistic Studies. *Usp. Khim.* **1996**, *65*, 125–139.

10. *Compendium of Chemical Terminology: IUPAC Recommendations* (Compiled by Gold, V.; Loening, K. L.; McNaught, A. D.; Sehmi, P.); Blackwell: Oxford, 1993.

11. Schmid, R.; Sapunov, V. N. *Non-Formal Kinetics*; Verlag-Chemie: Weinheim, 1982.

12. Kiperman, S. L. From the Kinetic Model to the Reaction Mechanism. In *Mechanism of Catalysis: The Nature of the Catalytic Action*; Boreskov, G. K.; Andrushkevich, T. V., Eds.; Nauka: Moscow, 1984, Vol. 1 (in Russian).

13. Trinajstić, N. On the Scientific Method. *Stud. Phys. Theor. Chem.* **1989**, *63*, 557–568.

14. See, for example, Barone, R.; Chanon, M. Computer-Aided Organic Synthesis. In *Computer Aids to Chemistry*; Vernin, G.; Chanon, M., Eds.; Ellis Horwood: Chichester, 1986, pp 19–102.

15. Bonchev, D.; Temkin, O. N.; Kamenski, D. On the Classification and Coding of Linear Reaction Mechanisms. *React. Kinet. Catal. Lett.* **1980**, *15*, 113–118.

16. Bonchev, D.; Temkin, O. N.; Kamenski, D. Graph-Theoretical Classification and Coding of Chemical Reactions with a Linear Mechanism. *J. Comput. Chem.* **1982**, *3*, 95-110.

17. Temkin, O. H.; Bonchev, D. Classification and Coding of Chemical Reaction Mechanisms. In *Graph Theory and Its Applications to Chemistry*; Tyutyulkov, N.; Bonchev, D., Eds.; Nauka Izkustvo: Sofia, 1987, pp 166–200 (in Bulgarian).

18. Temkin, O. N.; Bonchev, D. Classification and Coding of Chemical Reaction Mechanisms. In *Mathematical Chemistry. Chemical Graph Theory. Reactivity and Kinetics*; Bonchev, D.; Rouvray, D. H., Eds.; Gordon & Breach: Chichester, 1992; Vol. 2, Chapter 2.

19. Gordeeva, E.; Bonchev, D.; Kamenski, D.; Temkin, O. N. Enumeration, Coding, and Complexity of Linear Reaction Mechanisms. *J. Chem. Inf. Comput. Sci.* **1994**, *34*, 436–445.

20. Temkin, O. N.; Zeigarnik, A. V.; Bonchev, D. G. Graph-Theoretical Models of Complex Reaction Mechanisms. In *Graph Theoretical Approaches to Chemical Reactivity*; Bonchev, D.; Mekenyan O., Eds.; Kluwer: Dordrecht; 1994, pp 241–275.

21. Zeigarnik, A. V. A Graph-Theoretical Model of Complex Reaction Mechanisms: Special Graphs for Characterization of the Linkage between the Routes in Complex Reactions Having Linear Mechanisms. *Kinet. Katal.* **1994**, *35*, 711–713; *Kinet. Catal. Engl. Transl.* **1994**, *35*, 656–658.

22. For a review see Zeigarnik, A. V.; Temkin, O. N. A Graph-Theoretical Methods in the Theory of Complex Reaction Mechanisms: A Graphical Description of Complex Reaction Mechanisms. *Kinet. Katal.* **1994**, *35*, 691–701; *Kinet. Catal. Engl. Transl.* **1994**, *35*, 636–646.

23. Temkin, O. N.; Zeigarnik, A. V.; Bonchev, D. G. Application of Graph Theory to Chemical Kinetics. Part 2. Topological Specificity of Single-Route Reaction Mechanisms. *J. Chem. Inf. Comput. Sci.* **1995**, *35*, 729–737.

24. Zeigarnik, A. V.; Temkin, O. N.; Bonchev, D. G. Application of Graph Theory to Chemical Kinetics. Part 3. Topological Specificity of Multiroute Reaction Mechanisms. *J. Chem. Inf. Comput. Sci.* (in press).

25. Zeigarnik, A. V.; Temkin, O. N. A Graph-Theoretical Model of Complex Reaction Mechanisms: Bipartite Graphs and the Stoichiometry of Complex Reactions. *Kinet. Katal.* **1994**, *35*, 702–710; *Kinet. Catal. Engl. Transl.* **1994**, *35*, 647–655; Zeigarnik, A. V. A Graph-Theoretical Model of Complex Reaction Mechanisms: The Method for Classifying Complex Reaction Mechanisms. *Kinet. Katal.* **1995**, *36*, 653–657.

26. Bonchev, D.; Temkin, O. N.; Kamenski, D. On the Complexity Index for the Linear Reaction Mechanisms. *React. Kinet. Catal. Lett.* **1980**, *15*, 119–123.

27. Bonchev, D.; Kamenski, D.; Temkin, O. N. Complexity Index for the Linear Mechanisms of Chemical Reactions. *J. Math. Chem.* **1987**, *1*, 345–388.

28. Zeigarnik, A. V.; Temkin, O. N. A Graph-Theoretical Model of Complex Reaction Mechanisms: A New Complexity Index for Reaction Mechanisms. *Kinet. Katal.* **1996**, *37*, 372–385; *Kinet. Catal. Engl. Transl.* **1996**, *37*, 347–360.

29. Herges, R. Coarctate Transition States: The Discovery of a Reaction Principle. *J. Chem. Inf. Comut. Sci.* **1994**, *34*, 91–102.

30. For reviews see: Yablonskii, G.S.; Bykov, V.I.; Gorban', A. N. *Kinetic Models of Catalytic Reactions*; Nauka: Novosibirsk, 1983 (in Russian); Temkin, O. N.; Bonchev, D. G.

Application of Graph Theory to Chemical Kinetics. Part 1. Kinetics of Complex Reactions. *J. Chem. Educ.* **1992**, *69*, 544–550; Kuo-Chen Chou, Applications of Graph Theory to Enzyme Kinetics and Protein Folding Kinetics: Steady and Non-Steady-State Systems. *Biophys. Chem.* **1990**, *35*, 1–24; Clarke, B. L. Stability of Complex Reaction Networks. In *Advances in Chemical Physics*; Prigogine, I., Rice, S.A., Eds.; Wiley: New York, 1980, Vol. 43, pp 7–215.

Chapter 1

Graph Theory Assistance in Studies of Elementary Steps of Complex Reactions

Currently, numerous chemical reactions are widely believed to occur via several consecutive elementary reactions. This set of elementary reactions is usually called a reaction mechanism, or a mechanistic scheme. The study of reaction mechanism allows one to look inside a complex reaction and to predict its behavior under various conditions. Elementary reactions that are building blocks of a mechanism are usually the subject of special interest. Researchers involved in mechanistic studies carefully collect their experimental data to gain evidence for or to discard their conjectures related to elementary reactions. Proceeding along this avenue, in most cases, they intuitively use several paradigms related to chemical changes that can take place during one or another elementary reaction. The basis for these paradigms is analogies, heuristics, and generalizations. Among the latter are those that have a fundamental basis (e.g., the conservation of orbital symmetry, which is currently generally accepted). Others can stem from researcher's experience. In all cases such paradigms cannot be completely proved, although substantial evidence can exist. Many recent efforts of mathematical chemists were directed toward the search for new reaction rules and hidden principles of chemical reactions. The goal of this chapter is to show what has been achieved.

1.1. The Concept of an Elementary Step

An elementary reaction is defined as follows: elementary reaction is a repeated reproduction of a unit act that involve transformation of some (in most cases, two) species at the instant they collide or the transformation of a single species.

An elementary reaction is assumed to pass through a single transition state. No intermediates can be detected in the elementary reaction.[1]

Several additional ideas are associated with an *elementary reaction*. Thus, elementary reactions are usually termed *concerted* in order to highlight that forming or breaking of chemical bonds occurs via a single step. Sometimes a reaction is said to be *synchronous*. This term is often used as a synonym of *concerted*. However, these are not synonyms. According to Dewar,[2] a synchronous reaction is that in which all bond changes "take place in unison, having all proceeded to comparable extents in the transition states." An elementary reaction is usually suggested to have a certain direction; that is, the reverse reaction is assumed to be a separate process.

To determine whether a reaction is elementary or not, comprehensive studies are needed. However, sometimes a researcher can determine with a fair degree of certainty that a reaction is *not* elementary. What drives his reasoning? Analogies, intuition, and background knowledge. Heuristics and various reaction principles also play an important role. Sometimes new reaction principles can be discovered using the results of combinatorial analysis of reactions and the information stored in reaction databases. The typical scheme of this study is shown in Figure 1.1. As can be seen, there are two key steps in this study: combinatorial enumeration and generalization of reactions. These two steps are parts of a broader process called classification.

As was mentioned by Bawden, "*There are a number of clearly understood benefits of classification... They are involved in four distinct, though interrelated, processes. The first is 'straightforward' retrieval of information* [in information systems]. *The second is less-well-understood processes of information discovery, through analysis, correlation, and reasoning by analogy. The third is teaching and exposition of the variety and scope of chemical reactions. The fourth is the systematization of chemical reaction information...*"[3] Our primary interest here lies within this *less-well-understood process* of information discovery.

The pioneer of systematic enumeration and classification of reactions was Balandin.[4-7] However, his papers are not cited even by Russian authors. Until recently, his contribution to this area was largely forgotten.[8] There are several reasons for such an injustice. At that time, when his papers were published, researchers in Russia paid little attention to systematization of knowledge. They viewed classification and enumeration studies as unnecessary game of logic, which would never bring any practical results. The situation was redoubled by the policy of dictation in science. Suffice it to mention that somewhat later the information theory was marked as antiscience. Balandin's career was time and again interrupted by Stalinist repression. At that time, reading of his papers was strictly prohibited. Later, when his findings became accessible to the broad scientific audience, they appeared to be much ahead of his time. Not very many people were capable of comprehending his far-from-mainstream ideas or ever wished to try to understand his scientific language. When the reaction classification studies were resumed in many countries, Balandin's papers were already forgotten. Another reason is that publications in Russian are much less accessible to an international reader.

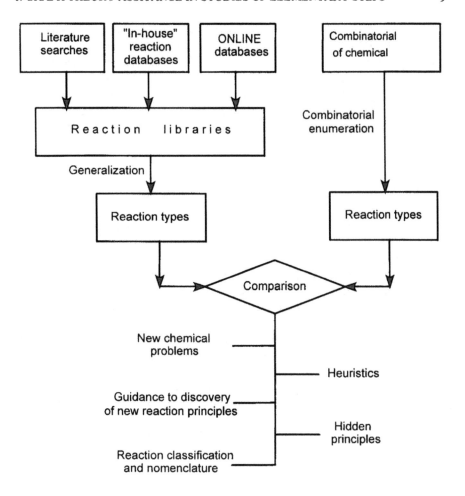

Figure 1.1. Combinatorial enumeration and generalization of reactions: a way to problems and solutions in chemistry of tomorrow.

In the 1960s, the study on enumeration and classification of chemical reactions was started de novo by Balaban,[9] who was familiar with Cayley,[10-17] Pólya,[18-22] and Harary's[23] findings. In the 1970s, the number of papers devoted to these problems began to increase exponentially.

Here we give a list (probably incomplete) of research groups and researchers whose efforts contributed to the development of this area (the order in which they are listed is not chronological): Vladutz (an approach to systematization of knowledge about chemical reactions),[24-26] Hendrickson (a systematic view of thermal 6-membered and 5-membered pericyclic reactions),[27] Arens (a "+–" formalism to reaction classification and coding),[28-30] Zefirov, Tratch, et al. (a formal–logical approach to the constructive enumeration and classification of organic reactions),[31-43] Brownscomb (enumeration of 4- and 6-membered pericyclic reactions),[44] Bart and

Garagnani (assessment of frequency of reaction types),[45] Ugi *et al.* (application of *be-* and *r*-matrices to reaction classification)[46,47] and the relevant approach to the classification and canonical representation of reaction classes,[48] Roberts[49] and Satchell[50] (a systematic approach to classification and nomenclature of organic reactions), Fujita (the concept of imaginary transition structures and enumeration of reaction classes using Pólya's enumerating theorem),[51-65] Kvasnička *et al.* (a graph-theoretical interpretation of the Dugundji–Ugi model of reactions),[66-68] and Herges (computer-assisted constructive enumeration and reaction design).[69-73]

1.2. A Reaction as a Combinatorial Object

Different approaches to the reaction classification and enumeration usually do not deal with reactions as they are recognized by chemists because it is quite obvious that the number of possible reactions is huge. Moreover, there is no need to enumerate all reactions, even if the upper limit of the number of atoms in initial and target molecules is specified. When conducting the combinatorial enumeration, reactions are described without superfluous details. In most cases valence-bond schemes are explicitly considered. There is also a general trend to describe a reaction as a graph from which subgraphs of reactants and products can be extracted or deduced. Clearly, at this level of generality, it is difficult to determine whether or not a graph represents an elementary reaction. All reactions (not only elementary ones) can be described by graphs and are subject to combinatorial enumeration. Let us trace how a reaction can be described as a combinatorial object.

Suppose we have the Diels–Alder reaction, written as a usual chemical equation:

$$(1.1)$$

In Herges' terms this is a *specific reaction*.[70-73] After deletion of atoms that do not participate in the bond change, we obtain a *transformation*:

$$(1.2)$$

The above procedure can be viewed as the removal of bonds whose bond order does not change during the reaction. Equation (1.2) can be termed a *reaction equation*[43] or a heteroreaction.[70–73] Then, let us replace symbols of atoms with vertexes thus arriving at a *symbolic equation*[43] or a *basic reaction*:[70–73]

$$G_1 \qquad\qquad G_2 \tag{1.3}$$

After deletion of edges that remain unchanged, we arrive at the following scheme (hereafter, double and triple bonds are regarded as two and three bonds, respectively):

$$G_3 \qquad\qquad G_4 \tag{1.4}$$

which is called a *reaction type* (our term) or a *skeleton of transformation* (Roberts' term[49]). Superposition of graphs G_3 and G_4 (Figure 1.2) produces a new graph G_5 that describes the topology of electron shifting (*topology of the reaction category,*[70–73] *topology identifier*[43]).

The topology identifier resembles the transition state structure. However, they are different, because the latter takes into account all molecular orbitals participating in the process. Conversely, the topology identifier is the idealized structure (graph), which takes into account only bonds that change their formal orders. For example, the theoretically studied hydrogen elimination from 1,4-cyclohexadiene should occur via the polycyclic transition state:[74]

From a formal standpoint, to construct the topology identifier, it is sufficient to include only one of the two cycles containing hydrogen atoms in the transition state. However, these two cycles are indistinguishable and, therefore, the transition state is regarded as polycyclic.

The three graphs below also have their specific names:

$$G_5 \qquad\qquad G_6 \qquad\qquad G_7$$

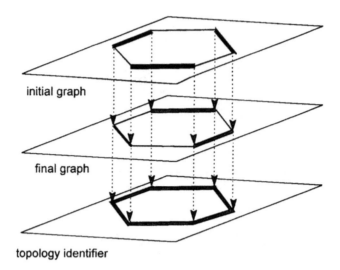

initial graph

final graph

topology identifier

Figure 1.2. Superposition of two graphs.

Graph G_6 denotes a skeleton of bonds that remain unchanged (*par-bond skeleton*,[52,53] σ-*frame*[70–73]). Graph G_7 shows that some bonds were broken (dashed lines), while others were made (this graph is called a *reaction category*[70–73]). The adjacency matrix of G_7 is homomorphic to the *r*-matrix in the Dugundji–Ugi model.[75–77]

Fujita proposed an alternative system of representation of chemical reactions based on the concept of imaginary transition structures.[51–65] The hierarchical system of generalized reaction representation is shown in Figure 1.3.

After all deletions and simplifying assumptions, we arrive at several types of graphs that describe a reaction as a combinatorial object. This can be either the topology identifier, or the graph that denotes a reaction category, or the imaginary transition structure, or one of its characteristic subgraphs, etc.[78] Suppose, this is a topology identifier. Then, if one conducts a search in the reaction database for reaction types, it will become clear that very few topology identifiers describe the overwhelming majority of the up-to-date known reactions. Then, a reasonable question arises: *How many reaction types have the same reaction "topology" (can belong to the same reaction category)?* To answer this question, a chemist has a tool called the enumeration technique.

1.3. Enumeration of Reaction Classes

1.3.1. Analytical Enumeration Using Pólya's Enumeration Theorem

The language of graph theory is quite natural to chemists because chemical structures are isomorphic to graphs. The number of chemical compounds increases with the number of atoms so rapidly that enumeration of possible chemical structures is the only way to develop their nomenclature. At present, many researchers employ combinatorial techniques in chemical problem solving. However, the terminology of combinatorial analysis is still difficult to understand for those who do not use combinatorics in everyday studies.

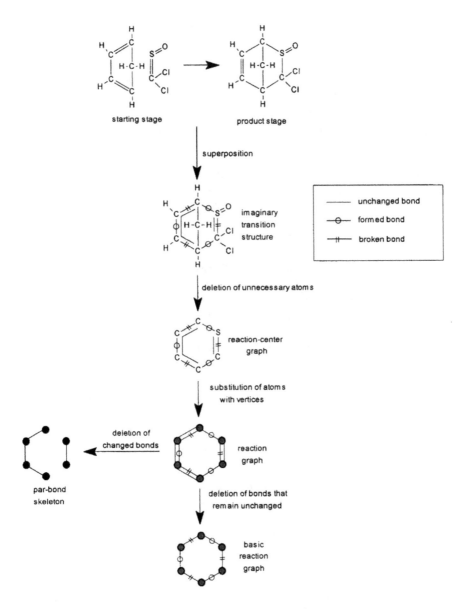

Figure 1.3. Fujita's terms and the description of reactions using the imaginary transition structures.

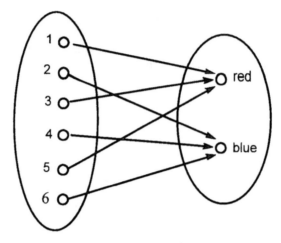

Figure 1.4. A mapping from the edge set to the color set for the Diels–Alder reaction.

Because this book is addressed not only to mathematical chemists, we will try to describe key ideas so that they will be easy to understand without special background in mathematics. If, nevertheless, something remains unclear, we direct the reader to several papers, some of which are written specifically for chemists.[79–87] An advanced reader can turn to Harary and Palmer's textbook[23] and to Fujita's recent textbook.[88]

The term *enumeration* means two things: analytical enumeration and constructive enumeration. Analytical enumeration answers the question, *How many objects (e.g., graphs, reaction classes, etc.) can exist?* Constructive enumeration is a procedure that allows one to derive the complete list of objects.

To introduce basic enumeration concepts, we will use the illustrative example of enumerating reaction types. Suppose one needs to know how many reaction typescan exist. Then, each particular reaction type is said to be a *configuration*.

The mathematical solution to the enumeration problem is usually given as a *configuration counting series*, a polynomial termed a *generating function*. For instance, a generating function of reaction types belonging to six-membered cyclic topology is as follows:

$$1 + x + 3x^2 + 3x^3 + 3x^4 + x^5 + x^6. \tag{1.5}$$

In this series, the exponent of a term represents the number of bonds broken during the reaction. Apparently, the number adding to the maximum exponent, 6, is the number of bonds formed during the reaction (otherwise, the topology will not hold). The coefficient is the number of isomeric reaction types. Thus, the term $3x^2$ indicates that there exist exactly three reaction types in which two bonds are broken and four bonds are formed.

The enumeration of reaction types can be reformulated as counting all possible ways of assigning two colors to the edges of the topology identifier. Say, if the bond is broken, we will color it with red. If the bond is formed, we will color it with blue. Because we need only two colors we can write a series

$$x^n + y^m \qquad (1.6)$$

in which x denotes red and y denotes blue. The series of this sort is called a *figure counting series* (or a *figure inventory*). The figure is always a thing to be permuted. If the exponent is 1 in the figure counting series, the figure (color) is present. If the exponent is 0, the figure (color) is absent. Further discussion will reveal the meaning of exponents in the figure counting series. If necessary, one can include more figures in the figure counting series. For instance, the enumeration of reaction graphs, which contain bonds remaining unchanged, requires the following figure counting series:

$$x^n + y^m + z^p. \qquad (1.7)$$

Suppose that we number the edges of the topology identifier as shown below:

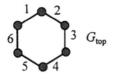

Thus, we have a set of edges E = {1, 2, 3, 4, 5, 6} and a set of colors C = {red, blue}. Then a Diels–Alder reaction belongs to the reaction type that corresponds to the mapping (function) shown in Figure 1.4.

Each mapping of this sort corresponds to a particular reaction type, but different mappings do not necessarily generate different reaction types. This is because G_{top} possesses a certain symmetry. The symmetry properties of the graph can be defined in terms of permutations. Let Π be a permutation group acting on edges of G_{top}, which consist of all conceivable ways to renumber edges of G_{top} (including the permutation that does not change edge numbers). There is a subgroup $\Gamma(G_{top})$ of the group Π which consists of all permutations that preserve adjacencies of G_{top}. The group $\Gamma(G_{top})$ is termed an *edge group* of G_{top} or *an automorphism group*. Similar group can be defined on vertexes of a graph. Permutations belonging to the group $\Gamma(G_{top})$ are said to be *automorphisms* of G_{top}.

For instance, the permutation

$$\begin{pmatrix} 1 & 2 & 3 & 4 & 5 & 6 \\ 1 & 6 & 5 & 4 & 3 & 2 \end{pmatrix}$$

means that edges 2 and 6 as well as edges 3 and 5 exchange their numbers, while edges 1 and 4 preserve their numbers. Then, after renumbering edges, we have:

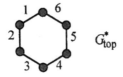

Note that after renumbering, all pairs of vertexes preserve their adjacencies, i.e., edges 1 and 2 remain adjacent, edges 2 and 3 also remain adjacent and so on. In other words, this permutation is an automorphism of G_{top}. We can describe the above permutation by its cycle representation:

$$\alpha^* = (1)(26)(4)(35).$$

This means that edges 1 and 4 remained on their sites, while edges 2 and 6 and edges 3 and 5 exchanged numbers without loss of adjacencies.

Next, we have to define the cycle index of a group Γ acting on the set of n elements. Let $j_k(\alpha)$ be the number of cycles of length k contained in the permutation a. For instance,

$$j_1(\alpha^*) = j_2(\alpha^*) = 2.$$

Then, the polynomial in variables $s_1, s_2, s_3,...s_n$, defined by the formula

$$Z(\Gamma) = |\Gamma|^{-1} \sum_{\alpha \in \Gamma} \sum_{k=1}^{n} s_k^{j_k(\alpha)} \tag{1.8}$$

is termed a *cycle index of the group* Γ.

Consider the irredundant system of permutations belonging to an automorphism group of G_{top} and respective terms of the cycle index:

$$
\begin{aligned}
\alpha_1 &= (1)(2)(3)(4)(5)(6), & s_1^6 \\
\alpha_2 &= (1)(4)(26)(35), & s_1^2 s_2^2 \\
\alpha_3 &= (2)(5)(13)(46), & s_1^2 s_2^2 \\
\alpha_4 &= (3)(6)(15)(24), & s_1^2 s_2^2 \\
\alpha_5 &= (12)(36)(45), & s_2^3 \\
\alpha_6 &= (14)(23)(56), & s_2^3 \\
\alpha_7 &= (14)(25)(36), & s_2^3 \\
\alpha_8 &= (16)(25)(34), & s_2^3 \\
\alpha_9 &= (135)(246), & s_3^2 \\
\alpha_{10} &= (153)(264), & s_3^2 \\
\alpha_{11} &= (123456), & s_6^1 \\
\alpha_{12} &= (165432). & s_6^1
\end{aligned}
\tag{1.9}
$$

The order of numbers in a permutation cycle shows the direction of number switches: permutation cycles (135) and (153) denote $(1\rightarrow3, 3\rightarrow5, 5\rightarrow1)$ and $(1\rightarrow5, 5\rightarrow3, 3\rightarrow1)$, respectively. Because we have 12 permutations belonging to this system, $|\Gamma(G_{top})| = 12$. After collecting all terms of the cycle index, we arrive at the following formula for the cycle index:

$$Z(\Gamma(G_{top})) = \frac{1}{12}(s_1^6 + 3s_1^2s_2^2 + 4s_2^3 + 2s_3^2 + s_6^1). \tag{1.10}$$

Note that the following rule holds for each term:

$$\sum_{k=1}^{n} kj_k(\alpha) = n. \tag{1.11}$$

Now we are able to introduce Pólya's enumeration theorem, which is of great utility in analytical enumeration of chemical reactions:

Theorem. *The configuration counting series $C(x)$ is obtained by substituting the figure counting series into the cycle index of group Γ so that each term s_k is replaced with a series $s(x^k, y^k,...)$:*

$$C(x) = Z(\Gamma, s(x^k, y^k, ...)). \tag{1.12}$$

Let us trace the procedure of obtaining the number of reaction types belonging to a cyclic 6-membered topology. First, it is necessary to define an irredundant system of permutations. This system uniquely determines the group Γ of the graph G_{top}. This group is called the dihedral group D_6. Second, collect terms and write the expression for the cycle index. Third, choose the form of the figure counting series. Fourth, substitute the figure counting series into the cycle index.

Let us choose the figure counting series as $s_n = x^n + y^n$. Then, substitution of s_n into the cycle index produces:

$$Z(D_6, x + y) = \frac{1}{12}([x + y]^6 + 3[x + y]^2[x^2 + y^2]^2 +$$
$$4[x^2 + y^2]^3 + 2[x^3 + y^3]^2 + [x^6 + y^6]) = \tag{1.13}$$
$$y^6 + y^5x + 3y^4x^2 + 3y^3x^3 + 3y^2x^4 + yx^5 + x^6.$$

This can be interpreted as follows. The term y^6 stands for a reaction in which 6 bonds are formed. The term y^5x stands for a reaction in which 5 bonds are formed and one bond is broken. The term $3y^4x^2$ means that there are three reactions in which four bonds are formed and two bonds are broken, etc.

Let us choose another figure counting series: $s_n = x^n + 1$. Indeed, because one of the exponents in each term can be deduced from the other it is sufficient to include only one of them:

$$Z(D_6, x + 1) = \frac{1}{12}([x + 1]^6 + 3[x + 1]^2[x^2 + 1]^2$$
$$+ 4[x^2 + 1]^3 + 2[x^3 + 1]^2 + [x^6 + 1]) = \tag{1.13'}$$
$$1 + x + 3x^2 + 3x^3 + 3x^4 + x^5 + x^6.$$

We arrived at expression (1.5).

Now, suppose that numbers of bonds broken and formed is unnecessary information for us and we need only to know the total number of possible graphs. Then the figure counting series should be chosen as $s = 1 + 1 = 2$:

$$Z(D_6, 2) = 1 + 2 + 3 \cdot 2^2 + 3 \cdot 2^3 + 3 \cdot 2^4 + 2^5 + 2^6 = 13. \qquad (1.13'')$$

Making use of Pólya's enumeration theorem, one can derive all reaction types (symbolic equations,[43] subgraphs of imaginary transition structures,[51-65] and other graphical representations of generalized reactions).

Figure 1.5 shows several characteristic reaction topologies discussed in the literature for which cycle indices are given in Table 1.1.

Fujita applied systematically Pólya's theorem to enumerate cyclic[54,55] and bicyclic[58,59] reaction graphs (i.e., transformations like $G_1 \rightarrow G_2$) and reaction center graphs (i.e., transformations like (1.2)). The general method was as follows. The procedure starts from a given type of the basic reaction graph, which is in one-to-one correspondence to a transformation like $G_3 \rightarrow G_4$ (reaction types) and r-matrix of the Dugundji–Ugi model.[75] Taking into account that edges of this graph are labeled as in-bonds or out-bonds (see Figure 1.3), one can derive the cycle index. The figure counting series (figure inventory) is

$$s_n = 1 + x^n + y^n, \qquad (1.14)$$

that is, three colors are used:

(1) the bond order is increased (decreased) from 2 to 3 (from 3 to 2);
(2) the bond order is increased (decreased) from 1 to 2 (from 2 to 1);
(3) a single bond is made (broken).

Thus, the procedure reduces to superposing all possible skeletons of bonds that remain unchanged upon the transformation (in Herges' terms, σ-frames[70-73]) on a given basic reaction graph.

For instance, consider a three-membered ring in which two bonds are broken and one bond is formed:

G_8

Then, if all edges were equivalent (Figure 1.5, graph 5), the cycle index would be $Z(D_3) = \frac{1}{6}(s_1^3 + 2s_1s_2 + 2s_3)$, where D_3 denotes the dihedral group. Because only edges 1 and 3 of G_8 are equivalent, its automorphism group is a cyclic group C_2. The cycle index $Z(C_2)$ is as follows:

$$Z(C_2) = \frac{1}{2}(s_1^3 + s_1s_2). \qquad (1.15)$$

In other words, G_8 contains only two of six permutations of the three-membered cyclic topology identifier. Substitution of (1.14) into (1.15) gives the following figure counting series:

$$Z(C_2, 1 + x + y) = 1 + 2x + 2y + 2x^2 + 3xy + 2y^2 + x^3 + 2x^2y + 2xy^2 + y^3. \quad (1.16)$$

The disadvantages of this procedure are as follows: one should have in advance the list of basic reaction graphs; second, the result contains a certain amount of chemically senseless results (small cycles containing triple and allene bonds). Fujita also reported the enumeration of graphs in which vertexes are replaced with heteroatoms.[63,89] This procedure is very similar to the above. He also showed that reaction types can be enumerated using several recent generalizations of Pólya's theorem and double cosets.[62]

1.3.2. Constructive Enumeration

Unlike analytical enumeration, the problem of constructive enumeration is somewhat more sophisticated and requires the use of computer programs.

Zefirov, Tratch, et al.[43,90] proposed an algorithm and the computer program SYMBEQ based on the edge labeling of the topology identifier G_{top}. One of the versions of the algorithm consists of two stages. At the first stage, edges of G_{top} are labeled according to the value of the bond multiplicity: 1, 2, 3. The result of this procedure is the set of multigraphs with up to three edges between two vertexes. The algorithm also imposes constraints on vertex degrees. Then, (multiple) edges are further labeled with *plus* and *minus* signs to denote the decrease or increase in the bond order. After this labeling, multigraphs transform to symbolic equations.

An alternative variant of this algorithm uses a one-step procedure: edges of the topology identifier are marked with unified labels from the set $M = \{m_1, m_2, ..., m_{12}\}$. Elements of this set correspond to the possible changes in the bond order: $^0/_1, ^0/_2, ^0/_3, ^1/_2, ^1/_3, ^2/_3, ^3/_2, ^3/_1, ^2/_1, ^3/_0, ^2/_0, ^1/_0$ (the first integer denotes the initial bond order and the second denotes the final bond order). Both algorithms use the principle of the earliest rejection of redundant solutions. Using a back-track procedure, the SYMBEQ program generates all automorphisms of G_{top} and stores them in memory.[91,92] Then, the program assigns labels so as to avoid noncanonical mappings from the edge set to the set of labels. The canon implies the use of the lexicographically smallest code for the string notation of mapping. The assignment of labels is also a back-track procedure. The program uses several constraints to match the requirements to the output equations. The SYMBEQ program can also generate all possible reaction equations in which vertexes are substituted with heteroatoms.

Another approach was proposed by Herges. He generated reaction types by his own algorithm[71] and the computer program IGOR developed by Bauer (Ugi's group).[73,93–95]

The Herges' algorithm is based on generating all possible σ-frames (skeletons that remain unchanged during the reaction) for a given reaction type, which is defined as the reaction matrix or as the corresponding graph with edges labeled with "+" and "–" signs. The complete list of σ-frames together with the equation like $G_3 \rightarrow G_4$ forms a complete list of symbolic equations (Zefirov's term[43]) or basic reactions (Herges' term[70–73]).

The procedure starts from the graph that contains p vertexes and does not contain edges. It consecutively adds edge by edge to graphs thus producing derivative graphs belonging to different "levels". Each level is characterized by the number of edges in graphs. On each nth level (n is greater by one than the number of edges in each graph of this level), the program finds classes of equivalent positions for a new edge to be added. It is likely that this procedure is much similar to that for finding automorphisms in SYMBEQ (details of the algorithm were not reported). A set of graphs within one level is generated by adding of an edge to each class of equivalent positions; that is, each equivalence class on nth level produces one graph on $(n + 1)$th level. This procedure consecutively generates a tree of spanning subgraphs of the topology identifier (or, in Herges' terms, σ-frames). If an edge was added to one of the positions of some equivalence class, all equivalent positions are further excluded as candidates for additions of edges to other graphs on this level and to their successors.

Table 1.1. Numbers of graphs (topology identifiers) shown in Figure 1.5 and their cycle indices

No.	Cycle index	No.	Cycle index
1	$\frac{1}{2}(s_1^2 + s_2)$	15	$\frac{1}{2}(s_1^6 + s_1^2 s_2^2)$
2	$\frac{1}{2}(s_1^3 + s_1 s_2)$	16	$\frac{1}{2}(s_1^6 + s_1^2 s_2^2)$
3	$\frac{1}{2}(s_1^4 + s_2^2)$	17	$\frac{1}{6}(s_1^6 + 2s_1^4 s_2 + s_1^2 s_2^2 + 2s_2^3)$
4	$\frac{1}{2}(s_1^5 + s_1 s_2^2)$	18	$\frac{1}{4}(s_1^7 + s_1^5 s_2 + s_1^3 s_2^2 + s_1 s_2^3)$
5	$\frac{1}{6}(s_1^3 + 3s_1 s_2 + 2s_3)$	19	$\frac{1}{6}(s_1^8 + 2s_1^4 s_2^2 + s_2^4)$
6	$\frac{1}{8}(s_1^4 + 2s_1^2 s_2 + 3s_2^2 + 2s_4)$	20	$\frac{1}{4}(s_1^8 + s_1^6 s_2 + s_1^4 s_2^2 + s_1^2 s_2^3)$
7	$\frac{1}{10}(s_1^5 + 5s_1 s_2^2 + 4s_5)$	21	$\frac{1}{4}(s_1^9 + 2s_1^5 s_2^2 + s_1 s_2^4)$
8	$\frac{1}{12}(s_1^6 + 3s_1^2 s_2^2 + 2s_2^3 + 2s_3^2 + 2s_6)$	22	$\frac{1}{6}(s_1^{10} + 2s_1^6 s_2^2 + s_1^2 s_2^4 + 2s_2^5)$
9	$\frac{1}{14}(s_1^7 + 7s_1 s_2^3 + 6s_7)$	23	$\frac{1}{6}(s_1^7 + s_1^5 s_2 + s_1^3 s_2^2 + s_1 s_2^3)$
10	$\frac{1}{16}(s_1^8 + 4s_1^2 s_2^3 + 5s_2^4 + 2s_4^2 + 4s_8)$	24	$\frac{1}{4}(s_1^5 + 3s_1 s_2^2)$
11	$\frac{1}{2}(s_1^4 + s_1^2 s_2)$	25	$\frac{1}{4}(s_1^7 + 2s_1^3 s_2^2 + s_1 s_2^3)$
12	$\frac{1}{2}(s_1^5 + s_1^3 s_2)$	26	$\frac{1}{4}(s_1^9 + 3s_1^1 s_2^4)$
13	$\frac{1}{2}(s_1^6 + s_1^4 s_2)$	27	$\frac{1}{4}(s_1^{10} + s_1^2 s_2^4 + 2s_2^5)$
14	$\frac{1}{2}(s_1^5 + s_1 s_2^2)$		

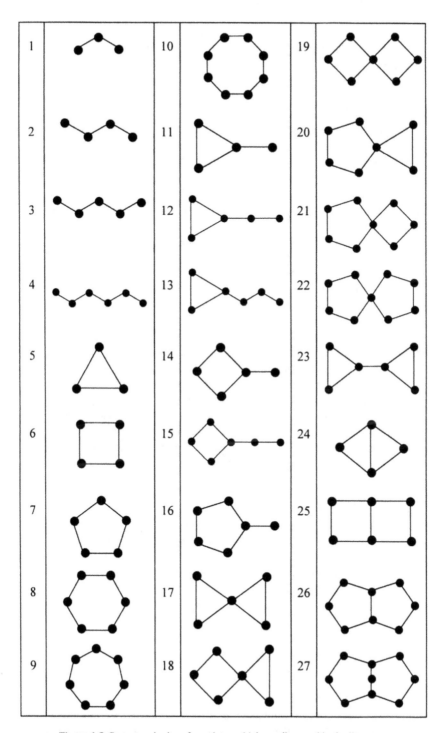

Figure 1.5. Some topologies of reactions, which are discussed in the literature.

Herges also applied the IGOR2 program. IGOR2 is a powerful device for generating reactions and structures. However, it is not directly applicable to generating reaction graphs. At the same time, Herges does not report details of how to generate reaction types (and basic reactions) with IGOR2. Unfortunately, it is a general trend to avoid the experimental section in theoretical papers.

Roughly, the procedure incorporated in IGOR is as follows. A user defines the transformation by inputting the reaction matrix and the so-called transition table. The transition table specifies allowable valence schemes of atoms and allowed (or forbidden) transitions between them. The user also imposes several constraints. For instance, "The product must contain exactly one 6-membered ring," "The educt consists of two reactants," etc. The reaction matrix, transition tables, and the constraints imposed by a user govern the procedure of reaction generation. The program generates all possible transformations that satisfy the conditions defined by the user. Then, all transformations are stored in an unambiguous manner to avoid redundant results.

Zeigarnik also proposed an algorithm for the generation of reaction types in which only changing bonds were taken into account.[96] The method is based on the application of the set of equations relating number of vertexes with different vertex degrees. Instead of searching for permutations belonging to the automorphism group and finding a representative and irredundant system of mappings from the set of edges to the set of labels, the algorithm implies a search for possible solutions to the set of equations using a number of reasonable constraints. All solutions are stored and further used to reconstruct the initial and final graphs in equations of reaction types. Apparently, this method generates all possible isomorphic copies. Because of this, some additional procedure for the rejection of redundant solutions is needed. However, because the number of solutions for topology identifiers of reasonable complexity appeared to be small, it was not difficult to delete copies. According to this scheme, atoms (reaction centers that participate in the bond change can be classified as follows (numbers in brackets denote the net change in the vertex degree and the desired minimum vertex degree in the initial graph, respectively): (i) breaking one bond and making another one $(0, 1)$; (ii) making one bond $(1, 0)$; (iii) making two bonds $(2, 0)$; (iv) breaking one bond $(-1, 1)$; (v) breaking two bonds $(-2, 2)$.

Let the following designations be used: symbols $\xi_{(i)}-\xi_{(v)}$ denote numbers of respective centers; q is the number of edges in the initial graph; Δq is the difference between numbers of edges in the initial and final graphs; N is the number of vertexes in the topology identifier (and in each graph). Then, the equations relating $\xi_{(i)}-\xi_{(v)}$ and Δq are as follows:

$$\xi_{(i)} + \xi_{(iv)} + 2\xi_{(v)} = 2q,$$

$$\xi_{(ii)} + 2\xi_{(iii)} - \xi_{(iv)} - 2\xi_{(v)} = 2\Delta q,$$

$$\xi_{(i)} + \xi_{(ii)} + \xi_{(iii)} + \xi_{(iv)} + \xi_{(v)} = N.$$

This approach does not contribute greatly to the theory of combinatorial algorithms. However, it can also be applied to the constructive enumeration of reaction types. To our knowledge, a similar approach was applied by Corio to the design of reaction mechanisms based on stoichiometric principles.[97] He applied material balance equations similar to those shown above.

Comparison of the above algorithms clearly demonstrates that generally one cannot avoid the combinatorial difficulties. Generation of reaction types with the aid of computer programs requires implementation of at least two time-expensive combinatorial modules.

Say, the first module detects automorphisms, while the second generates all possible combinatorial mappings from the edge set to the set of labels. If the first module is excluded, the program will generate a huge number of isomorphic copies. Then, it will be necessary to organize the procedure of checking for isomorphisms. Various user-defined constraints reduce the time of combinatorial operations. Apparently, the rejection of variants with user-defined constraints should be conducted at the earliest possible stage. Thus, every time the algorithm designer employs combinatorial modules, he has to decide between the generation of automorphisms and rejecting isomorphisms (i.e., he must decide which procedure requires less time). At the same time, one must anticipate that if an algorithm is exhaustive and irredundant, it can appear to be not so efficient.

1.4. Topological Heuristics

As mentioned in the beginning of this chapter, comparison of the reactions contained in reaction databases and the transformations that result from the formal enumeration procedures can reveal hidden reaction principles. As was already mentioned, on the combinatorial level, elementary and nonelementary reactions are indistinguishable. Garagnani and Bart[45] probably were the first to conduct the study according to the scheme described in Figure 1.1. They generalized the information contained in the Mathieu's compilation (1900 reactions)[98] and found that 92% of reactions can be described in terms of Ugi's r-matrices, i.e., in terms of valence-bond schemes. They showed that 30 reaction types can account for all reactions among this 92%. These reactions were not necessarily elementary ones.

Zeigarnik and Temkin created a "paper database" of elementary organometallic reactions (~3000 reactions), the major portion of which were elementary steps of organometallic catalysis, and compared them with results of the constructive enumeration of reaction types.[99] Reactions were taken from well-known textbooks,[100-176] reviews,[177] and original papers. All reactions were found to belong to linear or cyclic topology.[178] The only apparent exception was epoxidation reactions. For instance, the Mimoun's mechanism[179] of alkene epoxidation with peroxo complexes violates this principle:

$$\tag{1.17}$$

Comparison of the reactions contained in the database with reaction types derived by the enumeration technique also revealed a new heuristic principle, which currently has no apparent explanation. *All elementary reactions occur so that the number of bonds that decrease their order is not greater by one or less by one than the number of bonds that increase their order.* If a single bond is cleaved (formed), we say that the order is decreased (increased) from 1 to 0 (from 0 to 1). If a multiple bond is regarded as several single bonds, then it follows from the this principle that each bond that was made during the reaction must be compensated with a bond broken during the reaction. Only one bond can remain uncompensated. In graph-theoretical notation, the number of edges in the initial molecular graph is greater by one or less by one than the number of edges in the final molecular graph. We called this heuristic *the principle of bond compensation.*

The most comprehensive study along this line was reported by Herges.[72] He extracted 80 000 concerted organic reactions from the well-known "in-house" reaction databases ORAC[180] and REACCS[181] and several thousands from Houben Weyl[182] and found that ~90% of reactions belongs to linear or cyclic topologies. Others either cannot be classified[183] or belong to more complex topologies. Reactions possessing complex topologies were classified according to graphs, which are similar to topology identifiers (Figure 1.5). Figure 1.6 shows complex topologies discussed by Herges.[184] Reactions that belong to these categories were termed *coarctate* because in these reactions formal electron flow pattern resembles coarctate (constricted) circles. Thus, reaction (1.17) occurs via the coarctate transition state. Analysis of coarctate reactions served as a source of several heuristic rules related to concerted reactions:

(1) The more atoms and bonds involved in the transition state, the less probable the reaction. (This principle completely agrees with well-known intuitive principles of minimum structure change and minimum reaction participants.)
(2) Forming and breaking of single bonds is less favorable than the decrease or increase of the bond orders (also a quite obvious rule!).
(3) Hydrogen is the easiest group to be transferred (except for oxygen in epoxidation reactions).
(4) sp^3 hybridized atoms do not react with each other.
(5) All concerted reactions are thermodynamically highly favorable. (Indeed, this is quite clear without search in databases.)
(6) No examples of reactions having transition state like graph K in Figure 1.6 were found.

The search in REACCS and ORAC, which was conducted by Herges, substantiated the heuristic rule proposed by Zeigarnik and Temkin, which was called the bond compensation principle. The only exceptions found in databases were reactions belonging to the category B (Figure 1.6):

Ph-N=N-N=N-Ph ⟶ Ph-N̄ N≡N N̄-Ph

6 electrons

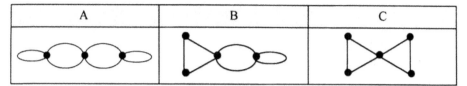

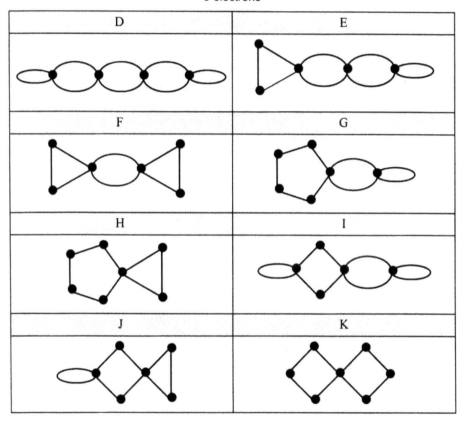

Figure 1.6. Graphs denoting complex topologies. These graphs are similar to topology identifiers except for multiple bonds (which denote an increase or decrease in the bond order by two) and loops (which denote electron lone pairs).

These exceptions, however, are not sound and convincing. It seems likely that these examples which violate the principle of bond compensation are not elementary ones; the more so because one of them is termolecular in reverse direction and thus cannot be elementary.

Sometimes, formulation of heuristic rules does not require the study according to the scheme shown in Figure 1.1. Many important and useful heuristics are known from times when databases were even not thought about. Thus, the classical *principle of minimum structure change* is known from the middle of the 19th century.[185] Currently, this principle is used in a number of computer programs for deductive solution of chemical problems (the synthesis planning, the mechanism elucidation, etc.). Dugundji and Ugi proposed a chemical metric, which made it possible to use the principle of minimum structure change in a quantitative manner.[75] In 1973, they published a paper,[75] which has become classical, and formulated the matrix model of constitutional chemistry. The fundamental equation of this model

$$B + R = E \qquad\qquad (1.18)$$

represents a chemical reaction. **B** and **E** are symmetric $n \times n$ be-matrices (bond and electron matrices), which describes adjacencies and valence electrons of n atoms in the reacting molecules and reaction products. In terms of the Dugundji–Ugi model, **B** and **E** describe the constitution of isomeric ensembles of molecules in the beginning and at the end of the reaction. Numbers of valence electrons are placed on the main diagonal of a be-matrix. The **R** matrix accounts for the electron redistribution during the reaction. The chemical metric, which was proposed by Dugundji and Ugi, was defined according to the formula:

$$D(\mathbf{B},\mathbf{E}) = \sum_{i,j} |b_{ij} - e_{ij}| = \sum_{i,j} |r_{ij}|, \qquad\qquad (1.19)$$

where b_{ij}, e_{ij}, and r_{ij} are entries of **B**, **E**, and **R** matrixes, respectively. This metric was called a *chemical distance*. Later Ugi and co-workers proposed *the principle of minimum chemical distance*,[75,76,186–120] which is a modern interpretation of the classical principle of minimum structure change. As follows from equation (1.19), chemical distance is nothing but the number of valence electrons that are redistributed during the reaction. The principle of minimum chemical distance says that a reaction normally occurs through the redistribution of the minimum number of valence electrons. The number of covalent bonds formed and broken is also minimal. This principle primarily concerns elementary reactions, but, to within a certain error (4 units of chemical distance), it is also applicable to multistep reactions.[75,76] In fact, the principle of minimum chemical distance implies that, at each stage of a complex reaction, an intermediate species (or an ensemble of molecules) has several alternative variants to be transformed. An intermediate undergoes those transformations which correspond to minimum (± 2 or 4) bonds to be formed or broken.

It is not surprising that the elementary steps of organometallic catalysis appeared to be very simple.[99] Indeed, many organic reactions do not occur without a catalyst. To a great extent, this happens because the catalyst divides a complicated transformation into simpler ones, which correspond to lower values of chemical distance between two adjacent intermediates. Surely, the chemical distance is not a driving force of a chemical reaction, but it is a very useful heuristic rule, which was successfully applied in computer programs for synthesis planning.[5,76,189–192]

Chemical distance has several interpretations:[192] (1) chemical (i.e., the number of valence electrons that are redistributed during the transformation or the double number of bonds that are made and broken during the reaction); (2) algebraic (see equation (1.19)); (3) geometrical; and (4) graph-theoretical.[66,193–195]

Another measure of similarity between two compounds was proposed by Hendrickson.[196,197] It was called a *reaction distance*. The reaction distance Δ between two molecules is calculated as

$$\Delta = \sum_i (|\Delta h_i| - |\Delta z_i|),$$

where Δh_i and Δz_i are the differences in numbers of hydrogen and heteroatom functional attachments, respectively, between two structures at carbon i. This similarity measure is mainly applicable to the computer-assisted solution of synthesis planning, but it can be used in mechanistic studies as well.

Kvasnička, Koča, *et al.* proposed a distance measure also called a *reaction distance*, which was defined differently from Hendrickson's reaction distance.[66,194,195,198] In terms of this approach a reaction distance between two molecular graphs is the minimum number of elementary chemical transformations that comprise the transformation of one molecular graph to another. This definition implies that there exists a fixed set of isomeric molecular graphs (isomeric ensembles of molecules) and a fixed set of elementary transformations (or a strong rule that defines which transformations are allowed).[199]

Bertz and Herndon showed that several similarity (distance) measures can be based on the list of subgraphs of molecular graphs.[200] Different interpretations of chemical similarity are due to a variety of their applications.[200]

1.5. Other Heuristics

In the preceding section we discussed topological heuristics, which stem from the application of combinatorics, graph theory and database searches. However, major portions of them are closely related to chemical principles, which are not topological heuristics and which are less formal although better understood and justified. We consider only those of them that characterize the possibility of an elementary step to occur at an appreciable rate within the overall process. Obviously, the rate of an elementary step is defined by the height of the potential energy barrier or, to be more precise, by the $\Delta G^{\ne}$ value (the free energy of activation, the Gibbs potential). However, the factors that determine this value (and related to $\Delta H^{\ne}$ and $\Delta S^{\ne}$ values) are used in different heuristic principles with a proper weight and thus reflect one or another aspect of the reagent interaction process (orbital energy and symmetry, topology of the transition state, etc.). In fact, each particular heuristic rule usually overestimates the significance of particular factors that contribute to the value of the potential energy barrier. By convention heuristic rules and reaction principles can be divided into several groups:

- Methodological (basic) principles
- Energy rules
- Symmetry rules
- Isolobality principle
- Electron rules
- Topological rules

Topological rules were already discussed. Others will be briefly reviewed in this section.

Methodological Principles. First, mention should be made of one of the most important postulates of chemical kinetics, the Ostwald *principle of independence of chemical reactions*. It is because of independence of each reaction that the forward reaction is always associated with a reverse one and equilibrium point can be reached. It is vital to note that the equilibrium in the system of several reactions is also a result of the equilibrium in each particular elementary step in the system by way of compensation of all forward reactions ("microprocesses") with reverse ones. This is known as the *principle of microscopic reversibility*.[201] The macroscopic manifestation of this principle is the *principle of detailed equilibrium* ($\vec{k} / \overline{k} = K$).[201,202] Interestingly, the symmetry of the Schrödinger equation with respect to inversion of time underlies the principle of microscopic reversibility.[202]

Another principle which is commonly regarded as a fundamental one is the paradigm that monomolecular and bimolecular elementary reactions are most probable (*principle of molecularity*). Most theories regard reactions that are monomolecular from the formal standpoint as complex reactions occurring via several bimolecular acts. Earlier, termolecular reactions were also considered (fast reactions of radical recombination with a third molecule in the gas phase).[203] However, up to now sufficient evidence for plausibility of termolecular reactions was not found. Currently, gas-phase reactions at high temperatures are considered to occur via bimolecular elementary steps.[202]

In the preceding section the *principle of minimum structure change* was mentioned, which was formulated in the 19th century.[185] In the beginning of the 20th century, this principle was rediscovered and reformulated several times: Müller and Peytral, *the principle of minimum molecular deformation* (1924);[204] Hückel, *the principle of minimum structure change;*[205] Rice and Teller, *principle of least nuclear motion*.[206] The principle of least nuclear motion postulates that the activation energy is low if the following conditions are met: (1) the motion of nuclei (that is, change in their coordinates) is minimal; (2) the motion of electrons is minimal and alterations of electron shells do not result in alterations of the valence states.

Clearly, the straightforward path of least motion rarely coincides with the minimum energy path.[204] Nevertheless, this principle and the principle of minimum molecular deformation can be categorized as fundamental principles of chemical reactivity. It is also obvious that fulfillment of both principles is defined by the closeness of orbital overlap energies when orbitals have the same symmetry. In this case, the motion along the reaction coordinate preserves maximum symmetry elements of final products.[207] A somewhat more rigorous approach was devised within the framework of *perturbation theory* (*vide infra*).

It seems likely that the least motion principle is associated with the classical *principle of least action* (Maupertuis, Lagrange, Hamilton), which is interpreted as a variational principle in modern science.[208] The principles of minimum structure change and minimum molecular deformation were interpreted as the principle of minimum chemical distance: "most chemical reactions proceed along a pathway of minimum chemical distance, i.e., with a redistribution of a minimum number of valence electrons."[187]

Energy Rules. The following energy principles are very common and widely used in theoretical organic chemistry: the linear free energy rules (*the Brønsted equation*),[201] the *Bell–Evans–Polanyi principle*,[204,209] and the *Hammond postulate.*[210] In particular, it follows from the Hammond postulate that as exothermicity of a step increases, $\Delta G^{\neq}$ is decreased and the transition state structure becomes closer to the structure of reagents. If a step is more endothermic, the transition state structure is closer to the reaction products.

One of the most general approaches to the assessments of energetic characteristics of elementary steps and the structure of transition states is perturbation theory,[204,211] which uses the approximation of frontier orbitals (FMO, Fukui).[211,212] This theory takes into account the topology of interacting orbitals (nodal structures and symmetry). This theory introduced the concepts of orbital-controlled and charge-controlled reactions and led to the justification of the Pearson *rules of soft and hard acid and bases*[134,213] as well as *Woodward–Hoffmann rules*[214] This theory was applied with success in homogeneous catalysis with metal complexes and heterogeneous catalysis.[160,215]

Symmetry Rules. The failure in the consistency of orbital symmetry of breaking and forming bonds, as well as the involvement of highly excited orbitals, reasonably affects the height of the potential energy barrier. At the same time, these rules should be discussed separately from other energy rules because they allow one to assess the reaction probability without sophisticated calculations of energy characteristics. Thus, for the concerted pericyclic reactions, *the principle of conservation of orbital symmetry* is as follows:[214] Concerted reactions in which occupied MO of reagents and products conform to each other in symmetry occur more readily than those in which there is no such conformity. Considering this principle, several rules were formulated for various kinds of concerted reactions, which depend on the total number of electrons on FMO and stereochemistry of drawing together. The extension of topological analysis to other reactions is discussed in several books.[134,211]

Isolobality Principle. In 1976, R. Hoffmann proposed the so-called isolobality principle or the *principle of isolobal analogy.*[216] Different functional groups or fragments of coordination compounds are called isolobal if (1) they have the same number of frontier orbitals, (2) these orbitals are similar in symmetry properties and spatial arrangement, (3) the same numbers of electrons occupy these orbitals, and (4) these orbitals are close to each other in energy. According to this principle, which allows the prediction of reactions producing various organometallic compounds and metal clusters, isolobal species should readily react with each other to produce stable molecules (intermediate or final). The conditions under which isolobal species react with each other via concerted synchronous mechanism are the subject of special analysis.

Electronic Rules. There are several heuristic rules, which are associated with the structure of electron shell or with the structure of reacting species. For example, according to the Tolman rules,[217] organometallic reactions (and reactions of coordination compounds) occur via steps, involving only complexes having 18 or 16 valence electrons. Correspondingly, diamagnetic organometallic intermediates can also occur at sufficient concentrations if their valence shells contain either 18 or 16 electrons.

An important rule in coordination chemistry is the Chernyaev effect of trans-influence in square planar complexes (1926).[134] The essence of this heuristic rule consists in that the ligand coordinated to Pt(II) (or to any other metal in a square planar or octahedral complex) notably affects the rate of substitution of the ligand that is in position *trans* to it. This effect is related to the symmetry and interaction energy of orbitals of the metal and two ligands, which are placed on a diagonal of the square.

* * *

As can be seen from the foregoing, the problem of constructive and analytical enumeration is generally solved. Although several open issues remain, solutions to them are primarily of theoretical interest. From a pragmatic standpoint, the knowledge gained is enough to classify the reaction types and to compare it to currently available information about concerted chemical reactions. Several attempts were made to perform such a "comparative" study. Several important (topological) heuristic rules were deduced. These rules add to the already known principles, which are usually considered when dealing with concerted reactions. To summarize the totality of the reaction principles, we can suggest that a reaction has a good chance of being an elementary one if the following conditions are met:

1. The molecularity of the reactions in both directions is less than or equal to 2.
2. The number of bonds that decrease their order is not greater by one than the number of bonds that increase their order and vice versa.
3. The topology of bond change (electron redistribution) is simple: linear or cyclic.
4. Frontier orbitals of reactants correlate in symmetry with those of products in the ground state.
5. Nodal structures of bonds to be formed and broken coincide with each other.
6. The are no extensive nuclear motion.

The use of these criteria does not exclude usual theoretical and experimental studies aimed at determining whether the step is elementary or not. But they are extremely useful because they recommend to the chemist who investigates a novel reaction to pay attention to factors that violate the above principles. In addition, the topological heuristic rules can be applied in the computer-assisted solution of chemical problems.

References and Notes

1. *Compendium of Chemical Nomenclature. IUPAC Recommendations*, Blackwell Scientific Publications: Oxford, 1987.
2. Dewar, M. J. S. Multibond Reactions Cannot Normally Be Synchronous. *J. Am. Chem. Soc.* **1984**, *106*, 209–219.
3. Bawden, D. Classification of Chemical Reactions: Potential, Possibilities, and Continuing Relevance. *J. Chem. Inf. Comput. Sci.* **1991**, *31*, 212–216.
4. Balandin, A. A. *Multiplet Theory of Catalysis: Theory of Hydrogenation. Classification of Organic Catalytic Reactions. Algebra Applied to Structural Chemistry*; Moscow State Univ.: Moscow, 1970; Vol. 3 (in Russian).
5. Balandin, A. A. An Experience of Natural Classification of Catalytic Reactions in Organic Chemistry. 1. Compounds of Carbon, Hydrogen, and Oxygen Having Constant Valencies and Double and Triple Bonds. *Zh. Org. Khim.* **1932**, *2*, 166–182.
6. Balandin, A. A. On the Classification of Catalytic Reactions in Organic Chemistry. 2. The Method for Deriving the Complete System of Doublet Reactions Involving Systems with H, C, N, O, S, and Cl Atoms and Zero to Quadruple Bonds. *Zh. Fiz. Khim.* **1934**, *5*, 679–706; Structural Theory of Chemical Change. Complete System of Doublet Reactions. *Acta Phys.-Chim. SSSR* **1935**, *2*, 177–202; On the Method for Deriving a Complete System of Triplet Reactions. *Zh. Fiz. Khim.* **1935**, *6*, 1145–1150.
7. Randić, M.; Trinajstić, N. Notes on Some Less Known Early Contributions to Chemical Graph Theory. *Croatica Chem. Acta* **1994**, *67*, 1–35.
8. Balandin, A. A. Structural Algebra in Chemistry. *Usp. Khim.* **1940**, *9*, 390–418.
9. Balaban, A. T. Chemical Graphs. III. Reactions with Cyclic Six-Membered Transition States. *Rev. Roum. Chim.* **1967**, *12*, 875–898.
10. Cayley, A. On the Theory of the Analytical Forms Called Trees. *Phil. Magazine* **1857**, *13*, 172–176; *Phil. Magazine* **1859**, *18*, 373–378.
11. Cayley, A. On the Mathematical Theory of Isomers. *Phil. Magazine* **1874**, *67*, 444.
12. Cayley, A. On the Analytical Forms Called Trees with Applications to the Theory of Chemical Compounds. *Rep. Brit. Ass. Adv. Sci.* **1887**, 257–305.
13. Cayley, A. On the Number of Univalent Radicals C_nH_{2n+1}. *Phil. Magazine, Ser. 5*, **1887**, *3*, 34–35.
14. Cayley, A. On the Analytical Forms Called Trees. *Amer. Math. J.* **1881**, *4*, 266–269.
15. Cayley, A. A Theorem on Trees, *Quar. J. Pure Appl. Math.* **1889**, *23*, 376–378.
16. Cayley, A. In *Collected Mathematical Papers*, Cambridge Univ. Press: Cambridge, 1889–1897, vols. 3, 9, 11, 13, etc.
17. For a review see also: Pólya, G.; Read, R. C. *Combinatorial Enumeration of Groups, Graphs, and Chemical Compounds*; Springer: New York, 1987.
18. Pólya, G. Un problème combinatoire général sur les groupes des permutations et le calcul du nombre des composés organiques. *Comp. Rend. Acad. Sci. Paris* **1935**, *201*, 1167–1169.
19. Pólya, G. Algebraische Berechnung der Isomeren einiger organischer Verbindungen. *Z. Krystallogr. A.* **1936**, *93*, 414.

20. Pólya, G. Uber das Anwachsen der Isomerenzahlen in den homologen Reihe der organischen Chemie. *Vierteljschr. Naturforsch. Ges.* (Zürich) **1936**, *81*, 243–258.

21. Pólya, G. Kombinatorische Anzahlbestimmungen für Gruppen, Graphen und chemische Verbindungen, *Acta Math.* **1937**, *68*, 145–254. (Later this paper was published in English with extensive Read's comments and additions as Pólya, G.; Read, R. C. *Combinatorial Enumeration of Groups, Graphs, and Chemical Compounds*; Springer: New York, 1987.)

22. Pólya, G. Sur le nombre des isomères de certains composées. *J. Symb. Logic,* **1940**, *5*, 98–103.

23. All Harary's findings available at that time are now included in textbooks: (a) Harary, F., *Graph Theory*; Addison-Wesley: Reading, MA, 1969; (b) Harary, F.; Palmer, E. M. *Graphical Enumeration*; Academic: New York, 1973.

24. Vladutz, G. E. Concerning One System of Classification and Codification of Organic Reactions. *Inf. Storage Retr.* **1963**, *1*, 117–116.

25. Vladutz, G. E., Geivadov, E. A. *Automated Information Systems for Chemistry*; Nauka: Moscow, 1974 (in Russian).

26. Vladutz, G. E. Do We Still Need a Classification of Reactions? In *Modern Approaches to Chemical Reaction Searching*; Willett, P., Ed.; Gover: Aldershot, 1986, pp 202–220.

27. Hendrickson, J. B. The Variety of Thermal Pericyclic Reactions. *Angew. Chem.* **1974**, *86*, 71–100; *Angew. Chem. Int. Ed. Engl.* **1974**, *13*, 47–76.

28. Arens, J. F. A Formalism for the Classification and Design of Organic Reactions. I. The Class of $(-+)_n$ Reactions. *Recl. Trav. Chim. Pays-Bas* **1979**, *98*, 155–161.

29. Arens, J. F. A Formalism for the Classification and Design of Organic Reactions. II. The Classes of $(+-)_n+$ and $(-+)_n-$ Reactions. *Recl. Trav. Chim. Pays-Bas* **1979**, *98*, 395–399.

30. Arens, J. F. A Formalism for the Classification and Design of Organic Reactions. I. The Class of $(+-)_n C$ Reactions. *Recl. Trav. Chim. Pays-Bas* **1979**, *98*, 471–483.

31. Zefirov, N. S.; Tratch, S. S. Rearrangements and Cyclizations. X. Formal-Logical Approach to Synchronous Reactions. Formal-Logical Method for Deriving the Complete Set of Possible Types of Multicentered Processes with Cyclic Electron Transfer. *Zh. Org. Khim.* **1975**, *11*, 225–231.

32. Zefirov, N. S.; Tratch, S. S. Rearrangements and Cyclizations. XI. Formal-Logical Approach to Synchronous Reactions. Classification of Multicentered Processes with Cyclic Electron Transfer, *Zh. Org. Khim.* **1975**, *11*, 1785–1800.

33. Zefirov, N. S.; Tratch, S. S. Rearrangements and Cyclizations. XII. Formal-Logical Approach to Synchronous Reactions. Complete Sets of Multicentered Processes with Cyclic Electron Transfer on Three, Four, Five, and Six Centers. *Zh. Org. Khim.* **1976**, *12*, 7–18.

34. Zefirov, N. S.; Tratch, S. S. Rearrangements and Cyclizations. XV. Tautomerism: General Problems, Classification, and Search for New Topological and Reaction Types. *Zh. Org. Khim.* **1976**, *12*, 697–718.

35. Zefirov, N. S.; Tratch, S. S. Systematization of Tautomeric Processes and Formal-Logical Approach to the Search for New Topological and Reaction Types of Tautomerism. *Chem. Scripta* **1980**, *15*, 4–12.

36. Tratch, S. S.; Zefirov, N. S. Problems of Molecular Design and Computers. VII. Formal-Logical Approach to Organic Reactions. Basic Concepts and Terminology, *Zh. Org. Khim.* **1982**, *18*, 1561–1583.

37. Zefirov, N. S.; Tratch, S. S. Problems of Molecular Design and Computers. VIII. Reaction Fragments and Classification Equations in the Formal-Logical Approach to Organic Reactions *Zh. Org. Khim.* **1984**, *20*, 1121–1142.

38. Zefirov, N. S.; Tratch, S. S.; Gamziani, G. A. Problems of Molecular Design and Computers. IX. Formal-Logical Approach to Organic Reactions. Characterization of the Processes with Linear Electron Transfer. *Zh. Org. Khim.* **1986**, *22*, 1341–1359.
39. Tratch, S. S.; Gamziani, G. A.; Zefirov, N. S. Problems of Molecular Design and Computers. X. Enumeration and Generation of Equations Characterizing Ionic, Radical, and Redox Processes with Linear Electron Transfer. *Zh. Org. Khim.* **1987**, *23*, 2488–2507.
40. Zefirov, N. S. An Approach to Systematization and Design of Organic Reactions. *Acc. Chem. Res.* **1987**, *20*, 237–243.
41. Tratch, S. S.; Baskin, I. I.; Zefirov, N. S. Problems of Molecular Design and Computers. XIII. Systematic Analysis of Organic Processes that Are Characterized by the Open Topologies of Bond Change, *Zh. Org. Khim.* **1988**, *24*, 1121–1133.
42. Tratch, S. S.; Baskin, I I.; Zefirov, N. S. Problems of Molecular Design and Computers. XIV. Systematic Analysis of Organic Processes that are Characterized by the Linear-Cyclic Topology of Bond Change, *Zh. Org. Khim.* **1989**, *25*, 1585–1606.
43. Zefirov, N. S.; Tratch, S. S. Symbolic Equations and Their Applications to Reaction Design. *Anal. Chim. Acta.* **1990**, *235*, 115–134.
44. Brownscomb, T. F. A Computer-Assisted Derivation of Possible Unimolecular Pericyclic Reactions. *Diss. Abstr. Int. B.* **1973**, *34*, 1035.
45. Garagnani, E.; Bart, J. C. J. Organic Reaction Schemes and General Reaction-Matrix Types. III. A Quantitative Analysis. *Z. Naturforsch., B: Chem. Sci.* **1976**, *32*, 465–468.
46. Dugundji, J; Gillespie, P.; Marquarding, D.; Ugi, I.; Ramirez, F. Metric Spaces and Graphs Representing the Logical Structure of Chemistry. In *Chemical Applications of Graph Theory*; Balaban, A. T. Ed.; Academic: London, **1976**, Chapter 6.
47. Bauer, J.; Ugi, I. Chemical Reactions and Structures without Precedent, Generated by Computer Program. *J. Chem. Res. (M)* **1982**, *11*, 3101–3260; *J. Chem. Res. (S)* **1982**, *11*, 289.
48. Brandt, J.; Bauer, J.; Frank, R. M.; von Scholley, A. Classification of Reactions by Electron Shift Patterns. *Chem. Scripta*, **1981**, *18*, 53–60.
49. Roberts, D. C. A Systematic Approach to the Classification and Nomenclature of Reaction Mechanisms. *J. Org. Chem.* **1978**, *43*, 1473–1480.
50. Satchell, D. P. N. The Classification of Chemical Reactions. *Naturwissenschaften*, **1977**, *64*, 113–121.
51. Fujita, S. "Structure–Reaction Type" Paradigm in the Conventional Methods of Describing Organic Reactions and the Concept of Imaginary Transition Structures Overcoming This Paradigm. *J. Chem. Inf. Comput. Sci.* **1987**, *27*, 120–126.
52. Fujita, S. A Novel Approach to Systematic Classification of Organic Reactions. Hierarchical Subgraphs of Imaginary Transition Structures. *J. Chem. Soc. Perkin Trans. 2* **1988**, 597–616.
53. Fujita, S. Description of Organic Reactions Based on Imaginary Transition Structures. 1. Introduction of New Concepts. *J. Chem. Inf. Comput. Sci.* **1986**, *26*, 205–212.
54. Fujita, S. Description of Organic Reactions Based on Imaginary Transition Structures. 2. Classification of One-String Reactions Having an Even-Membered Cyclic Reaction Graph. *J. Chem. Inf. Comput. Sci.* **1986**, *26*, 212–223.
55. Fujita, S. Description of Organic Reactions Based on Imaginary Transition Structures. 3. Classification of One-String Reactions Having an Odd-Membered Cyclic Reaction Graph. *J. Chem. Inf. Comput. Sci.* **1986**, *26*, 224–230.
56. Fujita, S. Description of Organic Reactions Based on Imaginary Transition Structures. 4. Three-Nodal and Four-Nodal Subgraphs for a Systematic Characterization of Reactions. *J. Chem. Inf. Comput. Sci.* **1986**, *26*, 231–237.
57. Fujita, S. Description of Organic Reactions Based on Imaginary Transition Structures. 5. Recombination of Reaction Strings in a Synthesis Space and Its Applications to

Description of Synthetic Pathways. *J. Chem. Inf. Comput. Sci.* **1986**, *26*, 238–242.

58. Fujita, S. Description of Organic Reactions Based on Imaginary Transition Structures. 6. Classification and Enumeration of Two-String Reactions with One Common Node. *J. Chem. Inf. Comput. Sci.* **1987**, *27*, 99–104.

59. Fujita, S. Description of Organic Reactions Based on Imaginary Transition Structures. 7. Classification and Enumeration of Two-String Reactions with Two or More Common Nodes. *J. Chem. Inf. Comput. Sci.* **1987**, *27*, 104–110.

60. Fujita, S. Description of Organic Reactions Based on Imaginary Transition Structures. 8. Syntheses Space Attached by Charge Space and Three-Dimensional Imaginary Transition Structures with Charges. *J. Chem. Inf. Comput. Sci.* **1987**, *27*, 111–115.

61. Fujita, S. Description of Organic Reactions Based on Imaginary Transition Structures. 9. Single-Access Perception of Rearrangement Reactions. *J. Chem. Inf. Comput. Sci.* **1987**, *27*, 115–120.

62. Fujita, S. Enumeration of Organic Reactions by Counting Substructures of Imaginary Transition Structures. Importance of Orbits Governed by Coset Representations. *J. Math. Chem.* **1991**, *7*, 111–133.

63. Fujita, S. Formulation of Isomeric Reaction Types and Systematic Enumeration of Six-Electron Pericyclic Reactions. *J. Chem. Inf. Comput. Sci.* **1989**, *29*, 22–30.

64. Fujita, S. The Description of Organic Reactions Based on Imaginary Transition Structures. A Novel Approach to the Computer-Oriented Taxonomy of Organic Reactions. *Pure Appl. Chem.* **1989**, *61*, 605–608.

65. Fujita, S. Graphic Representation and Taxonomy of Organic Reactions. *J. Chem. Educ.* **1990**, *67*, 290–293.

66. Koča, J.; Kratochvíl, M.; Kvasnička, V.; Matyska, L.; Pospíchal, J. *Synthon Model of Organic Chemistry and Synthesis Design*; Springer: Berlin, **1989** (Lecture Notes in Chemistry, no. 51).

67. Kvasnička, V. Classification Scheme of Chemical Reactions. *Collect. Czech. Chem. Commun.* **1984**, *49*, 1090–1097.

68. Kvasnička, V.; Kratochvíl, M.; Koča, J. Reaction Graphs. *Collect. Czech. Chem. Commun.* **1983**, *48*, 2284–2304.

69. Herges, R. Reaction Planning (Computer-Aided Reaction Design). In *Chemical Structures: The International Language of Chemistry*; Warr, W. A., Ed., Springer: Berlin, 1988, pp 385–398.

70. Herges, R. Reaction Planning: Computer-Aided Reaction Design. *Tetrahedron Computer Methodol.* **1988**, *1*, 15–25.

71. Herges, R. Reaction Planning: Prediction of New Organic Reactions. *J. Chem. Inf. Comut. Sci.* **1990**, *30*, 377–383.

72. Herges, R. Coarctate Transition States: The Discovery of a Reaction Principle. *J. Chem. Inf. Comut. Sci.* **1994**, *34*, 91–102.

73. Bauer, J.; Herges, R.; Fontain, E. Ugi, I. IGOR and Computer Assisted Innovation in Chemistry. *Chimia* **1985**, *39*, 43–53.

74. Rico, R. J.; Page, M.; Doubleday, C., Jr. Structure of the Transition State for Hydrogen Molecule Elimination from 1,4-Cyclohexadiene. *J. Am. Chem. Soc.* **1992**, *114*, 1131–1136.

75. Dugundji, J; Ugi, I. An Algebraic Model of Constitutional Chemistry as a Basis for Chemical Computer Programs. *Top. Curr. Chem.* **1973**, *39*, 19–64.

76. Ugi, I.; Bauer, J.; Bley, K.; Dengler, A.; Dietz, A.; Fontain, E.; Gruber, B.; Herges, R.; Knauer, M.; Reitsam, K.; Stein, N. Computer-Assisted Solution of Chemical Problems – The Historical Development and the Present State of the Art of a New Discipline of Chemistry. *Angew. Chem. Int. Ed. Engl.* **1993**, *32*, 201–207.

77. Ugi, I.; Bauer, J.; Blomberg, C.; Brandt, J.; Dietz, A.; Fontain, E.; Gruber, B.; v. Scholley-Pfab, A.; Senff, A.; Stein, N.; Models, Concepts, Theories, and Formal

Languages in Chemistry and Their Use as a Basis for Computer Assistance in Chemistry. *J. Chem. Inf. Comput. Sci.* **1994**, *34*, 3–16.

78. Recently, Hendrickson showed that most of the above systems of reaction description are essentially identical and equally well suited for different tasks in which generalized reactions are used. Hendrickson, J. B. Description of Reactions: Their Logic and Applications. *Recl. Trav. Chim. Pays-Bas* **1992**, *111*, 323–334.

79. Harary, F.; Palmer, E. M.; Robinson, R. W.; Read, R. C. Pólya's Contribution to Chemical Enumeration. In *Chemical Applications of Graph Theory*; Balaban, A. T., Ed.;. Academic: New York, **1976**, p. 11–24.

80. Balaban A. T. Enumeration of Chemical Isomers. In *Graph Theory and Its Application In Chemistry*; Sofia: Nauka i Izkustvo, 1987 (in Bulgarian). This chapter contains good introductory material to enumeration techniques.

81. Tratch, S. S.; Zefirov, N. S. Combinatorial Models and Algorithms in Chemistry: A Staircase of Combinatorial Objects and Its Application to the Formalization of Structural Problems in Organic Chemistry. In *Principles of Symmetry and Organization in Chemistry*; Stepanov, N. F., Ed.; Moscow State Univ.: Moscow, **1987**, pp 54–86 (in Russian).

82. Rouvray, D. H. Isomer Enumeration Methods. *Chem. Soc. Rev.* **1972**, *3*, 355–372.

83. Hansen, P. J.; Jurs, P. C. Chemical Applications of Graph Theory. Part II. Isomer Enumeration. *J. Chem. Educ.* **1988**, *65*, 661–664.

84. Balasubramanian, K. Applications of Combinatorics and Graph Theory to Spectroscopy and Quantum Chemistry. *Chem. Rev.* **1985**, *85*, 599–618. (This review contains a section with easy-to-understand introduction to the combinatorial enumeration.)

85. Parks, C. A.; Hendrickson, J. B. Enumeration of Monocyclic and Bicyclic Carbon Skeletons. *J. Chem. Inf. Comput. Sci.* **1991**, *31*, 334–339.

86. Pólya, G.; Tarjan, R. E.; Woods, D. R. *Notes on Introductory Combinatorics*; Birkhäuser: Boston, 1983 (Progress in Computer Science, no. 4).

87. Read, R. C. The Legacy of Pólya's Paper: Fifty Years of Pólya Theory. In *Combinatorial Enumeration of Groups, Graphs, and Chemical Compounds*; Springer: New York, 1987, pp 96–135.

88. Fujita, S. *Symmetry and Combinatorial Enumeration in Chemistry*; Springer: Berlin, 1991.

89. Fujita, S. Formulation and Enumeration of Five-Center Organic Reactions. *Bull. Chem. Soc. Jpn.* **1989**, *62*, 662–667.

90. Tratch, S. S. Logical and Combinatorial Methods for Design of Organic Structures, Reactions, and Configurations, Doctorate (Chemistry) Dissertation, Moscow State Univ., 1993. See also: Zefirov, N. S.; Baskin, I. I.; Palyulin, V. A. SYMBEQ Program and Its Application in Computer-Assisted Reaction Design. *J. Chem. Inf. Comput. Sci.* **1994**, *34*, 994–999.

91. Stankevich, M. I.; Tratch, S. S.; Zefirov, N. S. Combinatorial Models and Algorithms in Chemistry. Search for Isomorphism and Automorphisms of Molecular Graphs. *J. Comput. Chem.* **1988**, *9*, 303–314.

92. Zhidkov, N. P.; Zefirov, N. S.; Popov, A. I.; Smagin, S. B.; Tratch, S. S.; Shchedrin, B. M. An Algorithm to Search for All Permutations Belonging to an Automorphism Group. In *Mathematical Problems of Structural Analysis and Algorithms of Computer Experiments in Chemistry*; Zhidkov, N. P.; Shchedrin, B. M., Eds.; Moscow State Univ.: Moscow, 1979, pp 59–73.

93. Bauer, J. IGOR2: A PC-Program for generating New Reactions and Molecular Structures. *Tetrahedron Comput. Methodol.* **1989**, *2*, 269–280 (IGOR2 is available through *Tetrahedron Comput. Methodol.*, disk #21).

94. Fontain, E.; Bauer, J.; Ugi, I. Reaction Pathways on a PC. In *PCs for Chemists*; Zupan, J., Ed.; Elsevier: Amsterdam, 1990, 135–154.

95. Ugi, I. K.; Bauer, J.; Baumgartner, R., Fontain, E., Forstmeyer, D., Lohberger, S. Computer Assistance in the Design of Syntheses and a New Generation of Computer Programs for the Solution of Chemical Problems by Molecular Logic. *Pure Appl. Chem.* **1988**, *60*, 1573–1586.

96. Zeigarnik, A. V., unpublished paper.

97. Corio, P. L. Theory of Reaction Mechanisms. *Top. Curr. Chem..* **1989**, *50*, 249–283

98. Mathieu, J.; Weill-Raynal, J. *Formation of C–C Bonds. Vol. 1. Introduction of a Functional Carbon Atom*; Thieme: Stuttgart, 1973.

99. Zeigarnik, A. V.; Temkin, O. N. unpublished results.

100. Zeiss, H., Ed.; *Organometallic Chemistry*; Reinhold: New York, 1960.

101. Coates, G. E. *Organometallic Compounds*; Methuen: London, 1960.

102. Fischer, O. E.; Werner, H. *Metal π-Complexes*. Elsevier: Amsterdam, 1966, Vol. 1.

103. Falbe, J. *Synthesen mit Kohlenmonoxyd*. Springer: Berlin, 1967.

104. Basolo, F.; Pearson, R. G. *Mechanisms of Inorganic Reactions*. Wiley: New York, 1974, 2nd ed.

105. Pauson, P. L. *Organometallic Chemistry*. Edward Arnold: London, 1967.

106. Coates, G. E.; Wade, K. *Organometallic Compounds. The Main Group Elements*. Methuen: London, 1967, Vol. 1, 3rd ed.

107. Dolgoplosk, B. A.; Makovetskii, K. L.; Tinyakova, E. I.; Sharaev, O. K. *Diene Polymerization with π-Allyl Complexes*. Nauka: Moscow, 1968 (in Russian).

108. Temkin, O. N.; Flid, R. M. *Catalytic Transformations of Alkynes in Solutions of Metal Complexes*. Nauka: Moscow, 1968 (in Russian).

109. Candlin, J. P.; Taylor, K. A.; Thompson, D. T. *Reactivity of Transition Metal Complexes*. Elsevier: Amsterdam, 1968.

110. *Homogeneous Catalysis*; American Chemical Society: Washington, DC, (Adv. Chem. Ser., Vol. 70), 1978.

111. Coates, G. E.; Green, M. L. H.; Powell, P.; Wade, K. *Principles of Organometallic Chemistry*. Methuen: London, 1968.

112. Green, M. L. H. *Organometallic Compounds. Vol. 2. Transition Elements*. Methuen: London, 1968, 3rd ed.

113. Hagihara, N.; Kumada, M.; Okawara, R. *Handbook of Organometallic Chemistry*. Benjamin: New York, 1968.

114. Wender, I.; Pino, P. *Organic Syntheses via Metal Carbonyls*. Interscience: New York, 1968, Vol. 1; 1977, Vol. 2.

115. King, R. B. *Transition Metal Organometallic Chemistry*. Academic: New York, 1969.

116. Falbe, J. *Carbon Monoxide in Organic Synthesis*. Springer: Berlin, 1970.

117. Moiseev, I. I. *π-Complexes in the Liquid-Phase Alkene Oxidation*. Nauka: Moscow, 1970 (in Russian).

118. Tsutsui, M.; Levy, M. N.; Nakamura, A.; Ichikawa, M. Mori. K. *Introduction to Metal π-Complex Chemistry*. Plenum: New York, 1970.

119. Shrauzer, G. N., Ed.; *Transition Metals in Homogeneous Catalysis*. Dekker: New York, 1971.

120. Maitlis, P. M. *The Organic Chemistry of Palladium*. Academic: New York, 1971, Vols. 1, 2.

121. Becker, E. I., Tsutsui, M., Eds., *Organometallic Reactions*. Wiley: New York, 1972.

122. Herberhold, M. *Metal π-Complexes*. Elsevier: Amsterdam, 1972, Vol. 2.

123. Hancock, M.; Levy, M. N.; Tsutsui, M. *Organometallic Reactions*. Wiley: New York, 1972, Vol. 4.

124. Reutov, O. A.; Beletskaya, I. P.; Sokolov, V. I. *Reaction Mechanisms of Organometallic Compounds*. Khimiya: Moscow, 1972 (in Russian).

125. Martell, A. E.; Taqui Khan. *Homogeneous Catalysis*. Academic: New York, 1973, Vols. 1 and 2.

126. Shaw, B. L.; Tucker, N. I. *Organo-Transition Metal Compounds and Related Aspects of Heterogeneous Catalysis*. Pergamon: Oxford, 1973.
127. James, B. R. *Homogeneous Hydrogenation*. Wiley: New York, 1973.
128. Forster, D.; Roth, J. F., Eds.; *Homogeneous Catalysis II*; American Chemical Society: Washington, DC, (Adv. Chem. Ser., Vol. 132), 1974.
129. Heck, R. F. *Organotransition Metal Chemistry: A Mechanistic Approach*. Academic: New York, 1974.
130. Jolly, P. W.; Wilke, G. *Organic Chemistry of Nickel*. Academic: New York, 1974, 1975, Vols. 1, 2.
131. Ishii, Y.; Tsutsui, M., Eds.; *Organotransition Metal Chemistry*. Plenum: New York, 1974.
132. Hartley, F. R. *The Elements of Organometallic Chemistry*, The Chemical Society, 1974.
133. Matterson, D. S. *Organometallic Reaction Mechanisms*. Academic: New York, 1974.
134. Basolo, F.; Burwell, R. L., Eds.; *Catalysis: Progress in Research*. Plenum: New York, 1975.
135. Tsuji, J. *Organic Synthesis by Means of Transition Metal Complexes: A Systematic Approach*. Springer: Berlin, 1975.
136. Bonchev, P. R. *Complexation and the Catalytic Activity*. Mir: Moscow (in Russian), 1975; Nauka i Izkustvo: Sofia, 1972 (in Bulgarian).
137. Henrici-Olivé, G.; Olivé, S. *Coordination and Catalysis*. Chemie: Weinheim, 1977.
138. Rylender, P. N.; Greenfield, H., Eds.; *Catalysis in Organic Synthesis*. Academic: New York, 1976
139. Pracejus, H. *Koordinationschemie Katalyse Organischer Reaktionen*. Theodor Steinkopff: Dresden, 1977.
140. Kochi, J. K. *Organometallic Mechanism and Catalysis: The Role of Reactive Intermediates in Organic Reactions*; Academic: New York, 1978.
141. Fel'dblum, V. Sh. *Dimerization and Metathesis of Alkenes*. Khimiya: Moscow, 1978 (in Russian).
142. Nakamura, A.; Tsutsui, M. *Principles and Applications of Homogeneous Catalysis*. Wiley: New York, 1980.
143. Parshall, G. *Homogeneous Catalysis*. Wiley: New York, 1980.
144. Henry, P. M. *Palladium Catalyzed Oxidation of Hydrocarbons*. Reidel: Dordrecht, 1980.
145. Falbe, J. *New Synthesis with Carbon Monoxide*. Springer: New York, 1980.
146. Collman, J. P.; Hegedus, L. S.; Norton, J. R.; Finke, R. G. *Principles and Applications of Organotransition Metal Chemistry*; University Science Books: Mill Valley, CA, 1987, 2nd ed. (1980, 1st ed.).
147. Cotton, F. G. A., Wilkinson, G. *Advanced Inorganic Chemistry*; Wiley: New York, 1980.
148. Negishi, E. *Organometallics in Organic Synthesis*. Wiley: New York, 1980, Vol. 1.
149. Temkin O. N., *The Foundations of Catalysis with Metal Complexes: Catalysis and Coordination Chemistry*. Moscow Inst. of Fine Chem. Tech.: Moscow, 1980 (in Russian).
150. Temkin O. N., *Chemistry and Chemical Engineering of Catalysis with Metal Complexes: Catalysts and Reaction Mechanisms*. Moscow Inst. of Fine Chem. Tech.: Moscow, 1980 (in Russian).
151. Masters, C. *Homogeneous Transition–Metal Catalysis*; Chapman & Hall: London, 1981.
152. Reutov, O. A.; Beletskaya, I. P.; Artamkina, G. A.; Kashin, A. N. *Reactions of Organometallic Compounds as Redox Processes*. Nauka: Moscow, 1981 (in Russian).

153. Sheldon, R. A.; Kochi, J. K. *Metal-Catalyzed Oxidations of Organic Compounds.* Academic: New York, 1981.
154. Wilkinson, G.; Stone, F. G. A.; Abel, E. W. *Comprehensive Organometallic Chemistry.* Pergamon: Oxford, 1982.
155. Keim, W., Ed.; *Catalysis in C_1 Chemistry*; Reidel: Dordrecht, 1983.
156. Ivin, K. J. *Olefin Metathesis.* Academic: London, 1983.
157. Sheldon, R. A. *Chemicals from Synthesis Gas.* Reidel: Dordrecht, 1983.
158. Pignolet, L. H. *Homogeneous Catalysis with Metal Phosphine Complexes.* Plenum: New York, 1983.
159. Henrici-Olivé, G.; Olivé, S. *The Chemistry of the Catalyzed Hydrogenation of Carbon Monoxide.* Springer: Berlin, 1984.
160. Dorfman, Ya. A. *Catalysis and Mechanisms of Hydrogenation and Oxidation Reactions.* Nauka: Alma-Ata, 1984 (in Russian).
161. Gubin, S. P.; Shul'pin, G. B. *Chemistry of Complexes Having Metal–Carbon Bonds.* Nauka: Novosibirsk, 1984 (in Russian).
162. Metelitsa, D. I. *Modeling of Redox Ferments.* Nauka i Tekhnika: Minsk, 1984 (in Russian).
163. Pearson, A. J. *Metalo-Organic Chemistry.* Wiley: Chichester, 1985.
164. Rudakov, E. S. *Reactions of Alkanes with Oxidants, Metal Complexes, and Radicals.* Naukova Dumka: Kiev, 1985 (in Russian).
165. Lukehart, C. M. *Fundamental Transition Metal Organometallic Chemistry.* Brooks/Cole: Monterey, CA, 1985.
166. Yamamoto, A. *Organotransition Metal Chemistry: Fundamental Concepts and Applications*; Wiley: New York, 1986.
167. Chaloner, P. A. *Handbook of Coordination Catalysis in Organic Chemistry.* Butterworth: London, 1986.
168. Shmidt, F. K. *Catalysis by Complexes of Metals of the First Transition Row of Hydrogenation and Dimerization Reactions.* Irkutsk Univ.: Irkutsk, 1986 (in Russian).
169. Kukushkin, Yu. N. *Reactivity of Coordination Compounds.* Khimiya: Leningrad, 1987 (in Russian).
170. Shul'pin, G. B. *Organic Reactions Catalyzed by Metal Complexes*; Nauka: Moscow, 1988 (in Russian).
171. Crabtree, R. *The Organometallic Chemistry of the Transition Metals.* Wiley: New York, 1988.
172. Taube, R. *Homogene Katalyse.* Akademie: Berlin, 1988.
173. Hill, C. L., Ed.; *Activated Functionalization of Alkenes.* Wiley: New York, 1989.
174. Temkin, O. N.; Shestakov, G. K.; Treger, Yu. A. *Acetylene: Chemistry, Reaction Mechanisms, and Technology*; Khimiya: Moscow, 1991 (in Russian).
175. Ugo, R., Ed.; *Aspects of Homogeneous Catalysis*, Reidel: Dordrecht, 1970–1984, Vols. 1–5.
176. Books in the series *Fundamental Research in Homogeneous Catalysis.* Plenum: New York, 1977–1984, Vols. 1–4.
177. Mention should be made of the highly systematized information contained in the extensive review of Theodosiou et al.: Theodosiou, I.; Barone, R.; Chanon, M. Computer Aids for Organometallic Chemistry and Catalysis. In *Advances in Organometallic Chemistry*; Academic: New York, 1986, Vol. 26, pp 165–216.
178. Temkin, O. N.; Zeigarnik, A. V.; Bonchev, D. G., Graph-Theoretical Models of Complex Reaction Mechanisms. In: *Graph Theoretical Approaches to Chemical Reactivity*; Bonchev, D.; Mekenyan O., Eds.; Kluwer: Dordrecht; 1994, pp 241–275.
179. Jørgensen, K. A. Transition-Metal-Catalyzed Epoxidations. *Chem. Rev.* **1989**, *89*, 431–458.
180. ORAC (Organic Reaction Access by Computer) is a database from ORAC, Ltd.

181. REACCS is a reaction retrieval system from the Molecular Design Ltd., San Leonardo, CA.
182. Carbene (Carbenoid). *Houben Weyl E19b*; Thieme: Stuttgart, 1989.
183. A portion of reactions were found to be not stoichiometric and another portion were mistakenly categorized as concerted reactions.
184. Literature examples shows that topologies **11, 12, 16, 17, 23** (Figure 1.5) are also possible.
185. Kolbe, H. Ueber die chemische Constitution und Natur der organischen Radicale. *Ann. Chem. Pharm.* **1850**, *75*, 211–239; **1850**, *76*, 1–73.
186. Jochum, C.; Gastieger, J.; Ugi, I. The Principle of Minimum Chemical Distance (PMCD). *Angew. Chem. Int. Ed. Engl.* **1980**, *19*, 495–505.
187. Jochum C.; Gasteiger J.; Ugi I. The Principle of Minimum Chemical Distance and the Principle of Minimum Structure Change. *Z. Naturforsch., B: Chem. Sci.* **1982**, *37*, 1205–1215.
188. Wochner, M.; Brandt, J.; v. Scholley, A.; Ugi, I. Chemical Similarity, Chemical Distance, and Its Exact Determination. *Chimia* **1988**, *42*, 217–225.
189. Ugi, I.; Wochner, M. Molecular Logic and Computer Assistance in Chemistry. *J. Mol. Struct. (Teochem)* **1988**, *165*, 229–242.
190. Fontain, E. Application of Genetic Algorithms in the Field of Constitutional Similarity. *J. Chem. Inf. Comput. Sci.* **1992**, *32*, 748–752.
191. Ugi, I.; Fontain, E.; Bauer, J. Formal Methods for Reducing the Combinatorial Abundance of Conceivable solutions to a Chemical Problem — Computer-Assisted Elucidation of Complex Reaction Mechanisms. *Anal. Chim. Acta* **1990**, *235*, 151–161
192. For a comprehensive review see: Ugi, I.; Wochner, M.; Fontain, E.; Bauer, J.; Gruber, B.; Karl, R. Chemical Similarity, Chemical Distance, and Computer-Assisted Formalized Reasoning by Analogy. In *Concepts and Applications of Molecular Similarity*; Johnson, M. A.; Maggiora, G. M., Eds.; Wiley: New York, 1990, Chapter 2.
193. Kvasnička, V.; Pospíchal, J. Graph-Theoretical Interpretation of Ugi's Concept of the Reaction Network. *J. Math. Chem.* **1990**, *5*, 309–322.
194. Kvasnička, V.; Pospíchal, J. Two Metrics for a Graph-Theoretical Model of Organic Chemistry. *J. Math. Chem.* **1989**, *3*, 161–191.
195. Kvasnička, V.; Pospíchal J.; Baláž, V. Reaction and Chemical Distances and Reaction Graphs. *Theor. Chim. Acta* **1991**, *79*, 65–79.
196. Hendrickson, J. B.; Parks, C. A. A Program for the Forward Generation of Synthetic Routes. *J. Chem. Inf. Comput. Sci.* **1992**, *32*, 209–215.
197. Hendrickson, J. B. Descriptions of Reactions: Their Logic and Applications. *Recl. Trav. Chim. Pays-Bas* **1992**, *111*, 324–335.
198. Koča, J. The Reaction Distance. *Collect. Czech. Chem. Commun.* **1988**, 53, 3119–3130.
199. Here, we explicitly assume that a molecular graph belonging to a set of isomeric molecular graphs can match several molecules constituting an ensemble of molecules.
200. Bertz, S. H.; Herndon, W. C. The Similarity of Graphs and Molecules. In *Artificial Intelligence Applications in Chemistry*; Pierce, T. H., Hohne, B. A., Eds.; American Chemical Society: Washington, DC; **1986** (ACS Symp. Ser., no. 306), pp 169–175. For reviews on molecular similarity see: (a) Johnson, M. A.; Maggiora, G. M., Eds.; *Concept and Applications of Molecular Similarity*; Wiley: New York, 1990; (b) Johnson, M. A. A Review and Examination of the Mathematical Spaces Underlying Molecular Similarity Analysis. *J. Math. Chem.* **1989**, *3*, 117–145; (c) Rouvray, D. H. Definition and Role of Similarity Concepts in the Chemical and Physical Science. *J. Chem. Inf. Comput. Sci.* **1992**, *32*, 580–586.
201. Hammet, L. P. *Physical Organic Chemistry: Reaction Rates, Equilibria, and Mechanisms*. McGraw Hill: New York, 1970, 2nd ed.

202. Polak, L. S. *Nonequilibrium Chemical Kinetics and Its Applications*. Nauka: Moscow, 1979 (in Russian).
203. Benson, S. W. *The Fundamentals of Chemical Kinetics*. McGraw Hill: New York, 1960.
204. Salem, L. *Electrons in Chemical Reactions: First Principles*. Wiley: New York, 1982.
205. Hückel, W. *Theoretische Grundlagen der Organischen Chemie*. 2nd ed. Akad. Verlag: Leipzig, 1934.
206. Rice, F.O.; Teller, E. The Role of Free Radicals in Elementary Organic Reactions. *J. Chem. Phys.* **1938**, *8*, 489–496; **1939**, *9*, 199. See also: Hine, J. The Principle of Least Nuclear Motion. *Adv. Phys. Org. Chem.* **1977**, *15*, 1–61.
207. Pearson, R. G. *Symmetry Rules for Chemical Reactions*. Wiley: New York, 1976.
208. Polak, L. S. *Variational Principles of Mechanics*. Izdatel'stvo Fiz.-Mat. Literatury: Moscow, 1960 (in Russian).
209. Bell R. P. *The Proton in Chemistry*. Cornell Univ. Press: Ithaca, 1959.
210. Hammond, G.S. A Correlation of Reaction Paths. *J. Am. Chem. Soc.* **1955**, *77*, 334–338.
211. Klopman, G., Ed.; *Chemical Reactivity and Reaction Paths*. Wiley: New York, 1974.
212. For the review of FMO theory see: Fleming, I. *Frontier Orbitals and Organic Reactions*. Wiley: New York, **1976**.
213. Pearson, R. G. Hard and Soft Acids and Bases – The Evolution of a Chemical Concept. *Coord. Chem. Rev.* **1990**, *100*, 403–425.
214. Woodward, R. B.; Hoffmann, R. Conservation of Orbital Symmetry, *Angew. Chem. Int. Ed. Engl.* **1969**, *8*, 781–853; Hoffmann, R.; Woodward, R. B. Selection Rules for Concerted Cycloaddition Reactions. *J. Am. Chem. Soc.* **1965**, *87*, 2046–2048; Woodward, R. B.; Hoffmann, R. Selection Rules for Sigmatropic Reactions. *J. Am. Chem. Soc.*, **1965**, *87*, 2511–2513; Woodward, R. B.; Hoffmann, R. Stereochemistry of Electrocyclic Reactions. *J. Am. Chem. Soc.* **1965**, *87*, 395–397.
215. Dorfman, Ya. A. *Liquid-Phase Catalysis: Orbital Modeling*. Nauka: Alma-Ata, 1981 (in Russian).
216. Hoffmann, R.; Building Bridges Between Inorganic and Organic Chemistry. *Angew. Chem. Int. Ed. Engl.* **1982**, *21*, 711–724; Stone, F. G. An Extremely Fruitful Principle in Organometallic Chemistry. *Angew. Chem. Int. Ed. Engl.* **1984**, *23*, 89–99.
217. Tolman, C.A. The 16 and 18 Electron Rule in Organometallic Chemistry and Homogeneous Catalysis. *Chem. Soc. Rev.*, **1972**, *1*, 337–353.

Chapter 2

Reaction Mechanisms and Networks

2.1. Application of Graph Theory to Reaction Networks: An Overview of Different Methods and Events

In different papers, the concept of reaction network is treated differently. Most commonly, the reaction network is regarded as several consecutive reactions (not necessarily elementary ones). In others, a reaction network is said to be a graph that represents the reaction mechanism or several synthetic reactions that constitute different synthetic routes from starting molecule(s) to the target. In this chapter, we use this interpretation of reaction network with respect to reaction mechanisms. Thus, unless otherwise stated, we will assume that all reactions that constitute a reaction mechanism are elementary. Reaction network in many cases is a synonym of *reaction graph*, which represents different transformations of species. Vertexes of the reaction network (reaction graph) generally represent a species (or a set of species), which participate in the elementary reaction.

Before proceeding to the discussion of reaction networks, let us agree upon several terms. A reaction mechanism is said to be *linear* if it is expressed by the sequence of steps each containing at most one intermediate on the right-hand side and at most one intermediate on the left-hand side. A *nonlinear mechanism* contains at least one elementary step (reaction) in which the number of intermediates on one or both sides is greater than one.

A reaction network is *linear* if there is no need to introduce bipartition (or, generally, multipartition) of edges or vertexes. Thus, bipartite reaction graphs, which will be discussed below are not linear because the set of vertexes is divided into two proper subsets, and vertexes from different subsets are not equivalent. Vertexes and edges of reaction networks can have different labels. These labels denote that a particular vertex corresponds, say, to species X. The edge denotes the ordinal number that was assigned to the corresponding reaction step. These labels, however, do not make different vertexes or edges nonequivalent. In nonlinear networks, vertexes or edges must be differently colored or shaped. When two vertexes or edges are differently shaped or colored, they represent different objects. For instance, in bipartite networks, ●-vertexes denote species, while ■-vertexes denote elementary reactions (or transition states). The shape and color help to show that objects are qualitatively different.

Linear mechanisms can be described by both linear and nonlinear networks. However, nonlinear networks of linear mechanisms can be always reduced to linear networks without the loss of generality. This reduction always simplifies the structure of the network. Sometimes nonlinear mechanisms can also be described by linear networks, although such a linearization always requires special techniques which considerably change the mechanism description. Generally, when one deals with nonlinear mechanisms or with any mechanisms some of which are nonlinear, it is convenient to use nonlinear networks.

2.1.1. Linear Mechanisms in Terms of Linear Networks

The issue, *Who was the first?* has always been very controversial. The source of many ideas can be traced to ancient times. In particular, the history of mathematical chemistry can be regarded as a long story with many twists. Many things were rediscovered time and again, and now it is difficult to avoid errors in reconstructing the order of events. So, we will generally avoid the historical order, although dates will be mentioned.

In 1953, Christiansen proposed the classification of reactions using diagrams similar to graphs.[1] He noticed that the reaction networks are generally nonlinear because the corresponding reaction mechanisms can involve bimolecular elementary steps. Studying chemical kinetics he understood that steady-state concentrations can be expressed in terms of concentrations of initial and final products and rate constants. He used graphs to derive these formulas. Then, he introduced the concept of reaction probability ω. The probability of a linear elementary reaction

$$<reactant> + <intermediate_1> \rightarrow <products> + <intermediate_2>$$

is equal to the ratio $\omega = w/[intermediate1]$, where w is the rate of the elementary reaction expressed in terms of the mass action law. Because elementary reactions are normally either mono- or bimolecular their equations and respective probabilities can be written as follows:

$$<reactant> + <intermediate> \rightarrow ..., \qquad \omega = k[reactant],$$
$$<intermediate> \rightarrow ..., \qquad \omega = k,$$

where k is the rate constant.

Christiansen showed that steady-state concentrations can be derived if one will simply omit all initial and final products (but not intermediates!) in the mechanism and replace rate constants with respective probability values ω. This expedient allows one to obtain a reduced mechanism, and check whether it is linear or not.

Christiansen was the first to mention that even this expedient sometimes fails to linearize a network because an elementary step may contain two intermediates on one of the sides or on both. According to the Christiansen's approach to the classification of reaction mechanisms, vertexes in networks are regarded as intermediates, while arcs (directed edges) are regarded as elementary reactions. Thus, Christiansen's reaction networks are *directed graphs* (or *digraphs*). He also mentioned that catalytic and chain reactions must contain cycles in their networks.

In 1956, King and Altman proposed the method for the derivation of the rate laws of steady-state enzymatic reactions having linear mechanisms. This method was based on graph theoretical interpretation of the Cramer rules.[2]

In 1965, M.I. Temkin published a paper about application of graphs to the analysis of steady-state reactions.[3] He proposed graphs similar to those of Christiansen with the exception that, instead of an arc for each elementary reaction, he used an edge for each elementary step. If the elementary step is reversible (i.e., consists of forward and reverse elementary reactions), it is represented by an edge instead of two arcs. If a step is irreversible, it is represented by an arc, which shows the direction of a step. These graphs are usually termed *mixed graphs*. Thus, Temkin also linearized part of nonlinear networks by exclusion species that are not intermediates. In some instances, an elementary step may not involve an intermediate among the reactants or products. To overcome this, Temkin proposed to include fictitious intermediate 0-species. 0-Species allowed him to describe both catalytic and noncatalytic reactions by cyclic graphs without pendant vertexes. In derivation of the rate laws, the concentration of 0-species is taken to be unity. Somewhat more recently, Temkin's graphs were termed *kinetic graphs*.

As an illustration, consider the following reaction mechanism:

$$
\begin{aligned}
&(1) \ A_1 \rightleftarrows X_1 + P_1, \\
&(2) \ X_1 + A_2 \rightleftarrows X_2 + P_2, \\
&(3) \ X_2 + A_3 \rightleftarrows X_1 + P_3, \\
&(4) \ X_2 + A_4 \longrightarrow P_4,
\end{aligned}
\qquad (2.1)
$$

in which A_1, A_2, A_3, and A_4 are reactants; P_1, P_2, P_3, and P_4 are products; X_1 and X_2 are intermediate species. According to the Temkin's method, the reaction graph can be depicted as shown in Figure 2.1.

Kinetic graphs and their applications are discussed in detail in Section 2.2.

In 1966, Balaban *et al.*[4] published a paper in which graphs similar to kinetic graphs were used. In these graphs, a vertex depicts an isomer and an edge depicts an isomerization reaction. Because all reactions are monomolecular isomerizations, no additional linearization of reaction networks were needed. Balaban *et al.* discussed isomerizations of pentasubstituted ethyl cation via possible hydrogen 1,2-shifts. They found that the isomerization network can be described by the

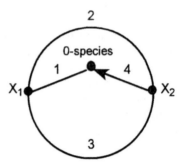

Figure 2.1. Kinetic graph of linear mechanism (2.1). Vertexes are intermediates. Edges 1, 2, and 3 represent reversible reactions. An arc (directed edge 4) denotes an irreversible reaction.

Desargues–Levi graph if carbon atoms in ethyl cation are distinguishable and by the Petersen graph if carbon atoms are indistinguishable.

Subsequent to the first paper on isomerization graphs, an extensive literature appeared that covers many types of isomerizations of organic compounds and metal complexes.[5-13] Graphical representation of a set of isomerization reactions, which is derived using different combinatorial methods, assisted in the elucidation of complex isomerization mechanisms and identification of possible intermediates.

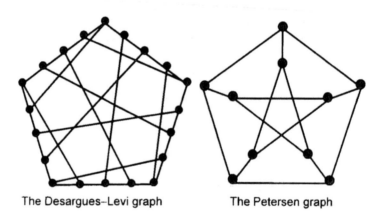

The Desargues–Levi graph   The Petersen graph

Figure 2.2. The Desargues–Levi graph and the Petersen graph of the hydrogen 1,2-shifts in pentasubstituted ethyl cations.

2.1.2. Any Mechanisms in Terms of Linear Networks

In 1976, another method to linearize nonlinear networks was proposed by Dugundji, Ugi *et al.*[14] They generalized the concept of isomerism by applying the mathematical model of constitutional chemistry,[15] which was briefly mentioned in Chapter 1. According to the generalized concept of isomerism, several molecules, which are constructed of a fixed set of atoms, can constitute an *ensemble of molecules* (EM), which can be isomeric to another ensemble. *A family of isomeric ensembles of molecules* (FIEM), each ensemble constructed from the same set of atoms, can be viewed as isomers. Thus, the reaction graph reduces to the graph of isomerizations, and each isomeric ensemble is viewed as an isomer. These graphs were called *graphs of FIEM.*[16]

A graph of FIEM for the two-route mechanism of NO_2 reduction by CO over the metal catalyst (Z denotes an active site) is shown in Figure 2.3. The mechanism is as follows:

$$(1) \ NO_2 + 2Z \rightarrow ZNO + ZO,$$

$$(2) \ ZNO \rightarrow NO + Z,$$

$$(3) \ CO + Z \rightarrow ZCO,$$

$$(4) \ ZCO + ZO \rightarrow 2Z + CO_2, \quad\quad (2.2)$$

$$(5) \ NO_2 + Z \rightarrow ZNO_2,$$

$$(6) \ ZCO + ZNO_2 \rightarrow 2Z + CO_2 + NO.$$

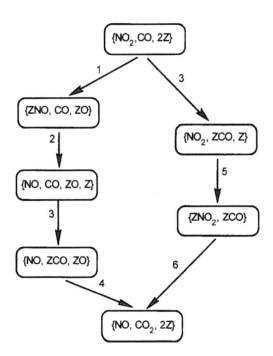

Figure 2.3. The graph of FIEM for the mechanism of NO_2 reduction by CO (2.2).

Several types of subgraphs of a graph of FIEM were shown to be applicable to a variety of chemical problems, such as analysis of multistep reaction mechanisms, mass spectrometry, formal-logical synthesis design, elucidation of synthetic pathways, etc.[14] Similar graphs as well as graphs of chemical distances were discussed in a series of papers of Kvasnička, Koča, Pospíchal, Kratochvíl, et al.[17-25] Another similar approach was proposed by Johnson.[26] He introduced the notion of metadigraphs. *Metadigraphs* contain vertexes, which are themselves graphs (molecular graphs). An arc of metadigraph joins two vertexes G_1 and G_2 if the molecular graph G_1 can transform to G_2. This formalism was shown to be useful in medicinal chemistry.[27,28]

In 1978, Willamowski and Rössler[29] showed that nonlinearity can be eliminated using Horn's idea of species complexes.[30-32] Reasoning from the fact that it is rare for the molecularity of an elementary reaction to exceed two, they considered pairs of species that constitute the left and right members of equations of steps. These pairs were completed with the fictitious intermediate X_0 if any member of any equation contained one (instead of two) species. A mechanism was described by the reaction network that consists of the following three sets: (1) the nonempty set of species $\mathcal{X} = \{X_0, X_1, ..., X_n\}$; (2) the set of complexes $\mathcal{C} = \{C_1, C_2, ..., C_k\}$; (3) the set of reactions that consists of the ordered pairs of complexes $\mathcal{R} = \{R_1, R_2, ...R_m\}$. To illustrate this idea, consider the reaction network of methanol carbonylation:[33]

$$CH_3OH + HI \rightleftharpoons CH_3I + H_2O,$$

$$CH_3I + [Rh] \longrightarrow CH_3[Rh]I,$$

$$CH_3[Rh]I + CO \longrightarrow CH_3CO[Rh]I, \qquad (2.3)$$

$$CH_3CO[Rh]I \longrightarrow CH_3COI + [Rh],$$

$$CH_3COI + H_2O \longrightarrow CH_3COOH + HI,$$

$$CH_3COI + CH_3OH \longrightarrow CH_3COOCH_3 + HI.$$

To take into account the material balance, Willamowski and Rössler proposed to include the supply and removal of species as pseudoreactions. However, because the conversion is generally less than 100%, all species (including a catalyst) can escape from the reactor. Recycling of several species is also possible. Then, let us simply agree upon which species are supplied and removed. Suppose CH_3OH and CO are supplied to the reactor, and CH_3COOCH_3 and CH_3COOH are removed. Using the designations $X_1 \equiv CH_3OH$, $X_2 \equiv CH_3I$, $X_3 \equiv CH_3[Rh]I$, $X_4 \equiv CH_3CO[Rh]I$, $X_5 \equiv CH_3COI$, $X_6 \equiv CH_3COOH$, and $X_7 \equiv CH_3COOCH_3$, we can rewrite the above reaction mechanism as follows:

$$X_0 + X_0 \longrightarrow X_1 + CO, \qquad R_1$$
$$X_6 + X_0 \longrightarrow X_0 + X_0, \qquad R_2$$

$$X_7 + X_0 \longrightarrow X_0 + X_0, \qquad R_3$$
$$X_1 + HI \longrightarrow X_2 + H_2O, \qquad R_4 \qquad\qquad (2.3')$$
$$X_2 + H_2O \longrightarrow X_1 + HI, \qquad R_5$$
$$X_2 + [Rh] \longrightarrow X_3 + X_0, \qquad R_6$$
$$X_3 + CO \longrightarrow X_4 + X_0, \qquad R_7$$
$$X_4 + X_0 \longrightarrow X_5 + [Rh], \qquad R_8$$
$$X_5 + H_2O \longrightarrow X_6 + HI, \qquad R_9$$
$$X_5 + X_1 \longrightarrow X_7 + HI. \qquad R_{10}$$

A set of species is $\{X_0, X_1, \ldots X_7, HI, H_2O, [Rh], CO\}$. The set of complexes is $\{(X_0, X_0), (X_1, CO), (X_6, X_0), (X_7, X_0), (X_1, HI), (X_2, H_2O), (X_2; [Rh]), (X_3, X_0), (X_3, CO), (X_4, X_0), (X_5; [Rh]), (X_5, H_2O), (X_6, HI), (X_5, X_1), (X_7, HI)\}$. The set of complexes can be depicted by the Horn graph in which a vertex stands for a species, and an edge connects two vertexes if the corresponding species are involved in the same complex (Figure 2.4). The reaction graph can be depicted in the form of the disconnected directed graph (Figure 2.5).

Using Pólya's enumeration theorem, Willamowski and Rössler managed to solve the problem of counting the number of sets of complexes for a given number of species and the number of reaction graphs for a given number of complexes. Proceeding from this method for the description of reaction networks, Othmer attempted to characterize the structural features of reaction graphs underlying the dynamic behavior of a multistep reaction.[34]

Veitsman[35] used the same idea to linearize networks and applied the methods of the network flow theory[36,37] to the study of the complex reaction mechanisms in which kinetic steps were combined with mass-transfer processes. He showed that the network flow theory can be applied to the identification of a set of stages of the process that determine the overall rate.[35]

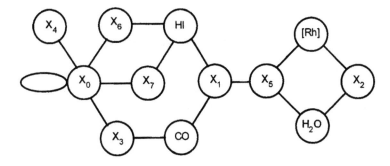

Figure 2.4. The Horn graph for the mechanism of ethanol carbonylation.

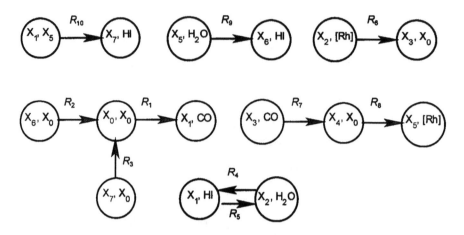

Figure 2.5. The reaction network for the mechanism of ethanol carbonylation.

Nemes *et al.*[38-40] studied nonlinear mechanisms in terms of linear networks as applied to oxidation of organic compounds. Reactions of this sort normally occur via a huge number of elementary steps. To systematize the information contained in these elementary steps, they proposed the following definition of the reaction network. Vertexes of the network are species that participate in elementary reactions of the mechanism. An arc $A_i \longrightarrow A_j$ joins species A_i and A_j if A_j contains at least one atom of A_i. In the latter case, one can state that $A_i \longrightarrow A_j$ is a one-step *kinetic communication*. Networks of kinetic communications, which are constructed on the basis of combinatorial methods or literature information, assist in experiment design for mechanistic studies of complex reactions.

A similar but somewhat more systematic procedure for the construction of reaction networks was proposed by Sokolov and Nikonorov.[41,42] These networks were termed *operator networks*, which were shown to be a useful tool in analysis of internal organization of chemical sets, formulation and interpretation of many heuristic principles based on similarity concepts and analogy (homology, isoelectronicity, isolobality, etc.), and the elucidation of reaction mechanisms.

2.1.3. Any Mechanisms in Terms of Nonlinear Networks

In parallel with studies of linear networks, numerous attempts were made to develop graph-theoretical models of nonlinear mechanisms. Balandin's graph-theoretical models were probably the earliest.

In the 1930s, Balandin started his work on the classification of multistep reactions.[43] The compilation of his findings was published as Volume 3 of the book *Multiplet Theory of Catalysis*.

In one of the models, he used directed *bicolored graphs*[44] with multiple edges in which vertexes of one class (color) were regarded as species, while vertexes of the other class (color) were regarded as transition states. Strictly speaking, it is difficult to define whether the reaction occurs via a single transition state or not (see Chapter 1). Some of reaction steps are pseudoelementary. Bearing this in mind, we say that vertexes of the second class represent reactions rather than transition states.

Consider the following mechanism of acetylene hydration:[45]

$$M + C_2H_2 \longrightarrow M(C_2H_2),$$

$$M(C_2H_2) + RSH \longrightarrow M(C_2H_2SR) + H^+,$$

$$M(C_2H_2SR) + H^+ \longrightarrow CH_2=CHSR + M, \qquad (2.4)$$

$$CH_2=CHSR + H^+ \longrightarrow (C_2H_3SHR)^+,$$

$$(C_2H_3SHR)^+ + H_2O \longrightarrow RSH + (C_2H_3 \cdot H_2O)^+,$$

$$(C_2H_3 \cdot H_2O)^+ \longrightarrow CH_3CHO + H^+.$$

For the sake of simplicity, all reactions are treated as irreversible. Note that if an elementary step is reversible, one should depict it by two elementary reactions (in the forward and reverse directions). The overall reaction for mechanism (2.4) is

$$C_2H_2 + H_2O \longrightarrow CH_3CHO.$$

Let us rewrite this scheme using the following designations: $X_1 \equiv M(C_2H_2)$, $X_2 \equiv M(C_2H_2SR)$, $X_3 \equiv CH_2=CHSR$, $X_4 \equiv (C_2H_3SHR)^+$, $X_5 \equiv (C_2H_3 \cdot H_2O)^+$, $X_6 \equiv RSH$, $X_7 \equiv H^+$, and $X_8 \equiv M$. Then, we arrive at the scheme:

$$(1)\ X_8 + C_2H_2 \longrightarrow X_1,$$

$$(2)\ X_1 + X_6 \longrightarrow X_2 + X_7,$$

$$(3)\ X_2 + X_7 \longrightarrow X_3 + X_8, \qquad (2.4')$$

$$(4)\ X_3 + X_7 \longrightarrow X_4,$$

$$(5)\ X_4 + H_2O \longrightarrow X_6 + X_5,$$

$$(6)\ X_5 \longrightarrow CH_3CHO + X_7.$$

This mechanism can be depicted in the form of a bipartite graph (Figure 2.6). In Figure 2.6, the graph is described so as to avoid intersection of edges. However, it is not always possible. In the general case, the image of a graph does not contain intersections of edges only in the three-dimensional Euclidean space.

Interestingly, that the classification of reaction mechanism proposed by Christiansen appeared to be much similar to that of Balandin, although Christiansen discussed only linear mechanisms.

Somewhat later, the method of bipartite graphs was used by Clarke,[46-53] Vol'pert,[54-57] and Ivanova,[57-62] who studied the qualitative dynamics of chemical systems. Although in papers cited designations are different, the idea to represent the reaction network as a bipartite graph is just the same. The work of Clarke[52] is a most comprehensive with the high level of generality. However, reading of this paper requires sufficient background in mathematics. Principal findings on stability of reaction networks are related to the analysis of cycles in bipartite graphs and several derivative graphs proposed by Clarke. The key concept in determining whether a reaction network is stable or unstable is the *current cycle*. He proved that current cycles are necessary for the instability. All networks that do not have current cycles are stable and thus cannot exhibit "exotic" dynamic behavior such as oscillations, chaos, etc. The exact definition of a current cycle is rather complex and will be given in Section 2.3. The results of Vol'pert and Ivanova are easier to understand and we briefly summarize them here.

In stability analysis of complex reaction networks, a key role belongs to cycles in bipartite graphs. Cycles may have regular and irregular orientation (see Figure 2.7). If a bipartite graph is cycle-free the dynamic behavior must be simple.[55] If a bipartite graph contains only irregular cycles the behavior is also simple. If a graph contains regular cycles, the reaction network can be (but not necessarily is) unstable. Note that all catalytic reactions must contain regular cycles in bipartite graphs.

Balandin applied also elements of topology to the description of complex reactions.[63,64] According to this model, classes of elementary reactions can be described by simplexes (see Figure 2.8). In the Balandin's simplexes, the species that enter the Horn complex are joined by a "double" edge, which is the boundary of a simplex. The reaction direction is shown with a dashed line. If an elementary step is reversible, an arrow on the dashed line is omitted. The stoichiometry is ignored; that is, reactions $A \rightarrow B + C$ and $2A \rightarrow B + C$ are depicted in like manner. The simplest mechanism of an enzymatic reaction following the Michaelis–Menten kinetics

$$E + S \rightleftharpoons ES,$$
$$ES \longrightarrow E + P, \tag{2.5}$$

can be described in the form of the simplicial complex shown in Figure 2.9.

Berreta, Solimano, et al.,[65–67] studied stability problems and proposed another model of nonlinear networks based on the theory of interactant-graphs and knot-graphs (i-graphs and k-graphs). This approach was based on ideas of Horn,[30,68,69] Delattre,[70] and Hyver[71] and was generalized in Clarke's review.[52] They proposed diagrams similar to ordinary graphs with the exception that vertexes in these diagrams (which denote chemical species) are joined into Horn's complexes and edges (which are counterparts of chemical reactions) are to be drawn between complexes. This diagrams were termed i-graphs, because the word *interactant* is treated in these papers as a synonym of the *Horn's complex*.

For instance, the graph of the Michaelis–Menten mechanism of the enzymatic reaction can be depicted as shown in Figure 2.10, and that for the Langmuir–Hinshelwood mechanism of the heterogeneous reaction

$$(1)\ A + Z \rightleftharpoons AZ,$$

$$(2)\ B + Z \rightleftharpoons BZ,$$

$$(3)\ AZ + BZ \rightleftharpoons PZ_2,$$

$$(4)\ PZ_2 \rightleftharpoons P + 2Z,$$

(2.6)

as in Figure 2.11. Figure 2.10 shows the two-step procedure for transforming the i-graph into the k-graph. First, complexes of species should be shaded (step 1). Then, each connected shaded region should be compressed into a point (vertex). Arrows should be replaced with edges (step 2). Knot-graphs were shown to be applicable to the analysis of dynamic behavior of complex reactions.

Perelson and Oster proposed the representation of reaction networks based on the circuit theory.[72] Within the framework of their approach, bond graphs were used to describe reaction networks. It was shown that there is a correspondence between bond graphs and bipartite graphs of reaction mechanisms.

Sinanoğlu proposed the method to depict reaction networks by graphs with two kinds of edges and two kinds of vertexes.[73–78] According to this method, a reaction itself is depicted in the form of a wavy line (with or without direction; in most cases, a direction is omitted). A wavy line is bound by two vertexes, which symbolize the beginning and the end of reaction (■). Each mole of species is

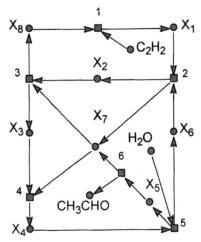

Figure 2.6. The bipartite graph for the mechanism of acetylene hydration.

shown in the form of a vertex (●).[79] If the species is a reactant or a product of reaction, the corresponding species-vertex is adjacent to a reaction-vertex. Sinanoğlu characterized the topological features of reaction networks[77] and proposed the method to generate all *a priori* possible reaction networks.[78] The topology of the reaction network can be described by the graph called a *skeleton*. To transform a reaction network into a skeleton, one should find all components of the network provided that wavy lines are not considered as edges. A component thus found is termed a *lineblock*. Then, each lineblock should be compressed into a single vertex and wavy lines should be displaced by solid lines. Figure 2.12 shows a reaction network of acetylene hydration. After the removal of wavy lines we arrive at lineblocks (Figure 2.13). In Figure 2.13, each line block is confined in a shaded region. A skeleton of this network is shown in Figure 2.14.

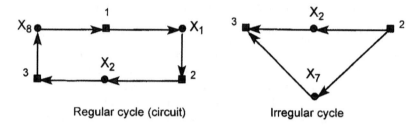

Regular cycle (circuit)   Irregular cycle

Figure 2.7. Two cycles which were cut off from the bipartite graph shown in Figure 2.6. In the regular cycle, one can access a vertex from any other vertex in a walk along edges in the direction shown by arrows. In the irregular cycle, two vertexes may be not accessible from each other.

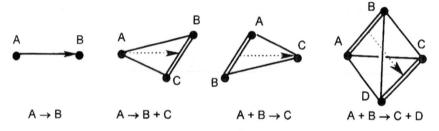

$A \rightarrow B$ $A \rightarrow B + C$ $A + B \rightarrow C$ $A + B \rightarrow C + D$

Figure 2.8. The description of elementary reactions using Balandin's method of simplexes.

Sellers proposed to depict complex (although single-route) reactions by diagrams which are closely related graphs.[80-83] Using the formalism of Abelian groups, he studied combinatorial relations among the sets of species reactions, reaction mechanisms, and catalytic routes. He proposed to depict the results of enumeration of reaction mechanisms by geometric figures (triangles, quadrangles, and other polygons). Reactions involving three species (reactants and products) are described in the form of a triangle and those involving four species are described in the form of a quadrangle. Reactions involving two species cannot be treated within the frame of this method (and this is one of its limitations). The overall reaction (equation) is described in the form of a polygon. The polygon sides correspond to reactants and products of the overall reaction. The reaction polygon is dissected into fields, triangles and quadrangles, the boundaries of which correspond to intermediates. An interior field of each interior polygon is a counterpart of an elementary step. If some elementary steps are irreversible, arrows are used to depict the direction of a step. Because the molecularity of an elementary reaction is

no more than two, the polygon should be dissected only into triangles and quadrangles. One more limitation of the method is that each elementary polygon should
be copied in the overall polygon several times if the stoichiometric number of any
step is greater than unity. Figure 2.15 shows the dissected triangle of acetylene
hydration mechanism.

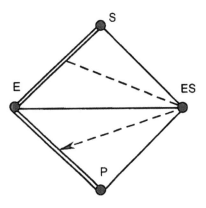

Figure 2.9. The simplicial complex that describes the Michaelis–Menten mechanism of the enzymatic
reaction.

2.1.4. Relationships between Different Graph-Theoretical Models[84]

As can be seen from the previous section, several models exist to graphically represent complex reaction mechanisms. These models make it possible to approach
several problems: (1) the classification or reactions according to their mechanisms;
(2) the computer representation of chemical reactions for synthesis design and
mechanism elucidation; (3) the study of dynamic behavior of multistep reactions
and stability analysis of complex reaction networks; (4) the enumeration of different reaction mechanisms; (5) the enumeration of possible mechanisms for a given
overall reaction; and many other important problems of theoretical chemistry. A
reasonable question arises: *Which model to choose?* In this section, we will show
that most of these models are closely related and fully compatible with each other
because they possess the same feature: they graphically represent the stoichiometry
of reaction networks. Simple methods exist to transform one model to another. In
particular, we will show that they all are transformable to the model based on bipartite graphs.

   Thus, we can join all species involved in a particular elementary reaction and
say that they form an edge of a hypergraph. The only problem that remains is the
direction of an edge in a hypergraph. Because species in an elementary reaction are
either reactants or products, one has to introduce edge directions. This can be done
as shown in Figures 2.10 and 2.11.

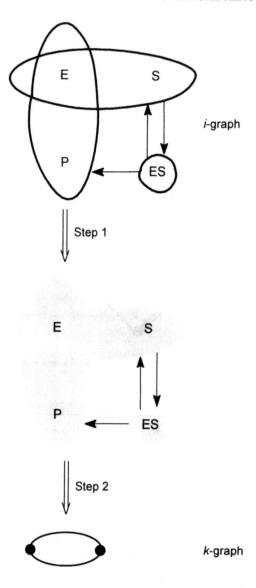

Figure 2.10. The procedure of obtaining the k-graph from the i-graph for the Michaelis–Menten mechanism of the enzymatic reaction.

Let us begin with the model of interactant-graphs (i-graphs). The key feature of an i-graph is that it contains two categories of edges. One of these represent elementary reactions. These edges were depicted as lines with arrows (if an elementary reaction is irreversible) or as two lines with contrarily directed arrows. Edges of another category were shown as fields bound by a closed line (see Figures 2.10 and 2.11). Indeed, an edge is a pair of vertexes no matter how it is depicted. These fields were termed *Horn's complexes* or *interactants*. Strictly speaking, an interactant may contain more than two different species if a reaction is pseudoelementary. In this case, the ordinary graph transforms to a *hypergraph*.[85,86] A hypergraph is a generalization of the graph such that edges can be not only two-element subsets of the vertex set, but subsets of greater and lower cardinality (that is, one-, two-, three-, four-element, and other substets). For instance, the hypergraph shown in Figure 2.16a contains five edges $e_1 = \{v_1, v_2\}$, $e_2 = \{v_2, v_3, v_4\}$, $e_3 = \{v_4, v_5, v_6, v_7\}$, $e_4 = \{v_6, v_7\}$, and $e_5 = \{v_8\}$.

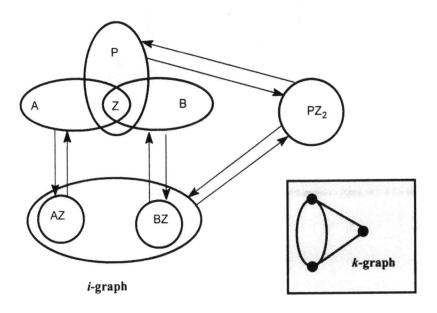

Figure 2.11. The *k*-graph and the *i*-graph for the Langmuir–Hinshelwood mechanism of the heterogeneous reaction.

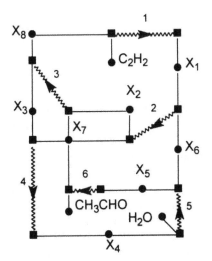

Figure 2.12. The graph of acetylene hydration.

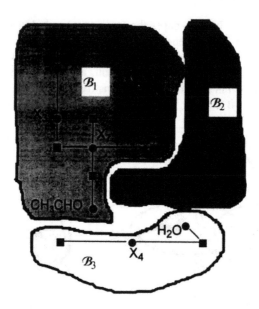

Figure 2.13. The graph obtained by the removal of wavy lines from the network of acetylene hydration.

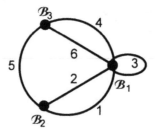

Figure 2.14. The skeleton of the network of acetylene hydration. $\mathcal{B}_1$, $\mathcal{B}_2$, and $\mathcal{B}_3$ are the components of the reaction network provided that wavy lines are removed.

Hypergraphs are closely related to bipartite graphs. A bipartite graph is often said to be a König's representation of a hypergraph.[86] Thus, one set of vertexes of a bipartite graph can be regarded as a set of vertexes of a hypergraph. The other set of vertexes of a bipartite graph can be regarded as a set of edges of a hypergraph. If a vertex v_i is included in the edge e_j, the respective vertexes v_i and e_j in the bipartite graph (König's representation) are joined by a line (edge). The König's representation shows incidence relations of a hypergraph. An example of a hypergraph and its König's representation are shown in Figure 2.16. Directed hypergraphs correspond to directed bipartite graphs.

Sinanoğlu's reaction networks are also transformable to bipartite graphs. Note that wavy lines are in principle unnecessary. Without loss of generality, one can assign directions to solid lines and shrink two bounds of each wavy line into a single vertex preserving all adjacencies of both bounding vertexes. Then, we arrive at a bipartite graph. The skeleton of a reaction network can be obtained by sectioning through ■-vertexes instead of sectioning through wavy lines. This procedure is shown in Figure 2.17. It can be also shown that skeletons of reaction networks and knot-graphs of the model proposed by Beretta *et al.* are just the same things although they are obtained by different methods.

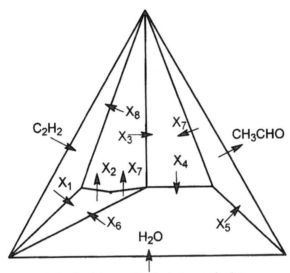

Figure 2.15. The dissected triangle of the acetylene hydration mechanism.

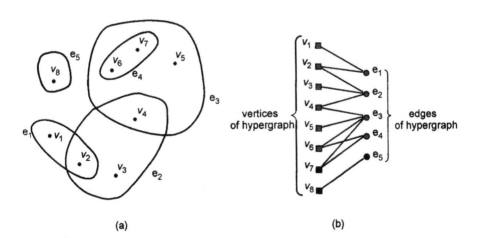

Figure 2.16. An example of a hypergraph (a) and its König's representation (b).

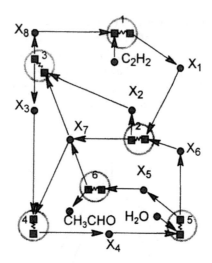

Figure 2.17. Transformation of the Sinanoğlu's reaction network for the mechanism (2.4) into the bipartite graph.

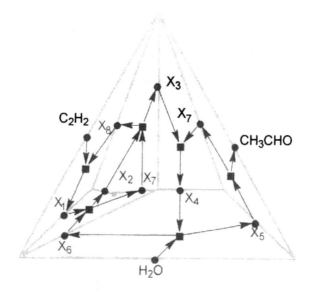

Figure 2.18. Transformation of a Sellers' map into the bipartite graph.

Geometric figures proposed by Sellers are also closely related to bipartite graphs. Note that these dissected polygons are maps. Harary, Prince, and Tutte mentioned a correspondence between maps and graphs.[87] A map also can be regarded as a dual form of reaction networks. Indeed, one can place a ■-vertex at the center of each interior face and a ●-vertex at the center of each line. Then, using arrows denoting directions of elementary reactions, one can reconstruct

●–■-adjacencies as shown in Figure 2.18. Thus, we arrive at a bipartite graph. After this procedure several vertexes can be found to possess similar labels. For instance, the bipartite graph shown in Figure 2.18 contains two vertexes labeled as X_7. It is necessary to shrink them. The resulting graph will be that shown in Figure 2.6.

Apparently, all reverse transformations are also possible. This fact implies that different graph-theoretical models, which represent a reaction mechanism and stoichiometric relations among species, within the frame of this mechanism are dual forms of each other and can be equally used in different studies. The results, which will be obtained, are invariant to a particular graph-theoretical model applied to reaction networks if these models are equally appropriate to describe the stoichiometry of reaction mechanisms. In our studies, we suggested that the model of bipartite graphs is most convenient and used it to characterize the topological features of reaction mechanisms (see Chapter 3).

2.2. Linear Reaction Networks

2.2.1. Basic Definitions

As can be seen from Section 2.2.1, the simplest graph-theoretical interpretation of linear mechanisms is based on kinetic graphs proposed by M.I. Temkin. Recall that mechanisms of this sort involve only linear steps. Kinetic graphs reflect the structure of a mechanism in the space of intermediates. As was already mentioned above, vertexes of kinetic graphs denote intermediates. Edges (arcs) of kinetic graphs denote elementary steps. *Intermediates* are species that are produced in some reaction steps and consumed in other steps; they do not appear in the overall equation(s) of the reaction (see Section 3.1 for more details). Species that are not intermediates (i.e., reagents and products) are termed *terminal*. The probability (weight) of the ith step ω_i is assigned to the respective edge of a kinetic graph. Complex reaction mechanisms also involve terminal species, which are of special sort. They are not involved in the overall stoichiometry of a reaction and do not participate in transformations of initial species, but they are involved in material (mass) balance of intermediate species. These species are involved in reversible activation of a catalyst precursor, deactivation of a catalyst, or binding intermediate species by reagents or products. To depict species of this sort, we use pendant vertexes (that is, terminal vertexes or vertexes of degree one).

Consider as an example the catalytic reaction

$$A + B \rightleftharpoons C + D \qquad (2.7)$$

with the three-step mechanism

$$(1)\ A + Z_1 \rightleftharpoons Z_2 + C,$$

$$(2)\ Z_2 + B \rightleftharpoons Z_3, \qquad\qquad (2.8)$$

$$(3)\ Z_3 \longrightarrow Z_1 + D,$$

where Z_1, Z_2, and Z_3 are intermediate species. The mechanism is depicted by the KG in Figure 2.19, in which the undirected edges 1 and 2 represent reversible reaction steps, while the directed edge (arc) 3 represents an irreversible step.

Suppose that two steps are added to mechanism (2.8):

$$(4)\ Z_1 + B \rightleftharpoons Z_1B,$$

$$(5)\ Z_2 + D \rightleftharpoons Z_2D.$$

Then, the graph of the mechanism complicated with two pendant vertexes is shown in Figure 2.20. The two inactive intermediates contribute to the mass balance of catalytically active intermediates:

$$Z_\Sigma = Z_1 + Z_2 + Z_3 + Z_1B + Z_2D. \qquad\qquad (2.9)$$

Stoichiometric analysis of reaction mechanisms and basic ideas of the theory of reaction routes will be discussed in Sections 2.2.4 and 2.3. At this point, we only note that the number of linearly independent cycles of kinetic graphs is equal to the number of linearly independent routes.

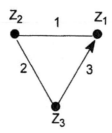

Figure 2.19. The kinetic graph used to depict the catalytic reaction (2.7) whose mechanism is given by equations (2.8).

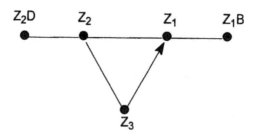

Figure 2.20. Kinetic graph with two pendant vertexes added to the kinetic graph in Figure 2.19 to depict the presence of two inactive intermediates.

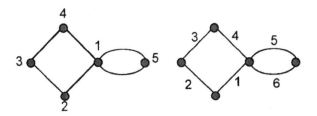

Figure 2.21. The kinetic graph (KG) of the reaction of methane conversion by water vapor on nickel whose mechanism is described by equations (2.10). Vertexes 1–5 correspond to Z, ZCH$_2$, ZCHOH, ZCO, and ZO, respectively, whereas edges 1–6 stand for the respective reaction steps 1–6.

The overall stoichiometric equation of a reaction is obtained by summing the steps of a mechanism, upon which all intermediates vanish. Thus, the stoichiometric equation (2.7) results from summing steps (2.8).

An example of a linear reaction network containing two independent routes, five intermediates, and six steps, is the mechanism of methane conversion by water vapor on nickel.[88] It is presented below by equations (2.10).

$$(1) \ CH_4 + Z \rightleftharpoons ZCH_2 + H_2,$$

$$(2) \ ZCH_2 + H_2O \rightleftharpoons ZCHOH + H_2,$$

$$(3) \ ZCHOH \rightleftharpoons ZCO + H_2,$$

$$(4) \ ZCO \rightleftharpoons Z + CO, \qquad\qquad (2.10)$$

$$(5) \ Z + H_2O \rightleftharpoons ZO + H_2,$$

$$(6) \ ZO + CO \rightleftharpoons Z + CO_2.$$

$$CH_4 + H_2O = CO + 3H_2, \qquad\qquad (2.11)$$

$$CO + H_2O = CO_2 + H_2. \qquad\qquad (2.12)$$

The intermediates included in equations (2.10) are Z (a bivalent reaction site on the nickel surface), ZCH$_2$, ZCHOH, ZCO, and ZO (chemisorbed species). The two independent reaction routes given by equations (2.11) and (2.12) incorporate steps 1–4 and 5 and 6, respectively. In Figure 2.21, they are represented by the two KG cycles.

The property of kinetic graphs to reflect the entire information about the mechanism structure was used for the classification and coding of reaction mechanisms.[89-93]

The manner in which linearly independent routes are interrelated in the reaction network is of utmost importance for the classification and coding of the networks. This task is greatly facilitated by using *kinetic face graphs (KFGs)*[93] or *cycle graphs,*[94] introduced in our earlier papers as *kinetic supergraphs (KSGs)*.[92] Kinetic face graphs are defined on the basis of cycle (face) adjacency in the initial kinetic graph. Let us introduce some terms.

Two cycles are adjacent if they have a common subgraph or joined by a bridge. An *n*-bridge between two cycles contains *n* edges and $(n + 1)$ vertexes and meets the following conditions: (1) Numbers 1 to *n* can be assigned to the edges of the bridge and numbers 0 to *n* can be assigned to the vertexes of the bridge so that the ends of the *j*th edge are vertexes v_{j-1} and v_j, $j = 1 \ldots n$. (2) If one considers the graph constructed of these two cycles and the bridge between them, the removal of a vertex or an edge of a bridge produces a disconnected graph. (3) There is no edge of the bridge, which belongs to a "third" cycle.

Let us define a cycle graph. Let *KG* be a kinetic graph with the cyclomatic number ϕ. Then, ϕ independent cycles of *KG* induce another graph *CG*. *CG* describes the adjacency of ϕ cycles of the cycle basis $B(KG)$ and two vertexes of *CG* are adjacent if the corresponding cycles of *KG* are adjacent. Obviously, in the general case if *KG* contains ξ cycles, $\xi > \phi$, several cycle graphs can exist for *KG*.

In terms of the graph theory, the kinetic graph *KG* is a connected multigraph without pendant vertexes. The loops of *KG* are treated as cycles. An edge is said to be a *diagonal* of a cycle if it joins two nonadjacent vertexes of a cycle. A *linking point* is the end of the bridge between two cycles of $B(KG)$ or a vertex of a common subgraph linking a pair of cycles of $B(KG)$.

Theorem:[94] *No cycle graph of KG contains an n-cycle without a diagonal, $n \geq 4$.*

It is enough to prove that no *CG* contains a 4-cycle without a diagonal, because an *n*-cycle ($n \geq 5$) can be broken into cycles of smaller lengths: $n - 1$, $n - 2, \ldots, 4, 3$ (the variant $n = 3$ does not contradict the condition of the theorem). Therefore, for $n \geq 5$ this theorem can be proved by induction.

Proof: Let *CG* contains an arbitrary sequence of vertexes v_1, v_2, v_3, and v_4 constituting a cycle. Pairs (v_1, v_2), (v_2, v_3), (v_3, v_4), and (v_1, v_4) are pairs of adjacent vertexes, whereas (v_1, v_3) and (v_2, v_4) are pairs of nonadjacent vertexes. The graph *KG* contains the respective cycles C_1, C_2, C_3, and C_4, which induce analogous pairs of adjacent and nonadjacent cycles. Any two points of *KG* linking a given cycle with two other cycles are accessible from each other (that is, joined by a path) because any two vertexes of a connected graph are accessible from each other. Consequently, *KG* contains at least one additional cycle C_5 such that (1) part of its vertexes are the linking points or belong to a bridge between two cycles in the graph, (2) each edge of the cycle belongs to only one or none of the cycles C_1, C_2, C_3, and C_4, (3) C_5 has a common edge with each of these four cycles. Then, because of the second property, the fifth cycle is independent of C_1, C_2, C_3, and C_4 and *CG* contains the corresponding vertex v_5, which, in view of the third property, is adjacent to the vertexes v_1, v_2, v_3, and v_4. These five vertexes produce a wheel-graph:

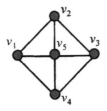

In a wheel-graph, each 4-cycle contains a diagonal. The theorem is proved.☐

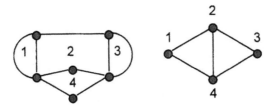

Figure 2.22. A kinetic graph with four cycles and the derivative kinetic face graph with four vertexes. The KFG edges {1,2}, {2,3}, and {2,4} stand for the C-class of linkage of these pairs of cycles (common KG edge), whereas KFG edges {1,4} and {3,4} represent a B-class of cycle linkage (a common KG vertex).

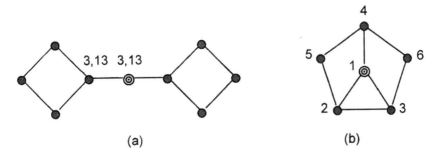

Figure 2.23. (a) A graph with two central vertexes, according to criteria 1 and 2 (minimum eccentricities and overall distances) and one pseudocenter (◉); (b) A graph with all vertexes being classical centers, vertexes 1, 2, 3, and 4 being a pseudocenter, and vertex 1 being an IVEC center.

Thus, each vertex in the KFGs represents a cycle (a face) in the initial KG, while a KFG edge represents the *adjacency* of a pair of KG cycles. Two KG cycles are regarded to be adjacent when they share either a common edge or a common vertex (i.e., when two routes share a common step or a common intermediate, respectively), as well as when they are connected by a bridge. By definition, a *bridge* contains two ends in cycles it connects. All other vertexes between the ends are of degree two. Three types of pairwise linkage of the KG cycles express major features of the *reaction network topology*. In the next subsection, they are shown (see Figure 2.24) to determine the three basis classes **A**, **B**, and **C** of linear reaction networks. The KFGs also make it possible the canonical numbering of KG vertexes, edges, and cycles, which is a prerequisite for the coding and enumeration of linear networks.

An example of a kinetic face graph is presented in Figure 2.22, along with its parent kinetic graph.

The computer storage and retrieval of linear reaction networks requires a unique network coding. In perceiving this goal, *a canonical numbering* (a numbering that does not depend on the manner the graph is presented) of the KG cycles, vertexes, and edges is needed. We developed a method for numbering, which is based on the centric description of the graph elements.[89]

The classical definition of a graph center[95,96] requires this to be the vertex with the minimum *eccentricity*, i.e., to have the minimum distance (number of edges along the shortest path) to its most distant vertex in the graph. For cyclic graphs this definition is insufficient, providing often several nonequivalent vertexes, and in some cases all vertexes, as graph centers. This definition was generalized during the last 15 years.[97-100] It incorporates the classical definition as a first stage in a hierarchical procedure, the second and third conditions in which require the minimum overall distance to the other graph vertexes, and the minimum *distance degree sequence d_1, d_2, ..., d_{max}* (d_L is the number of distances of length L for the vertex), respectively. The procedure *IVEC* (Iterative Vertex–Edge Centricity) then repeats the operations for finding the central graph edge. All vertexes and edges are thus ordered centrically into equivalence classes and ranked according to their centricity, assigning the lowest rank to the resulting classes of pseudocentral vertexes and edges. Then, an iterative redefining the centricity of each vertex and edge proceeds by using the vertex-edge incidence until the central vertex(es) is ultimately found. Examples of graphs with their classical centers, pseudocenters and IVEC centers are shown in Figure 2.23. Vertexes 1 and 2 in the first graph are centers according to criteria 1 and 2: they have the minimum eccentricity equal to three and the minimum overall distance equal to 13. However, vertex 1 has the minimum distance degree sequence of 2, 4, 1, whereas that of vertex 2 is 3, 2, 2. The second graph is a paradoxical case when all graph vertexes (though nonequivalent) appear as classical centers (eccentricity is 2). After applying criteria 2 and 3 vertexes 1–4 remain as pseudocenter. However, the IVEC procedure eventually determines vertex 1 as a single center.

2.2.2. Classification, Coding, and Enumeration of Linear Reaction Networks

Graph-theoretical studies of multiroute reactions with linear mechanisms have shown[89-94] that, proceeding from graph theory, one can build a mechanistic classification system (classification based on the topological structure of reaction mechanisms), and enumerate and code all classes of mechanisms with any number of reaction routes. Solving these problems was essential for the development of computer-assisted methods for the generation of mechanistic hypotheses and for their discrimination.[101,102] These problems will be discussed below in more detail.

The complete matching of linear mechanisms by the KGs made possible the classification of linear reaction networks proceeding from their topological structure.[103] In 1980–1982, we proposed an early version of this classification.[89,92] In the later development of these ideas, the enumeration of linear mechanisms and their computer storage and retrieval indicated the need for some changes in both the classification and coding systems. The resulting classification[93] is based on a hierarchical set of topological criteria.

Categories and Subcategories. The principal factors determining the network category and subcategory are the number of cycles and vertexes in KG. The cycles in any graph are the basic feature that determines the graph complexity. In dealing with kinetic graphs, one thus classifies linear reaction networks according to their basic structural characteristics, the number of their independent routes, into *single-route* networks, *two-route* networks, etc.

The next classificational criterion could be either the number of graph vertexes or that of graph edges. Both are relevant to reaction network classification because they correspond to the number of intermediates and the number of elementary reactions (steps). Indeed, according to the well-known relation between the number of cycles, vertexes and edges in a graph, only one of these two important network characteristics is independent. We preferred to make use of the KG vertexes, because in multiroute networks their number is smaller than that of edges (the number of steps exceeds the number intermediates). The subcategories of *one-intermediate* network, *two-intermediate* network, etc., were thus defined.

Types. The types and classes of linear networks are determined by the manner in which the KG cycles are connected. This was the major challenge because there were no standard methods to describe the complex cycle linkages.

Our approach was based on constructing the kinetic face graph, KFG,[89,93] described in Section 2.2.1. The larger the number of KFG edges, the more interconnected the reaction routes, the more complex type of reaction networks results in the hierarchical classification.

In the initial version of our classification,[89] serial numbers that increase with the number of KFG edges were assigned to KFGs to determine the type of a linear network. That was a satisfactory solution of the problem for the most common linear networks, those that have one to four routes. However, proceeding to the five-route networks, there are at least several groups of KFGs with the same number of edges L which requires additional classificational criteria. On the other hand, computer handling of multiroute reaction networks would require the storage and retrieval of too many KFG serial numbers. Due to this fact, only the number of KFG edges was preserved as a classificational criterion. *Types* of networks with more strongly interrelated reaction routes appear when the number of edges is increased. Table 2.1 and Table 2.2 illustrate the steadily increasing number and complexity of linear reaction networks having 1–5 routes with the increase in the number of KFG edges from 0 to 10.[94]

Classes. Within each type, classes of linear reaction networks were defined depending on the manner in which pairs of reaction routes are interrelated.

Table 2.1. Enumeration of the types L of the linear reaction networks having 1–5 routes (KFG vertexes)

L M	0	1	2	3	4	5	6	7	8	9	10	Total
1	1											1
2		1										1
3			1	1								2
4				2	1	1	1					5
5					3	3	3	3	2	1	1	16

Table 2.2. Types $L = 0$–10 of linear reaction networks having 1–5 routes (KFG vertexes)

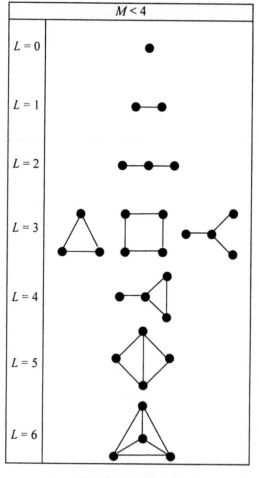

Table 2.2. (*Continued*)

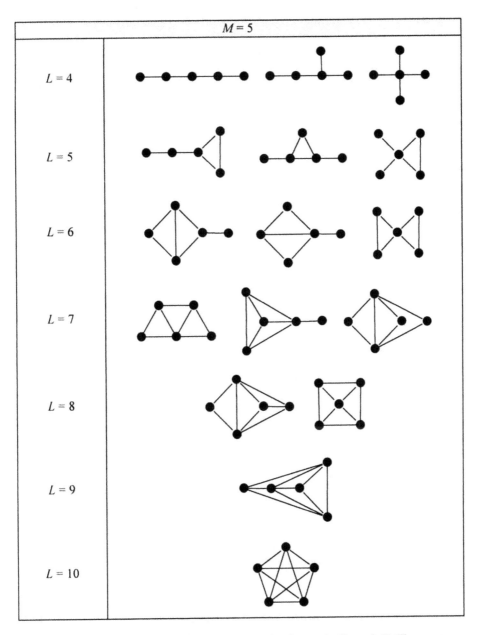

There are only three classes of two-route mechanisms: **A**, **B**, and **C**. They correspond to a pair of KG cycles connected via bridge, a common vertex, and a common edge, respectively (Figure 2.24). A bridge is the weakest kind of connection
between two KG cycles. The cycle interrelation is stronger in class **B**
(corresponding to the so-called spiro-linkage in the structural formulas of chemical
compounds) where the two routes share a common intermediate. It is particularly
strong in class **C** (corresponding to the fusion of two atomic rings in molecules)

where two routes share one or more steps. When dealing with multiroute networks a fourth two-route class **Z** is formally introduced to mark the absence of linkages **A**, **B** or **C** between the pair of corresponding KG cycles.

The multiroute classes are combinations of the two-route classes and they describe exhaustively all pairs of KG cycles. It is useful to introduce a larger classificational unit termed *generalized class* and denoted by $\mathbf{A}^a\mathbf{B}^b\mathbf{C}^c$ where a, b, and c stand for the total number of cycle adjacencies of types **A**, **B**, and **C**, respectively. Some examples of generalized classes are $\mathbf{AB}^2$, $\mathbf{C}^3$, $\mathbf{A}^2\mathbf{BC}$, $\mathbf{B}^3\mathbf{C}$, $\mathbf{A}^3\mathbf{BC}^2$, etc. The sum of the three superscripts equals the total number of KFG edges:

$$a + b + c = L, \tag{2.13}$$

which determines the network type. The **Z** type of indirect cycle linkage is omitted for brevity. Their number z can be retrieved from the relation

$$a + b + c + z = M(M-1)/2 \tag{2.14}$$

which shows that the sum of all four superscripts equals the total number of edges in the complete KFG with the same number of routes M.

The notation used for the *individual classes*, $\prod_{ijkd}\mathbf{A}^i\mathbf{B}^j\mathbf{C}^k\mathbf{Z}^l$, lists successively the types of cycle connections for all pairs of KG cycles, according to the canonical numbers assigned to each cycle; e.g., $\mathbf{A}^2\mathbf{ZCB}^2$, $\mathbf{B}^2\mathbf{CZCZ}$, $\mathbf{ABCZ}^2\mathbf{C}$, etc. Exponents are used for brevity in case of successive cycle connections of the same kind.

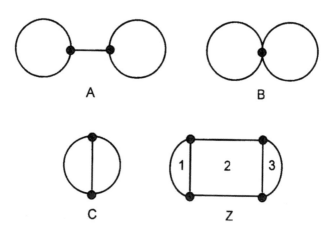

Figure 2.24. Four basic classes of linear two-route reaction networks. Class **Z** refers to the nonadjacent pair of cycles 1 and 3. Substituting any loop for a cycle of arbitrary size preserves the class. (Reprinted with permission from *J. Chem. Inf. Comput. Sci.* **1994**, *34*, 436–445 © 1994 American Chemical Society.)

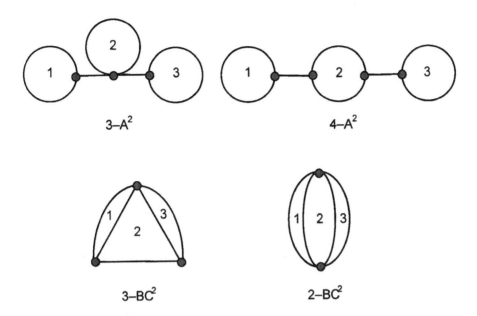

Figure 2.25. The two pairs of classes of three-route linear networks that can be discriminated only by using the class numerical prefix. The class notation follows the increasing numbers of each pair of cycles. Substituting any loop for a cycle of arbitrary size preserves the class.

Despite the variety of the above classificational criteria, some classes still need additional characterization. It seems reasonable to assume that all networks belonging to the same class have the same *basic topology* or, in other words, to assume that their KGs must have the same *smallest homomorphic image SHI* (the latter is obtained by omitting all vertexes of degree one and two). Hence, the complete class discrimination is reached when the number of branched KG vertexes (those with vertex degree $a_i \geq 3$) in the SHIs of all KGs of the class under consideration is added as classificational criterion. In the class notation this number is written as a general prefix, e.g., **4-B^{4}C** or **4-B^{2}CZB^2**. The need for the class numerical prefix n is exemplified in Figure 2.25 with the two pairs of such classes, **A^2** (or **AZA**) and **BC^2**, existing for three-route linear networks.

All classes of linear networks having one, two, and three routes are shown in Table 2.3 with their smallest homomorphic image KGs.

Computer-Assisted Generation and Enumeration of Linear Networks. The classes of networks presented in Table 2.3 were constructed manually.[89] The generation and enumeration of the classes of networks with four or more routes, as well as the generation and enumeration of all linear networks with an arbitrary number of reaction routes could be done only with computer assistance.

Table 2.3. Classes of linear reaction networks having one, two, and three routes and their notation

Table 2.4. All classes of linear reaction networks having four routes (Reprinted with permission from *Graph Theoretical Approaches to Chemical Reactivity*; Bonchev, D.; Mekenyan, O., Eds.; Kluwer: Dordrecht © Kluwer Academic Publishers)

$L = 3$	(29) $4\text{-}B^2Z^2CZ$	(57) $4\text{-}A^2C^2Z^2$	(85) $4\text{-}B^2CZC^2$
	(30) $4\text{-}B^2CZ^3$	(58) $5\text{-}ABCZ^2C$	(86) $4\text{-}BC^3ZB$
	(31) $4\text{-}BCZ^3B$	(59) $5\text{-}AC^2Z^2B$	(87) $5\text{-}BC^3ZC$
(1) $4\text{-}A^3Z^3$	(32) $5\text{-}BCZ^3C$	(60) $4\text{-}AC^2Z^2C$	(88) $6\text{-}C^5Z$
(2) $5\text{-}A^3Z^3$	(33) $5\text{-}BCZ^2CZ$	(61) $6\text{-}AC^2Z^2C$	
(3) $6\text{-}A^3Z^3$	(34) $5\text{-}BC^2Z^3$	(62) $2\text{-}B^4Z^2$	$L = 6$
(4) $4\text{-}A^2Z^2AZ$	(35) $6\text{-}C^2Z^2CZ$	(63) $3\text{-}B^3CZ^2$	
(5) $5\text{-}A^2Z^2AZ$	(36) $6\text{-}C^3Z^3$	(64) $3\text{-}B^2CBZ^2$	(89) $5\text{-}A^6$
(6) $6\text{-}A^2Z^2AZ$		(65) $3\text{-}B^2CZBZ$	(90) $6\text{-}A^6$
(7) $4\text{-}A^2BZ^3$	$L = 4$	(66) $4\text{-}B^2Z_2C^2$	(91) $4\text{-}A^5B$
(8) $5\text{-}A^2BZ^3$		(67) $4\text{-}B^2CZCZ$	(92) $5\text{-}A^5C$
(9) $4\text{-}A^2Z^2BZ$	(37) $5\text{-}A^4Z^2$	(68) $4\text{-}B^2C^2Z^2$	(93) $2\text{-}A^2B^2C^2$
(10) $5\text{-}A^2Z^2BZ$	(38) $6\text{-}A^4Z^2$	(69) $4\text{-}BC^2Z^2B$	(94) $2\text{-}A^3B^3$
(11) $5\text{-}ABZ^3A$	(39) $3\text{-}A^3BZ^2$	(70) $5\text{-}BCZCZC$	(95) $3\text{-}A^2BCA^2$
(12) $4\text{-}AB^2Z^3$	(40) $4\text{-}A^3BZ^2$	(71) $3\text{-}BC^2Z^2C$	(96) $4\text{-}A^2C^2A^2$
(13) $4\text{-}ABZ^2B$	(41) $5\text{-}A^2BAZ^2$	(72) $5\text{-}BC^2Z^2C$	(97) $3\text{-}A^3B^2C$
(14) $4\text{-}ABZ^3B$	(42) $4\text{-}A^2BZAZ$	(73) $5\text{-}BC^3Z^2$	(98) $4\text{-}A^3BC^2$
(15) $5\text{-}ABZ^3C$	(43) $3\text{-}A^2B^2Z^2$	(74) $4\text{-}C^4Z^2$	(99) $3\text{-}A^3C^3$
(16) $5\text{-}ABZ^2CZ$	(44) $3\text{-}A^2ZB^2Z$	(75) $6\text{-}C^4Z^2$	(100) $1\text{-}B^6$
(17) $5\text{-}ABCZ^3$	(45) $4\text{-}A^3CZ^2$		(101) $2\text{-}B^5C$
(18) $5\text{-}ACZ^3B$	(46) $5\text{-}A^3CZ^2$	$L = 5$	(102) $3\text{-}B^4C^2$
(19) $5\text{-}A^2Z^2CZ$	(47) $6\text{-}A^2CAZ^2$		(103) $3\text{-}B^2C^2B^2$
(20) $6\text{-}A^2Z^2CZ$	(48) $5\text{-}A^2CZAZ$	(76) $3\text{-}A^2BZA^2$	(104) $3\text{-}B^3C^3$
(21) $5\text{-}A^2CZ^3$	(49) $3\text{-}AB^2Z^2B$	(77) $4\text{-}A^2CZA^2$	(105) $3\text{-}B^2CBC^2$
(22) $6\text{-}A^2CZ^3$	(50) $3\text{-}A^2BCZ^2$	(78) $4\text{-}A^2CZ_2A^2$	(106) $4\text{-}B^2C^2BC$
(23) $6\text{-}ACZ^3A$	(51) $4\text{-}A^2ZBCZ$	(79) $3\text{-}A^2ZCB^2$	(107) $3\text{-}B^2C^4$
(24) $6\text{-}ACZ^3C$	(52) $4\text{-}A^2ZCBZ$	(80) $4\text{-}A^2ZCBC$	(108) $4\text{-}B^2C^4$
(25) $6\text{-}ACZ^2CZ$	(53) $4\text{-}A^2CBZ^2$	(81) $5\text{-}A^2ZC^3$	(109) $5\text{-}BC^4B$
(26) $6\text{-}AC^2Z^3$	(54) $4\text{-}AB^2Z^2C$	(82) $2\text{-}B^2CZB^2$	(110) $2\text{-}C^6$
(27) $3\text{-}B^2Z^2BZ$	(55) $4\text{-}ABCZ^2B$	(83) $3\text{-}B^2CZBC$	(111) $6\text{-}C^6$
(28) $3\text{-}B^3Z^3$	(56) $5\text{-}A^2ZC^2Z$	(84) $4\text{-}B^2ZC^3$	

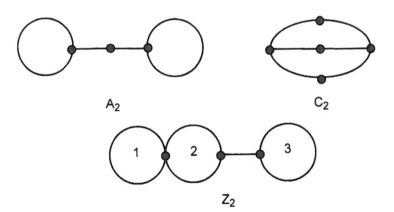

Figure 2.26. Illustration of the A_I, C_K, and Z_V subclasses of linear reaction networks: there is a two-edge bridge in A_2, two edges shared by the two cycles in C_2, and two edge distance between cycles 1 and 3 in Z_2.

Recently, the first large-scale enumeration of the topologically distinct linear reaction networks and their classes was reported.[93] An original program, KING (KINetic Graphs), was employed. It generates all nonredundant KGs for a given number of cycles and vertexes. The algorithm used is similar to that of the GENESIS program[104] and employs an approach to the enumeration of graphs developed by Faradzhev et al.[22]

In Table 2.4, we show the detailed list of all 111 generated classes of linear networks having four routes. They are ordered in four types, corresponding to the number of pairwise cycle adjacencies $L = 3, 4, 5$, and 6. In Table 2.5, we present the enumeration results for the classes of linear networks having up to six routes and up to 12 intermediates. It is seen that the number of classes of linear mechanisms increases exponentially with the number of routes and intermediates. We have also found that, at a constant number of reaction routes and an increasing number of intermediates, the number of classes passes through a maximum and behaves close to the normal distribution.

Table 2.5. Total number of classes of linear reaction networks having $M = 2-6$ routes and $N = 2-12$ intermediates[†] (Reprinted with permission from *Graph Theoretical Approaches to Chemical Reactivity*; Bonchev, D.; Mekenyan O., Eds.; Kluwer: Dordrecht © Kluwer Academic Publishers)

N / M	2	3	4	5	6	7
2	1	1	1	0	0	0
3	1	2	6	3	2	1
4	1	4	14	24	33	19
5	1	5	30	85	192	249
6	1	7	55	239	798	1746
M \ N	8	9	10	11	12	
2	0	0	0	0	0	
3	0	0	0	0	0	
4	11	4	1	0	0	
5	250	153	77	26	7	
6	2800	3082	2576	CE[§]	CE[§]	

[†] N for a class includes vertexes with $a_i \geq 2$, as well as all loops.
[§] Combinatorial explosion.

Subclasses. Within each class, subclasses of linear reaction networks of increasing complexity were specified according to the number of graph edges taking part in the linkage of the pairs of KG cycles. For classes **A** and **C** this is the number of edges I in the bridge, and the number of shared edges K, respectively. No subclasses exist for class *B* because the presence of two or more common KG vertexes requires changing the class from **B** to **C**. For class **Z**, this criterion is the number of edges V along the shortest path linking the pair of nonadjacent cycles, lacking the closer connections of type **A**, **B** or **C** (Figure 2.25). The case $V = 0$

corresponds to a KG in which two disjoint cycles are actually connected by a "bridge" that contain a vertex belonging a third cycle. This is illustrated by the first KG in Figure 2.26 whose generalized notation $3\text{-}A^2$ can be written as individual class notation $3\text{-}AZ_0A$. Generally, the subclass symbol (I, K or V) is denoted as a subscript to the class symbol, e.g., the subclasses A_2, A_3, C_2, C_3, Z_1, Z_2, Z_3, etc. If a subscript equals unity, it can be omitted for brevity.

Classification Summary. The hierarchical classification of linear reaction networks which thus resulted is summarized below:
(i)   **Category.** *Criterion:* Number of linearly independent reaction routes (KG cycles), $M = 1, 2, 3, ...$
(ii) **Subcategory.** *Criterion:* Number of intermediates (KG vertexes), $N = 2, 3, 4, ...$
(iii) **Type.** *Criterion:* Number of adjacent pairs of routes (KFG edges), $L = 0, 1, 2, 3, ...$
(iv) **Classes.** *Criterion:* Type of adjacency for a pair of routes (KG cycles)

Two-route networks:

Classes	Criteria
Class A	bridging of cycles
Class B	cycles sharing a common vertex (routes sharing a common intermediate)
Class C	cycles sharing a common edge (routes sharing a common elementary reaction)
Class Z	nonadjacent cycles (linkage via other cycles)

Prefix n denotes the number of KG vertexes with degree $a \geq 3$.
Multiroute networks:
Generalized classes: $A^a B^b C^c$

Individual classes: $\Pi_{ijkl} A^i B^j C^k Z^l$
(v)   **Subclasses.** *Criterion:* Number of elements connecting a pair of KG cycles:

Subclasses	Criteria
Subclasses A, $A_2, A_3,..., A_I$	length of a bridge, I
Subclasses C, $C_2, C_3,..., C_K$	number of common edges, K
Subclasses Z_0, $Z_1, Z_2,..., Z_V$	number of edges V along the shortest path linking a pair of nonadjacent cycles

(vi) Number of vertexes in each cycle, N_i.

The individual networks within classes and subclasses thus defined are uniquely discriminated by the number of vertexes in their cycles, as shown above in item (vi).

Coding of Linear Reaction Networks. The alphanumerical notation used in our hierarchical classification produces a unique linear code which can be used in computer storage and retrieval

$$M-N-n-\mathbf{A}_I^{\ i}\,\mathbf{B}^j\mathbf{C}_K^{\ k}\mathbf{Z}_V^{\ v}-N_1, N_2, ..., N_M. \qquad (2.15)$$

Based on simple (nondirected) kinetic graphs, it describes linear reaction networks incorporating reversible steps only. The code does not contain the symbol L for the mechanism type because it is redundant ($L = i + j + k$). However, preserving the mechanism type makes sense from the viewpoint of classification as shown, for example, in Table 2.4. Our code is more detailed than the Balaban code for cyclic graphs,[106] which classifies graphs mainly by vertex degrees and the number of cycles of different size. This is explained by specific purposes of our code.

The network code described above can be conveniently converted into the *complexity index* by using spanning trees of the KGs and some of their subgraphs (a spanning tree is an acyclic subgraph that contains all vertexes of the initial graph; for exact definitions see Section 2.2.4).[107,108] The mechanism complexity thus evaluated parallels the mechanism hierarchical ordering in types, classes, and subclasses.[89,107,108] Complexity of reaction networks will be discussed in detail in Chapter 4.

A basic requirement for any code is to be in one-to-one correspondence with the object described. This condition is not satisfied straightforwardly in coding linear mechanisms by using kinetic graphs, because KGs can be often depicted differently. In some instances, it is difficult to say that two graphs are isomorphic, while different images of the same graph correspond to different codes. To overcome these difficulties, we introduced two requirements that standardize the pictorial representation of KGs:

(1) The graph must be depicted by its minimal cycles;

(2) Among different pictorial representations of the same graph, the image whose code is minimal should be preferred.

Condition (1) is illustrated in Figure 2.27. It can be seen that the standardized pictorial representation of a graph is that with the highest feasible symmetry; whenever possible the cycles are depicted as regular *n*-angles.

Condition (2) for the minimal alphanumerical code is as follows. The sequence of symbols $a_1, b_1, c_1, ...$, is regarded being lesser that the sequence $a_2, b_2, c_2, ...$ if the following conditions are met:

$$a_1, b_1, c_1, ... < a_2, b_2, c_2, ... \qquad (2.16)$$

for

$$a_1 < a_2$$

or for

$$a_1 = a_2, \; b_1 < b_2$$

or for

$$a_1 = a_2, \; b_1 = b_2, \; c_1 < c_2, \text{ etc.}$$

When different individual class notations appear, due to the different nature of the edges joining equivalent face-graph vertexes, as well as due to the different choice of the points representing the cycles in the face graph, the minimal code condition is applied keeping the priority order $A < B < C < Z$. This is illustrated in Figure 2.28.

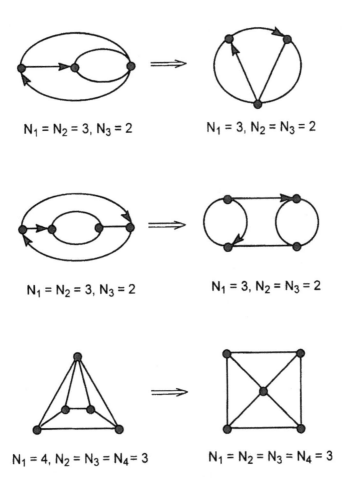

$N_1 = N_2 = 3, \; N_3 = 2$ $N_1 = 3, \; N_2 = N_3 = 2$

$N_1 = N_2 = 3, \; N_3 = 2$ $N_1 = 3, \; N_2 = N_3 = 2$

$N_1 = 4, \; N_2 = N_3 = N_4 = 3$ $N_1 = N_2 = N_3 = N_4 = 3$

Figure 2.27. Converting KGs into a standard pictorial representation having minimal cycles.

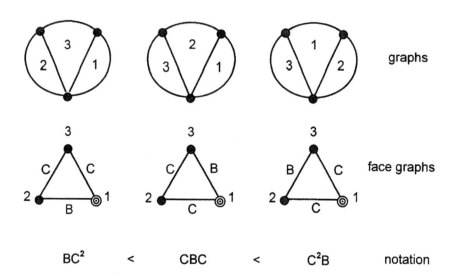

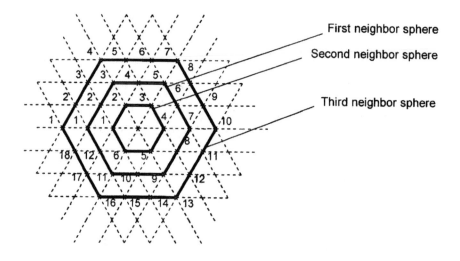

Figure 2.28. Three different pictorial presentations of a three-route network. The minimal code condition favors the BC^2 individual class notation.

Figure 2.29. The standard hexagonal lattice used in the canonical numbering of the kinetic face graph vertexes.

The uniqueness of the linear network code requires also the canonical numbering of the KG cycles, vertexes, and edges. Many approaches to the canonical numbering of graph vertexes are discussed in the literature.[109-120] Because the the KG cycles are of importance for our classification, we developed an original procedure which first canonically number the graph cycles and, only then, the graph vertexes and edges.[89] Indeed, the canonical numbering of the KG cycles means to number uniquely the KFG vertexes. With that purpose, one needs first to unambiguously determine the KFG central vertex (see Section 2.2.1).

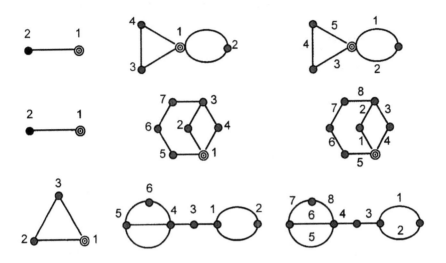

Figure 2.30. Examples of canonical numbering of the cycles, vertexes, and edges in three linear reaction networks. The cycles are numbered in the first column as vertexes of the kinetic face graph.

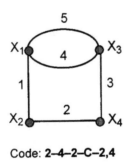

Code: **2–4–2–C–2,4**

Figure 2.31. The kinetic graph and linear code of the reaction, described by the stoichiometric equations (2.18) and (2.19), and by mechanistic steps (2.17).

Once the KFG central vertex is defined, the corresponding central ring is numbered by number 1. Because all KFG cycles can are always dissected into three-membered ones,[89,94] they can be embedded onto a *standard hexagonal lattice*[121,122] (Figure 2.29) so as to impose the central KFG vertex on the lattice center. That of the several possible orientations of the KFG in the lattice is selected, which provides the minimal vertex numbering in the hexagonal lattice. As pointed out earlier, the canonical numbering of the KFG vertexes with more than four vertexes faces some difficulties. More specifically, the embedding of the KFGs on the hexagonal lattice is generally possible in a way that all KFG vertexes are embedded on the standard lattice vertexes, but some KFG edges cannot be embedded on the standard lattice edges. Thus, such KFG edges have to be described by two or more standard lattice edges, connecting the respective two standard lattice vertexes along the shortest path and, if more than one shortest path is available, the vertex numerical priority is used to produce a minimal code.

Having thus all KG cycles canonically numbered, the KG vertexes and edges are numbered consecutively clockwise in the first cycle, then in the second, third, etc. cycles. The procedure is illustrated in Figure 2.30.

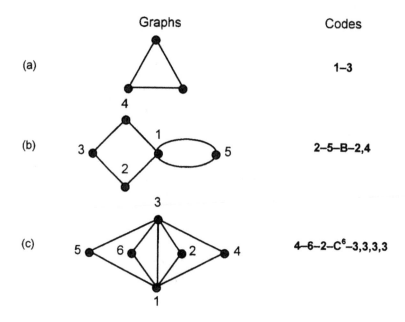

Figure 2.32. Three linear reaction networks and their codes: (a) the formic acid decomposition on metals,[124] (b) the methane conversion by water vapors on nickel (equations (2.10));[88] (c) the amination of alcohols and carbonyl-containing compounds in the presence of melted iron catalyst.[125]

Examples of Linear Reaction Networks and Their Codes. To illustrate the approach, we present below one of the mechanisms proposed for the catalytic reaction of methanol synthesis.[123] It includes two routes, five reaction steps and four intermediates. Correspondingly, its KG contains two cycles, four vertexes, and five edges (Figure 2.31). The network code starts with the number of routes and intermediates, then includes the class 2-C (the prefix $n = 2$ shows the presence of two vertexes of degree higher than 2), and ends with the number of intermediates in the two routes.

$$(1) \ Z{\cdot}H_2O + CO_2 \ \rightleftharpoons \ Z{\cdot}H_2O{\cdot}CO,$$

$$(2) \ Z{\cdot}H_2O{\cdot}CO_2 \ \rightleftharpoons \ Z{\cdot}CO_2 + H_2O,$$

$$(3) \ Z{\cdot}CO_2 + H_2 \ \rightleftharpoons \ Z{\cdot}CO_2{\cdot}H_2, \qquad\qquad (2.17)$$

$$(4) \ Z{\cdot}CO_2{\cdot}H_2 + 2H_2 \ \rightleftharpoons \ Z{\cdot}H_2O + CH_3OH,$$

$$(5) \ Z{\cdot}CO_2{\cdot}H_2 \ \rightleftharpoons \ Z{\cdot}H_2O + CO.$$

$$CO_2 + 3H_2 \ \rightleftharpoons \ CH_3OH + H_2O, \qquad\qquad (2.18)$$

$$CO + 2H_2 \ \rightleftharpoons \ CH_3OH. \qquad\qquad (2.19)$$

More examples are given in Figure 2.32.

The classification and codes of all topologically distinct linear networks having 1–3 routes and 2–6 intermediates are given in Table 2.6. Complexity indices are also presented in this table, related to the discussion of network complexity in Chapter 4. One can find in Chapter 4 the classification and codes of all four-route networks with 2–6 intermediates as well.

The Z class notation is omitted for brevity in the codes of three-route networks except for four pairs of networks (30, 31), (45, 46), (53, 54), and (63, 64), which cannot be otherwise discriminated.

Further Network Enumeration Results. All mechanisms having up to six reaction routes and up to 12 vertexes were enumerated by the program KING, except in the case of $M = 6$ for $N = 11$, and $N = 12$, for which the computational time was too long (Table 2.7).[93] The table evidences the extremely large variety of topologically distinct linear mechanisms. This theoretical prediction disagrees with other estimates based on detailed chemical (but not topological) description of the mechanisms, which produce rather limited mechanistic variety.[82,83] The pitfalls in this purely chemical approach do not account for all possible interrelations of reactants, elementary steps, and reaction routes. However, such comparisons may also indicate that some mechanisms that are topologically allowed might be chemically forbidden. Work is under way to elucidate this intriguing question.

The program KING allowed us also to generate all topologically distinct linear networks having 1–4 routes and 2–6 intermediates.[93] One hundred and thirty six networks having one, two, and three routes are shown in Figure 2.33. Their classification and codes are given in Table 2.6. (In Figure 2.33, KGs 134 and 136 are KGs 56' and 37' from Table 2.6. Graph 135 is omitted in the table.) The 388 generated networks with four routes are given in Chapter 4.

Networks Containing Irreversible Steps and Pendant Vertexes. The classification and coding given in the foregoing refer to reaction networks containing reversible steps only. A number of networks containing both reversible and irreversible steps can be generated for each linear network with reversible steps. In these cases, the networks are depicted by directed KGs and their code (2.15) is supplemented by listing successively the edge types E_i following their canonical numbering.

A step has forward, reverse, or both directions. These three types of steps are denoted by **i**, **î** and **e**, respectively. For brevity, exponents may be used in case of identical successive edge symbols, i.e., e^2, i^3, etc. To make the code unique, the priority order **i** < **î** < **e** and the minimum code criterion are used. The general code is as follows:

$$M\text{–}N\text{–}n\text{–}\mathbf{A}_I^i\mathbf{B}^j\mathbf{C}_K^k\mathbf{Z}_V^v\text{–}N_1, N_2, \dots , N_M\text{–}\mathbf{E}_1, \mathbf{E}_2, \dots, \mathbf{E}_E. \qquad (2.20)$$

The linear networks containing irreversible steps are common in catalytic reactions. Thus, the four most important mechanisms in homogeneous catalytic redox processes[126] are presented by monocyclic networks having one or two irreversible steps (Figure 2.34). Several more examples are presented in Figure 2.35.

The problem for the enumeration of linear networks containing both reversible and irreversible steps reduces to counting the digraphs, graphs containing both arcs and edges (*mixed graphs*),[95] that correspond to a certain nondirected graph. This is a well-known edge coloring problem in graph theory. In dealing with linear networks, however, this problem is complicated by the fact that some digraphs that result from edge coloring do not correspond to any mechanism.

Table 2.6. Classification, codes and complexity indices of linear reaction networks having one, two, and three routes

I. One-Route Networks

No.	Code	Index	No.	Code	Index
1	1-2	8	4	1-5	110
2	1-3	24	5	1-6	192
3	1-4	56			

II. Two-Route Networks

No.	Code	Index	No.	Code	Index
	Class A		18	2-6-B-3,4	804
6	2-4-A-2,2	128			
7	2-5-A-2,3	290		Class C	
8	2-6-A-2,4	552	19	2-2-C-2,2	28
9	2-6-A-3,3	612	20	2-3-C-2,3	90
10	2-5-A$_2$-2,2	200	21	2-4-C-2,4	216
11	2-6-A$_2$-2,3	420	22	2-4-C-3,3	240
12	2-6-A$_3$-2,2	288	23	2-5-C-2,5	430
			24	2-5-C-3,4	510
	Class B		25	2-6-C-2,6	756
13	2-3-B-2,2	72	26	2-6-C-3,5	936
14	2-4-B-2,3	184	27	2-6-C-4,4	996
15	2-5-B-2,4	380	28	2-5-C$_2$-4,4	560
16	2-5-B-3,3	420	29	2-6-C$_2$-4,5	1068
17	2-6-B-2,5	684			

Table 2.6. (*Continued*)

III. Three-Route Networks

No.	Code	Index	No.	Code	Index
	TYPE L = 2		56′	3-6-3-BC-3,2,4	2220
	Class 3-A^2		57	3-6-3-BC-4,2,3	2280
30	3-6-3-A^2-2,2,2	864	58	3-6-3-BC-3,3,3	2472
			59	3-6-3-BC_2-4,2,4	2496
	Class 4-A^2				
31	3-6-4-A^2-2,2,2	864		*Class 4-C^3*	
			60	3-4-4-C^2-4,2,2	576
	Class 3-AB		61	3-5-4-C^2-5,2,2	1180
32	3-5-3-AB-2,2,2	600	62	3-5-4-C^2-4,2,3	1380
33	3-6-3-AB-2,3,2	1272	63	3-6-4-C^2Z_1-6,2,2	2112
34	3-6-3-AB-2,2,3	1272	64	3-6-4-C^2Z_2-6,2,2	2112
35	3-6-3-AB-3,2,2	1272	65	3-6-4-C^2-5,2,3	2598
36	3-6-3-A_2B-2,2,2	864	66	3-6-4-C^2-4,2,4	2700
			67	3-6-4-C^2-4,3,3	3072
	Class 4-AC		68	3-6-4-CC_2-5,2,4	2916
37	3-5-4-AC-3,2,2	750			
37′	3-6-4-AC-4,2,2	1488		*TYPE L = 3*	
38	3-6-4-AC-3,3,2	1590		*Class 2-A^2B*	
39	3-6-4-AC-3,2,3	1680	69	3-5-2-A^2B-2,2,2	600
40	3-6-4-A_2C-3,2,2	1080	70	3-6-2-A^2B-3,2,2	1272
			71	3-6-2-A^2B-2,2,3	1272
	Class 2-B^2		72	3-6-2-A^2_2B-2,2,2	864
41	3-4-2-B^2-2,2,2	384			
42	3-5-2-B^2-2,2,3	880		*Class 3-A^2C*	
43	3-5-2-B^2-3,2,2	880	73	3-4-3-A^2C-2,2,2	304
44	3-6-2-B^2-2,2,4	1680	74	3-5-3-A^2C-3,2,2	690
45	3-6-2-B^2Z_1-4,2,2	1680	75	3-5-3-A^2C-2,2,3	750
46	3-6-2-B^2Z_2-4,2,2	1680	77	3-6-3-A^2C-3,2,3	1590
47	3-6-2-B^2-2,3,3	1872	76	3-6-3-A^2C-4,2,2	1308
48	3-6-2-B^2-3,2,3	1872	78	3-6-3-A^2C-2,3,3	1680
			79	3-5-3-A^2_2C-2,2,2	470
	Class 3-BC		80	3-6-3-A^2_2C-3,2,2	990
49	3-4-3-BC-3,2,2	480	81	3-6-3-A^2_2C-2,2,3	1080
50	3-5-3-BC-4,2,2	1030	82	3-6-3-A^2_3C-2,2,2	672
51	3-5-3-BC-3,3,2	1100			
52	3-5-3-BC-3,2,3	1160		*Class 1-B^3*	
53	3-6-3-BCZ_1-5,2,2	1896	83	3-4-1-B^3-2,2,2	384
54	3-6-3-BCZ_2-5,2,2	1896	84	3-5-1-B^3-2,2,3	880
55	3-6-3-BC-3,4,2	2100	85	3-6-1-B^3-2,2,4	1680
56	3-6-3-BC-4,3,2	2190	86	3-6-1-B^3-2,3,3	1872

Table 2.6. (Continued)

	Class 2-B^2C			110	3-6-2-BC^2-4,4,2	2520
87	3-3-2-B^2C-2,2,2	174		111	3-6-2-BCC_2-3,5,3	2490
88	3-4-2-B^2C-3,2,2	444		112	3-6-2-BC^2_2-4,4,4	3360
89	3-4-2-B^2C-2,2,3	480		113	3-3-3-BC^2-2,2,3	228
90	3-5-2-B^2C-4,2,2	910		114	3-4-3-BC^2-2,2,4	576
91	3-5-2-B^2C-2,2,4	1030		115	3-4-3-BC^2-2,3,3	620
92	3-5-2-B^2C-3,2,3	1100		116	3-5-3-BC^2-2,2,5	1180
93	3-5-2-B^2C-2,3,3	1160		117	3-5-3-BC^2-2,4,3	1320
94	3-6-2-B^2C-5,2,2	1626		118	3-5-3-BC^2-2,3,4	1380
95	3-6-2-B^2C-2,2,5	1896		119	3-5-3-BC^2-3,3,3	1510
96	3-6-2-B^2C-4,2,3	2100		120	3-6-3-BC^2-2,2,6	2112
97	3-6-2-B^2C-3,2,4	2190		121	3-6-3-BC^2-2,5,3	2418
98	3-6-2-B^2C-2,2,4	2280		122	3-6-3-BC^2-2,3,5	2598
99	3-6-2-B^2C-3,2,4	2190		123	3-6-3-BC^2-2,4,4	2700
100	3-6-2-B^2C-2,2,4	2280		124	3-6-3-BC^2-3,4,3	2982
				125	3-6-3-BC^2-3,3,4	3072
	Class 2-BC^2			126	3-5-3-BCC_2-2,4,4	1490
101	3-2-2-BC^2-2,2,2	64		127	3-6-3-BCC_2-2,4,5	2916
102	3-3-2-BC^2-2,3,2	210		128	3-6-3-BCC_2-3,4,4	3300
103	3-4-2-BC^2-2,4,2	504				
104	3-4-2-BC^2-3,3,2	584			Class 4-C^3	
105	3-5-2-BC^2-2,5,2	1000		129	3-4-4-C^3-3,3,3	768
106	3-5-2-BC^2-3,4,2	1260		130	3-5-1-C^3-3,3,4	1740
107	3-5-2-BCC_2-3,4,3	1480		131	3-6-1-C^3-3,3,5	3312
108	3-6-2-BC^2-2,6,2	1752		132	3-6-1-C^3-3,4,4	3594
109	3-6-2-BC^2-3,5,2	2328		133	3-6-1-C^2_2-4,4,4	3744

When a kinetic graph contains pendant vertexes (see Section 2.2.1), the code is supplemented by the total number of these vertexes, N_p, and in an increasing order, the numbers n_i of the vertexes with which P pendant vertexes are adjacent. The resulting code of a linear reaction network having irreversible steps and pendant vertexes is as follows:

$$M-N-n-A_I^i B^j C_K^k Z_V^v -N_1, N_2, \dots , N_M -E_1, E_2, \dots , E_E: n_1, n_2, \dots, n_p. \qquad (2.21)$$

As an example, consider the code 1–3–3–e^2i–2: 1, 2, corresponding to the linear network having one irreversible step and two pendant vertexes, depicted in Figure 2.20. More examples can be found in Table 2.6, which contains codes of 135 linear networks involving 1–3 routes and 2–6 intermediates.

When p_i pendant vertexes are attached to the same "root" vertex n_i, the number p_i appears in the code in parentheses immediately after the root vertex number. The last part of the code is then written in the following form (omitting for brevity all $p_i = 1$):

$$-N_p: n_1(p_1), n_2(p_2), \dots, n_p(p_p).$$

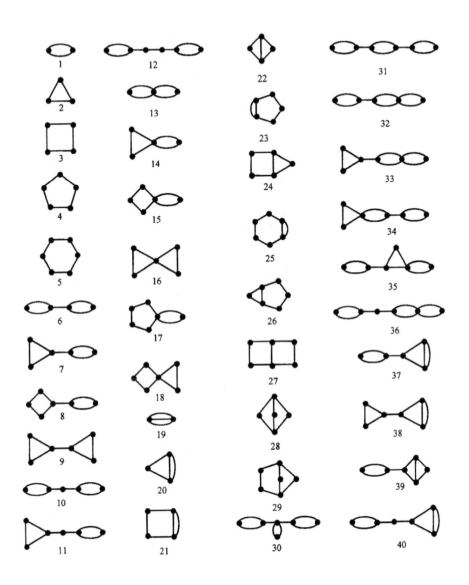

Figure 2.33. The topologically distinct linear networks having 1-3 routes and 2-6 intermediates.[108]

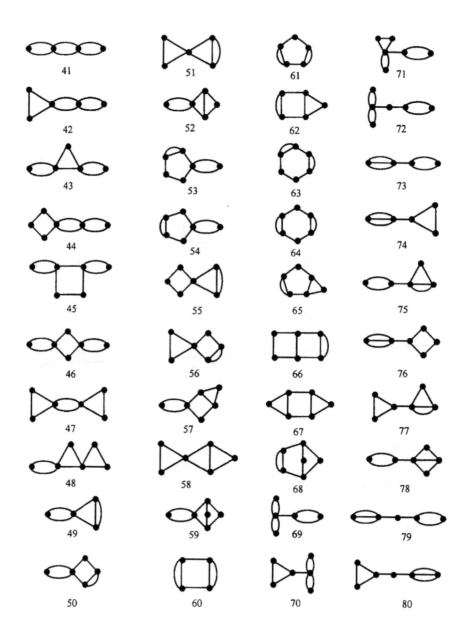

Figure 2.33. (*Continued*).

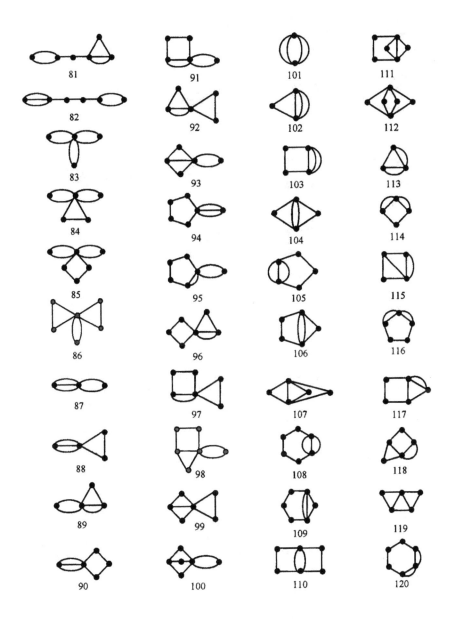

Figure 2.33. (*Continued*).

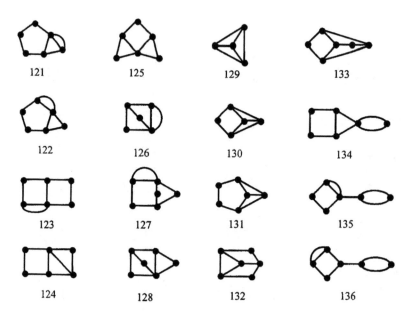

Figure 2.33. (*Continued*).

Table 2.7. The total number of linear reaction networks having $M = 2-6$ routes and $N = 2-12$ intermediates (Reprinted with permission from *Graph Theoretical Approaches to Chemical Reactivity*; Bonchev, D.; Mekenyan, O., Eds.; Kluwer: Dordrecht © Kluwer Academic Publishers)

N / M	2	3	4	5	6	7
2	1	2	4	7	10	14
3	1	3	12	27	65	129
4	1	5	23	85	276	764
5	1	6	43	210	924	3403
6	1	8	72	469	2652	12644

N / M	8	9	10	11	12
2	19	24	30	37	44
3	245	422	710	1113	1710
4	1935	4466	9583	19291	36859
5	11242	33156	89789	224621	526346
6	52727	194909	651008	CE*	CE*

*CE = combinatorial explosion.

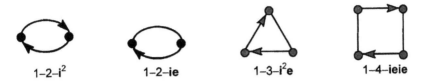

$1-2-i^2$ $1-2-ie$ $1-3-i^2e$ $1-4-ieie$

Figure 2.34. The linear networks of four basic mechanisms in homogeneous catalytic redox processes: (I) alternate oxidation and reduction of the catalyst, (II) activation by a complex formation, (III) complex formation before the oxidation step (or reduction step), (IV) complex formation before both steps of the redox conversion of the catalyst.

Branches with more than one pendant vertex are not taken into account because their edges represent equilibrium reaction steps. Therefore, two such steps 1 and 2 can always be replaced by a third step 3 = 1 + 2, or, alternatively, by two other steps directly attached to the root vertex (Figure 2.36). The latter case occurs when two products depicted by pendant vertexes have contributions to the mass balance.

In terms of graph theory, the enumeration of linear networks having equilibrium steps can be formulated as enumeration of graphs having pendant vertexes. The exhaustive topological enumeration of linear reaction networks, described above in several stages, could be expanded even further by accounting for network specificity. The latter implies both particular reactions (e.g., synthesis or decomposition or molecular rearrangement) and/or particular intermediates (e.g., different catalysts that fit the same mechanism), and the problem can be reformulated as enumeration of graphs with weighted edges and/or weighted vertexes, respectively.

2.2.3. Computer-Assisted Generation of Specific Linear Networks

Background. One can summarize the main idea of the graph-theoretical approach to reaction networks by stating that networks are associated with two types of information: chemical and topological information.[101–103] Chemical information includes the type, composition, and structure of the intermediates and transition states of the reaction steps, whereas the topological information describes relationships of intermediates and routes in the space of intermediates.

Usually, researchers in chemical kinetics avoid a closer look at the topology of a reaction mechanism. They proceed from the traditional strategy: from kinetic data to kinetic law models, and from these models to hypotheses concerning the mechanism of the process. On the other hand, for complicated multiroute chemical reactions (in disagreement with some existing points of view)[130–132] it is impossible to determine the topological structure of the mechanism without knowing chemical information. Moreover, it is either difficult or even impossible to obtain information on the composition of intermediates, as well as on the kinetic constants of their transformations not specifying their locations in the set of reaction steps and the sequences of species conversion.

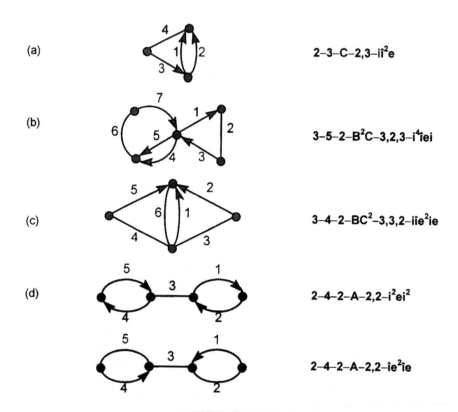

(a)   2–3–C–2,3–ii²e

(b)   3–5–2–B²C–3,2,3–i⁴iei

(c)   3–4–2–BC²–3,3,2–iie²ie

(d)   2–4–2–A–2,2–i²ei²

                                                      2–4–2–A–2,2–ie²ie

Figure 2.35. Linear networks of some catalytic processes containing irreversible steps: (a) syntheses of vinylacetylene and 2-chlorovinyl-acetylene in the presence of CuCl-CuCl₂-NH₄Cl-H₂O;[127] (b) catalytic oxidation of CO to CO₂ in PdCl₂ solutions;[128] (c) oxidative chlorination of acetylene;[129] and (d) two cases of homogeneous catalyzed redox process[126] in which the complex formed with the catalyst is partially inhibited.

We advocate a more efficient strategy for kinetic modeling:[101–103]

$$\begin{aligned} &\textit{generating } \textbf{\textit{hypotheses}} \textit{ for reaction network} \rightarrow \\ &\rightarrow \textit{mechanistic analysis and reaction rate model deriving} \rightarrow \\ &\rightarrow \textit{design of discriminating kinetic experiments} \rightarrow \\ &\rightarrow \textit{experiments} \rightarrow \textit{discrimination of hypotheses} \rightarrow \\ &\rightarrow \textbf{\textit{kinetic model(s)}}. \end{aligned} \qquad (2.22)$$

The topological approach based on graph theoretical methods is applicable at almost every step of this strategy. In what follows, we will discuss fruitful topological methods to conduct two stages of scheme (2.22). These stages are the generation of mechanistic hypotheses and derivation of rate laws.

The approach is exemplified by the reaction of alkene epoxidation with organic hydroperoxides in the presence of homogeneous or heterogeneous molybdenum catalysts. This reaction is not very complicated but offers new paths for the synthesis of organic compounds and might be of great importance for synthetic organic chemistry and pharmaceutical industry. The extensive publications on the catalytic epoxidation of alkenes with organic hydroperoxides,[131-142] do not shed enough light on the mechanism of this reaction, while offering a large variety of kinetic models in the form of fractional-rational functions. Some difficulties in elucidating the mechanism of this reaction stem from a great diversity of catalysts. However, we suppose principal problems are the researchers' unwillingness to examine more than one or two hypothetical reaction networks and the fact that they usually choose the first model that matches the experimental data.

During the last 25 years, several formal methods for generating mechanistic hypotheses were proposed.[102] Mention should be made of the approach based on the formalism of Abelian groups proposed by Sellers.[80-83] The Dugundji-Ugi model of constitutional chemistry was used in computer programs RAIN (Reactions And Intermediates Networks) designed by Fontain,[143-149] GRACE (Generalized Reaction Analysis for Creation and Estimation) designed by Yoneda,[150] and COMSICAT designed by the Novosibirsk group in Russia.[151-155] Theodosiou, Baron, and Chanon advanced the computer program TAMREAC (TrAnsition Metal REACtivity) for the generation of reaction networks in organometallic catalysis based on the use of reaction library.[156-158] Similar program ChemNet was designed by Temkin et al.[102] Stoichiometric relations in the step-species space were applied to hypothesis generation independently by several authors.[159,160] Recently, Valdés-Pérez designed the computer program MECHEM, which is a combination of combinatorial algorithms and stoichiometric analysis.[161-168] Alternatively, we recently developed a method that combines the use of reaction library (proceeding from expert evaluation) with graph-theoretical methods for hypotheses generation.[169]

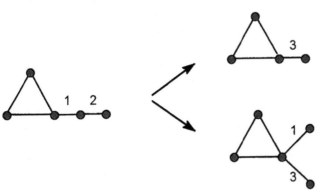

Figure 2.36. Pendant branches with two or more vertexes are not allowed in the linear networks. They are transformed via two paths shown.

Potential Intermediates and Reaction Steps. The epoxidation of alkenes by hydroperoxides in the presence of molybdenum or vanadium catalysts is described by the equation:

$$\text{C}{=}\text{C} + ROOH \longrightarrow \underset{O}{\text{C}{=}\text{C}} + ROH \qquad (2.23)$$

To account for the mechanism of this reaction, the formation of catalyst-hydroperoxide complex is quite frequently postulated. However, some experimental points to the formation of a binary alkene–catalyst complex, and a ternary hydroperoxide–catalyst–alkene complex.

The inhibiting effect of reaction products also suggests the formation of complexes between the catalyst and reaction products. Some more complexes are also thought of as taking part in this reaction, although strong evidence was not found. All potential intermediates of this reaction are collected in Table 2.8. Some of them are inactive; they do not take part in steps yielding reaction products. Table 2.9 lists reaction steps that can be generated from the 14 complexes of Table 2.8.

Variants of Single-Route Networks. For this type of networks only complexes X_1, X_2, X_9 and X_{14} are regarded as active intermediates. Another limitation imposed is the generation of only two-step, three-step and four-step networks (larger single-route networks seem unlikely for this reaction). The number of possible networks is thus reduced to six. They are shown in Table 2.10 with their simple numbering which starts with the symbols showing the network category (1 for single-route) and subcategory (2, 3, and 4 for the number of intermediates). A serial number is then added to distinguish the specific networks that belong to the same classificational unit and are depicted by the same KG. Otherwise, the complete code for these six networks is **1-2-ei** for 1.2.1 and 1.2.2; **1-3-e²i** for 1.3.1 and 1.3.2; and **1-4-e²i²** for 1.4.1 and 1.4.2.

Networks 1.2.1 and 1.2.2 describe the cases in which the rate determining step is the reaction of alkene with the X_1 complex (hydroperoxide–catalyst) and that of hydroperoxide with the X_2 complex (alkene–catalyst), respectively. Networks 1.3.1 and 1.3.2 differently include the formation of the ternary complex X_9 (peroxide-catalyst–alkene). Steps of formation the ternary complex X_{14} (epoxide-catalyst–alcohol) are added to networks 1.4.1 and 1.4.2.

The addition of inactive intermediates to six basic networks in all possible combinations produces thousands of networks depicted by KGs with pendant vertexes. Twelve such KGs with 1–12 pendant vertexes are generated for network 1.2.1 (Table 2.11). By replacing the pendant vertexes by specific complexes a total of 2047 networks result. One can generate in much similar manner all networks with pendant vertexes for the remaining five basic networks from Table 2.10.

Two-Route Mechanisms. As shown in (Section 2.2.2), there are only three classes **A**, **B**, and **C** of two-route networks. However, a bridging step that connects the two routes (cycles) in class **A** (Figure 2.24) is an equilibrium step. This reduces class **A** to class **B**, therefore, only cases related to classes **B** and **C** will be considered.

Thirteen such two-route networks devoid of equilibrium steps (KGs without pendant vertexes) can be generated for the alkene epoxidation reaction proceeding from six basic single-route networks (Table 2.12). The completeness of the generation was checked by a comparison with the standard Table 2.6. The networks of class **B** are assembled from single-route ones by sharing a common KG vertex (e.g., vertex X_0 for networks 1.2.1 and 1.2.2; 1.3.2 and 1.4.2, etc.), whereas those of class **C**, by sharing a graph edge (e.g., 1.2.1 and 1.3.1; 1.2.2 and 1.4.2. etc.) or two edges (e.g., edges 1 and 9 are common for networks 1.3.1 and 1.4.1; edges 2 and 12 are common for networks 1.3.2 and 1.4.2 ; and edges 20, 21 are common for networks 1.4.1 and 1.4.2).

The last column of Table 2.12 contains the linear codes of the networks generated. It can be seen that for three pairs of networks the code is degenerate. As a matter of fact, the two networks in each of these pairs have identical topology. These isomorphic networks differ only in labeling of some KG vertexes and edges, i.e., by the nature of some intermediates and reaction steps they involve.

The last KG in the table is the only representative of the C_2 subclass of networks, in which two routes share two common reaction steps, 20 and 21. The two other cases mentioned above with sharing pairs of steps 1,9 and 2,12 do not actually belong to this subclass but to the standard **C** (or C_1) class. This is due to the requirements for a canonical KG representation according to which graphs should be presented by their minimal cycles. As exemplified below in Figure 2.37, in combining single-route networks 1.3.1 and 1.4.1, the KG-I obtained has cycles with three and four vertexes, whereas its canonical representation KG-II has two three-membered cycles.

An inspection of the two-route networks in Table 2.12 indicates that most of the cases of class **C** contain the competitive transformation of one and the same intermediate into reaction products with or without involving other intermediates (e.g., $X_1 \rightarrow X_0$ and $X_1 \rightarrow X_9 \rightarrow X_{14} \rightarrow X_0$ in 2.4.3). Other networks seem less probable from the kinetic standpoint, although they are preserved for completeness.

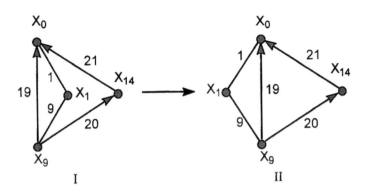

Figure 2.37. The two-route network **I**, resulting from the combining of the single-route networks 1.3.1 and 1.4.1 by sharing steps 1 and 9, is transformed into its canonical representation **II** which has minimal cycles ($X_0X_1X_9$ and $X_0X_9X_{14}$).

Table 2.8. Possible complexes between the catalyst, reactants, and reaction products ([Alk] = alkene, [Ep] = epoxide) [Reprinted with permission from *Appl. Catal.* 1992, *88*, 1–22 © 1992 Elsevier Science Publishers B.V.]

No.	Symbol	Complex	No.	Symbol	Complex
1	X_1	$X_0[ROOH]$	8	X_8	$X_0[ROH]_2$
2	X_2	$X_0[Alk]$	9	X_9	$X_0[ROOH][Alk]$
3	X_3	$X_0[Ep]$	10	X_{10}	$X_0[ROOH][Ep]$
4	X_4	$X_0[ROH]$	11	X_{11}	$X_0[ROOH][ROH]$
5	X_5	$X_0[ROOH]_2$	12	X_{12}	$X_0[Alk][Ep]$
6	X_6	$X_0[Alk]_2$	13	X_{13}	$X_0[Alk][ROH]$
7	X_7	$X_0[Ep]_2$	14	X_{14}	$X_0[Ep][ROH]$

Table 2.9. Possible reaction steps in the alkene epoxidation reaction (Reprinted with permission from *Appl. Catal.* 1992, *88*, 1–22 © 1992 Elsevier Science Publishers B.V.)

No.	Reaction step	No.	Reaction step
1	$X_0 + ROOH \rightleftharpoons X_1$	12	$X_2 + ROOH \rightleftharpoons X_9$
2	$X_0 + Alk \rightleftharpoons X_2$	13	$X_0 + Alk + Ep \rightleftharpoons X_{12}$
3	$X_0 + Ep \rightleftharpoons X_3$	14	$X_0 + Alk + ROH \rightleftharpoons X_{13}$
4	$X_0 + ROH \rightleftharpoons X_4$	15	$X_0 + Ep + ROH \rightleftharpoons X_{14}$
5	$X_0 + 2ROOH \rightleftharpoons X_5$	16	$X_6 + ROOH \longrightarrow X_9 + Alk$
6	$X_0 + 2Alk \rightleftharpoons X_6$	17	$X_1 + Alk \longrightarrow X_0 + Ep + ROH$
7	$X_0 + 2Ep \rightleftharpoons X_7$	18	$X_2 + ROOH \longrightarrow X_0 + Ep + ROH$
8	$X_0 + 2ROH \rightleftharpoons X_8$	19	$X_9 \longrightarrow X_0 + Ep + ROH$
9	$X_0 + Alk \rightleftharpoons X_9$	20	$X_9 \longrightarrow X_{14}$
10	$X_0 + ROOH + Ep \rightleftharpoons X_{10}$	21	$X_{14} \longrightarrow X_0 + Ep + ROH$
11	$X_0 + ROOH + ROH \rightleftharpoons X_{11}$	22	$X_9 + ROOH \longrightarrow X_1 + Ep + ROOH$

Table 2.10. Generated single-route networks for alkene epoxidation (Reprinted with permission from *Appl. Catal.* **1992**, *88*, 1–22 © 1992 Elsevier Science Publishers B.V.)

Network numbering	Networks*	Kinetic graphs*
1.2.1	(1) $X_0 + ROOH \rightleftharpoons X_1$ (17) $X_1 + Alk \longrightarrow X_0 + Ep + ROH$	$X_0 \overset{1}{\underset{17}{\rightleftharpoons}} X_1$
1.2.2	(2) $X_0 + Alk \rightleftharpoons X_2$ (18) $X_2 + ROOH \longrightarrow X_0 + Ep + ROH$	$X_0 \overset{2}{\underset{18}{\rightleftharpoons}} X_2$
1.3.1	(1) $X_0 + ROOH \rightleftharpoons X_1$ (9) $X_1 + Alk \rightleftharpoons X_9$ (19) $X_9 \longrightarrow X_0 + Ep + ROH$	triangle: X_0 —1→ X_1; X_0 —19→ X_9; X_9 —9→ X_1
1.3.2	(2) $X_0 + Alk \rightleftharpoons X_2$ (12) $X_2 + ROOH \rightleftharpoons X_9$ (19) $X_9 \longrightarrow X_0 + Ep + ROH$	triangle: X_0 —2→ X_2; X_0 —19→ X_9; X_9 —12→ X_2
1.4.1	(1) $X_0 + ROOH \rightleftharpoons X_1$ (9) $X_1 + Alk \rightleftharpoons X_9$ (20) $X_9 \longrightarrow X_{14}$ (21) $X_{14} \longrightarrow X_0 + Ep + ROH$	square: X_0 —1→ X_1; X_1 —9→ X_9; X_9 —20→ X_{14}; X_{14} —21→ X_0
1.4.2	(2) $X_0 + Alk \rightleftharpoons X_2$ (12) $X_2 + ROOH \rightleftharpoons X_9$ (20) $X_9 \longrightarrow X_{14}$ (21) $X_{14} \longrightarrow X_0 + Ep + ROH$	square: X_0 —2→ X_2; X_2 —12→ X_9; X_9 —20→ X_{14}; X_{14} —21→ X_0

* The notation for intermediates and the reaction step numbering is as in Table 2.8.

At the next stage of the procedure, 13 basic two-route networks of the catalytic alkene epoxidation can be combined with numerous equilibrium steps producing again thousands of kinetic graphs with pendant vertexes. Those of them having the maximum number of pendant vertexes are shown in Table 2.13.

The generation procedure generally produces networks with more than two routes. Thus, for example, the same three-route network can be designed for the reaction under study by combining the pairs of single-route mechanisms; e.g mechanisms 1.3.1 and 1.4.2 (Figure 2.38).

Table 2.11. Total number of networks for the basic network 1.2.1 (Reprinted with permission from *Appl. Catal.* 1992, *88*, 1–22 © 1992 Elsevier Science Publishers B.V.)

No.	Kinetic graph	Number of networks
KG1		12
KG2		66
KG3		220
KG4		495
KG5		792
KG6		924
KG7		792
KG8		495
KG9		220

Table 2.11. (*Continued*)

No.	Kinetic graph	Number of networks
KG10		66
KG11		12
KG12		1

Such networks are not discussed here, because they are hardly possible from a kinetic standpoint. Indeed, direct transformation of an intermediate into a reaction product and three indirect transformations involving another intermediate occurring simultaneously within the single system seem unlikely. On the other hand, we do have the aim to present here an exhaustive generation of all theoretically possible networks but rather to illustrate our computer-assisted methodology. We suppose that combining the direct network generation from a library (or database) of reaction intermediates and elementary steps with the search in our standard tables of networks with a given number of routes and intermediates, plus the elimination of the kinetically low probable networks, is an effective approach to mechanistic studies in chemical kinetics.

1.3.1 + 1.4.2:

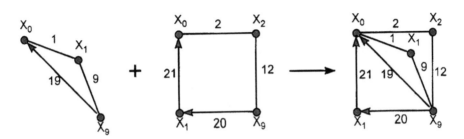

Figure 2.38. Example of a three-route network for the catalytic alkene epoxidation, obtained by combining two single-route networks: 1.3.1 and 1.4.2.

Table 2.12. Kinetic graphs of all two-route networks generated from the single-route mechanisms given in Table 2.10

Kinetic pairs of single-route networks	Resulting two-route networks	Resulting kinetic graphs	Linear code
1.2.1 + 1.2.2	2.3.1		**2-3-B-2,2-eiei**
1.2.1 + 1.3.2	2.4.1		**2-4-B-2,3-eie^{2}i**
1.2.2 + 1.3.1	2.4.2		**2-4-B-2,4-eie^{2}i**
1.2.1 + 1.4.2	2.5.1		**2-5-B-2,5-eie^{2}i^2**
1.2.2 + 1.4.1	2.5.2		**2-5-B-2,5-eie^{2}i^2**
1.2.1 + 1.3.1	2.3.2		**2-3-C-2,3-eiîe**

Table 2.12. (Continued)

Kinetic pairs of single-route networks	Resulting two-route networks	Resulting kinetic graphs	Linear code
1.2.2 + 1.3.2	2.3.3		2-3-C-2,3-eiîe
1.2.1 + 1.4.1	2.4.3		2-4-C-2,4-eiî^{2}e
1.2.2 + 2.4.4	2.4.4		2-4-C-2,4-i^{2}î^{2}e
1.3.1 + 1.3.2	2.4.5		2-4-C-3,3-ie^4
1.3.1 + 1.4.1	2.4.6		2-4-C-3,3-ie^{2}i^2

Table 2.12. (*Continued*)

Kinetic pairs of single-route networks	Resulting two-route networks	Resulting kinetic graphs	Linear code
1.3.2 + 1.4.2	2.4.7		2-4-C-3,3-ie^{2}i^2
1.4.1 + 1.4.2	2.5.3		2-5-C$_2$-4,4-i^{2}e^4

2.2.4. Graphical Methods for the Derivation of Rate Laws

The methods for the derivation of steady-state and pseudo-steady-state rate laws of linear mechanisms based on the graph theory were discussed in a number of review papers and textbooks.[126,170–179] In this section, we provide readers with the discussion and analysis of the most important ideas and algorithms used in the derivation of rate laws. Several recent results are discussed in more detail.

Stoichiometric Relation in Reaction Mechanisms. It is common knowledge that the mass action law, the principle of detailed equilibrium ($K = k_+/k_-$), and all derivative principles are valid only if one is concerned with elementary steps. The stoichiometric equation of a complex (multistep) reaction describes only balance of elements. For example, the equation

$$\tfrac{1}{2}SO_2 + CO \underset{k_{-1}}{\overset{k_{+1}}{\rightleftharpoons}} \tfrac{1}{4}S_2 + CO_2 \tag{2.24}$$

does **not** imply that the rate can be expressed according to the mass action law:

$$r = k_{+1} P_{SO_2}^{\frac{1}{2}} P_{CO} - k_{-1} P_{S_2}^{\frac{1}{4}} P_{CO_2}. \tag{2.25}$$

Table 2.13. Kinetic graphs with the maximum number of pendant vertexes used to depict the two-route networks of the catalytic alkene epoxidation (Reprinted with permission from *Appl. Catal.* 1992, *88*, 1–22 © 1992 Elsevier Science Publishers B.V.)

Network numbering	Kinetic graphs
2.3.1	
2.3.2	
2.3.3	
2.4.1	

Table 2.13. (*Continued*)

Network numbering	Kinetic graphs
2.4.2	
2.4.3	
2.4.4	
2.4.5	

Table 2.13. (*Continued*)

Network numbering	Kinetic graphs
2.4.6	
2.4.7	
2.5.1	
2.5.2	

Table 2.13. (*Continued*)

Network numbering	Kinetic graphs
2.5.3	

Equation (2.24) can be multiplied by two, four, or any other integer to produce equations like

$$SO_2 + 2CO \;\rightleftharpoons\; \tfrac{1}{2}S_2 + 2CO_2, \qquad (2.24')$$

$$2SO_2 + 4CO \;\rightleftharpoons\; S_2 + 4CO_2, \text{ etc.} \qquad (2.24'')$$

Equations (2.24) and equations (2.24'), (2.24''), etc., are equally correct unless one of them is proved to match the actual mechanism. At the same time, none of them can be used to write the expression for the reaction rate based on the mass action law.

Consider an intriguing example of a reversible catalytic reaction, which was discussed by Yablonskii:[171]

$$A \;\rightleftharpoons\; B. \qquad (2.26)$$

Suppose that the mechanism of reaction (2.26) is as follows:

$$A + Z_1 \;\underset{k_{-1}}{\overset{k_{+1}}{\rightleftharpoons}}\; Z_2, \qquad (2.27)$$

$$Z_2 + A \;\underset{k_{-2}}{\overset{k_{+2}}{\rightleftharpoons}}\; Z_3 + B, \qquad (2.28)$$

$$Z_3 \;\underset{k_{-3}}{\overset{k_{+3}}{\rightleftharpoons}}\; Z_1 + B, \qquad (2.29)$$

where Z_1, Z_2, and Z_3 are intermediate species. Upon adding up equations (2.27)–(2.29), the stoichiometric equation of the overall reaction becomes

$$2A \rightleftarrows 2B. \qquad (2.30)$$

It is equation (2.30) which corresponds to mechanism (2.27)–(2.29), although, from the stoichiometry standpoint, equations (2.26) and (2.30) are equally correct.

On the assumption that

$$[A], [B] \gg [Z]_\Sigma \sim [Z_1] \gg [Z_2], [Z_3],$$

the steady-state rate law can be written as follows:

$$r = r_+ - r_- = -\frac{1}{2}\frac{d[A]}{dt} = \frac{1}{2}\frac{d[B]}{dt} = \frac{(k_{+1}k_{+2}k_{+3}[A]^2 - k_{-1}k_{-2}k_{-3}[B]^2)[Z]_\Sigma}{k_{+2}k_{+3}[A] + k_{-1}k_{+3} + k_{-1}k_{-2}[B]}. \qquad (2.31)$$

To express the rate of the reverse reaction in terms of the rate of the forward reaction, one should determine the equilibrium constant K derived using equation (2.30) rather than (2.26): $K = [A]^2/[B]^2$. Note that the numerator of equation (2.31) also corresponds to the stoichiometry of equation (2.30): $k_+[A]^2 - k_-[B]^2$, where k_+ and k_- are some constants.

It is clear from this example that to derive the steady-state rate law of the overall equation, it is necessary first to elucidate the correct stoichiometry of a mechanism (i.e., correct stoichiometric numbers of steps). This problem was solved by Horiuti, who proposed the so-called *theory of reaction routes*.[180–182] More recently, this theory was revisited by M.I. Temkin, who proposed the *theory of steady-state reactions*.[183–185]

For an arbitrary complex linear mechanism, all intermediates to be eliminated in the overall equation, each step should be multiplied by a *stoichiometric number* γ_i. The set of stoichiometric numbers of steps is termed a *reaction route*, because this set indicates which steps and to which extent are involved on the pathway from reagents to products. Sometimes, elementary steps that correspond to nonzero reaction routes are also called a reaction route, although in this sense this term is somewhat arbitrary. The number of the overall equations corresponding to different reaction routes is infinite, because the number of sets of stoichiometric numbers is also infinite. To describe the reaction kinetics, it is sufficient to obtain one of the possible sets of linearly independent reaction routes. As it was mentioned in Section 2.1, the multiroute linear mechanisms can be described by polycyclic reaction networks (kinetic graphs). In kinetic graphs, the number of linearly independent cycles (the cyclomatic number ϕ) is equal to the number of linearly independent reaction routes M, which can be determined by the Horiuti–Temkin rule

$$M = s - I, \qquad (2.32)$$

where s is the number of steps in the mechanism; I is the number of linearly independent intermediates (or, in other words, the rank of the $\mathbf{B}_X$ matrix):

$$M = s - \text{rank}(\mathbf{B}_X). \qquad (2.32')$$

The matrix $\mathbf{B_X}$ of the stoichiometric coefficients of intermediates has s rows and I columns. Its entries b_{ji} represent the stoichiometric coefficients of the ith intermediate in the jth elementary step. Their signs are taken "+" and "–" for intermediates from the right-hand side and left-hand side of the step equation, respectively.

Intermediates

Steps	X_1	X_2	X_3	...	X_I
1	b_{11}	b_{12}	b_{13}	...	b_{1I}
2	b_{21}	b_{22}	b_{23}	...	b_{2I}
3	b_{31}	b_{32}	b_{33}	...	b_{3I}
...	...	...	...	...	...
s	b_{s1}	b_{s2}	b_{s3}	...	b_{sI}

$$\mathbf{B_X} = \qquad \qquad \qquad \qquad \qquad \qquad \qquad \qquad (2.33)$$

Equation (2.32) for the number of linearly independent reaction routes can be compared to the Euler's formula for the cyclomatic number ϕ (the number of independent cycles) of a graph

$$\phi = q - p + c, \qquad\qquad (2.34)$$

where q and p are the number of edges and vertexes, respectively; c is the number of graph components. In a kinetic graph, the number of linearly independent reaction routes is equal to the cyclomatic number ($M = \phi$); the number of steps is equal to the number of edges ($s = q$); the total number of intermediates plus 0-species (if any) is equal to the number of vertexes p; the number of components is always 1. Then, equations (2.32′) and (2.34) taken together can be rewritten as follows:

$$s - \text{rank}(\mathbf{B_X}) = s - p + 1.$$

Then,

$$p = \text{rank}(\mathbf{B_X}) + 1 = I + 1.$$

In the case of a noncatalytic reaction, this equation implies that all intermediates are linearly independent, and the total number of vertexes is equal to the number intermediates plus 0-species (trivial result). In the case of a catalytic reaction, the number of vertexes is equal to the total number of intermediates, which is greater by one than the number of linearly independent intermediates. That is, a material balance of a catalyst does exist, and there is a linear dependence relation acting on the set of intermediates (catalyst entities).

The sets of stoichiometric numbers γ_i can be found from the linear algebraic equations:

$$\sum_j b_{ji}\gamma_i = 0; \quad i = 1,\dots,I \tag{2.35}$$

or, more generally, from the matrix equation

$$\mathbf{B_X}^T \Gamma = 0, \tag{2.36}$$

where $\mathbf{B_X}^T$ is the transposed stoichiometric matrix for the linearly independent intermediates, and Γ is the matrix of the stoichiometric numbers γ of elementary steps.

Each independent route from the set of M independent routes (columns of the matrix Γ) has its overall equation. Stoichiometric (overall) equations of routes can be derived from the expression:

$$\mathbf{B_M} = \Gamma^T \mathbf{B_N}, \tag{2.37}$$

where $\mathbf{B_M}$ is the stoichiometric matrix of the overall equations and $\mathbf{B_N}$ is the s by N submatrix of the stoichiometric matrix for nonintermediate (initial and final) species. The stoichiometric matrix $\mathbf{B}$ can be obtained by the concatenation of $\mathbf{B_N}$ and $\mathbf{B_X}$.

Obviously, $\mathrm{rank}(\mathbf{B_M}) \leq M$, because not necessarily all overall equations of routes are linearly independent. In this case, a basis of routes can be found such that part of overall equations will be $0 = 0$. Overall equations of this sort are ordinary routes and correspond to cycles in kinetic graphs as well. They are distinct in that intermediates are transformed to each other without the consumption of reagents and formation of products.

Note that the number of linearly independent overall equations $Q^* = \mathrm{rank}(\mathbf{B_M})$ is less than or equal to the value Q_{max} determined by the Gibbs relation:[186]

$$Q^* = \mathrm{rank}(\mathbf{B_M}) \leq Q_{max} = N - \mathrm{rank}(\mathbf{H}), \tag{2.38}$$

where N is the total number of reagents and products, $\mathbf{H}$ is the N by I atomic matrix of reagents and products; I is the number of different chemical elements from which reagents and products are constructed.

Consider the mechanism of coal conversion by water vapor, simplified by assuming that all reaction steps are reversible:

$$(1) \; C + H_2O + Z \rightleftarrows H_2 + COZ,$$

$$(2) \; COZ \rightleftarrows CO + Z, \tag{2.39}$$

$$(3) \; COZ + CO \rightleftarrows CO_2 + C + Z.$$

Z is an active site on the coal surface, and COZ is a chemisorbed CO. The linear network containing two intermediates and two independent reaction routes (KG cycles) is shown in Figure 2.39.

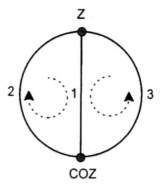

Figure 2.39. The kinetic graph depicting the mechanism of coal conversion by water vapors, equations (2.39). There are only two independent reaction routes. (Reprinted with permission from Temkin, O. N.; Bonchev, D. G. Application of Graph Theory to Chemical Kinetics. Part 1. Kinetics of Complex Reactions. *J. Chem. Educ.* **1992,** *69*, 544–550 © 1994 American Chemical Society.)

The sets of stoichiometric numbers for the network under consideration can be found by applying equation (2.36). Only one of the two intermediates is independent and the $\mathbf{B}_X$ matrix is reduced to a row-vector of stoichiometric coefficients

$$\mathbf{b}_{si} = [1 \quad -1 \quad -1].\tag{2.40}$$

This vector is multiplied by the column-vector of stoichiometric numbers of the steps:

$$\begin{Vmatrix} \gamma_1 \\ \gamma_2 \\ \gamma_3 \end{Vmatrix}.$$

Taking the vector product equal to zero, one arrives at

$$\gamma_1 = \gamma_2 + \gamma_3.\tag{2.41}$$

Selecting integers that satisfy equation (2.41), one can construct several sets of γ_s vectors; for example, the following four vectors can be obtained:

$$\gamma_s^{I} = [1 \quad 1 \quad 0],$$

$$\gamma_s^{II} = [1 \quad 0 \quad 1],$$

$$\gamma_s^{III} = \gamma_s^{I} - \gamma_s^{II} = [0 \quad 1 \quad -1],$$

$$\gamma_s^{IV} = \gamma_s^{I} + \gamma_s^{II} = [2 \quad 1 \quad 1].$$

They define four reaction routes and respective overall equations:

(I) $C + H_2O \rightleftharpoons H_2 + CO,$ (2.42)

(II) $H_2O + CO \rightleftharpoons H_2 + CO_2,$ (2.43)

(III) $CO_2 + C \rightleftharpoons 2CO,$ (2.44)

(IV) $C + 2H_2O \rightleftharpoons 2H_2 + CO_2.$ (2.45)

These four routes can be traced in Figure 2.39. Routes I and II correspond to the left and right cycle, respectively (steps 1 and 2 and steps 1 and 3). Route III is the enveloping cycle which includes step 2 in the forward direction and step 3 in the reverse direction. Route IV is a combined traverse along steps 1, 2, 1, and 3. Only two of these four cycles are independent. This follows from equation (2.32), which indicates that $M = 3 - 1 = 2$ (three steps minus one independent intermediate). Hence, one can find the Γ matrix of the stoichiometric numbers:

$$\Gamma = \begin{Vmatrix} 1 & 1 \\ 1 & 0 \\ 0 & 1 \end{Vmatrix}. \tag{2.46}$$

Indeed, this choice of the route basis is not unique. Any other pair of cycles in Figure 2.39, as well as any other pair of γ_s sets (or stoichiometric equations), is equally valid. In this example, $Q_{max} = 2$, $Q^* = 2$ and $M = 2$.

Within the method of reaction routes, an important quantity is the rate r_m of a route m. This is the number of turnovers per unit time in a unit volume. In fact, the rate of a route is the rate corresponding to the overall equation under steady-state or pseudo-steady-state conditions.

Note also that the rates of routes are related to the rates of steps by the Horiuti–Temkin equation:[175,185]

$$w_i = \sum_{m=1}^{M} \gamma_i^{m} r_m; \quad i = 1,...,I; \tag{2.47}$$

$$\mathbf{w} = \Gamma \times \mathbf{r}, \tag{2.48}$$

where r_m is the rate of route m, w is the column-vector of the rates of steps w ($w_i = w_{+i} - w_{-i}$ is the rate of ith step), and r is the column-vector of the rates of routes.

The system (2.47) incorporates s equations and the same number unknown rates of routes and concentration of linearly independent intermediates. This fact makes it possible to derive the steady-state rate law for the linear mechanism.

Basic Concepts of Graph Theory Applied in Chemical Kinetics. The problem of the derivation of steady-state rate laws is reduced to solving the system of linear inhomogeneous algebraic equations in which unknown are concentrations of linearly independent intermediates. Equations of this sort can be solved using Cramer rule:

$$x_i = d_i / d,$$

where x_i is the unknown steady-state concentration of an intermediate X_i; d is the determinant made up of the coefficients of the unknowns of the system; and d_i is the determinant obtained by replacing the ith column in d by a column of constant terms.

When the concentration x_i is obtained, one can substitute it where appropriate into the rate law of the elementary step. Thus, one can arrive at the formula for the rate of a product formation or the rate of a route.

As was mentioned above, King and Altman were first to apply the graphical method for the derivation of determinants d and d_i and expressions for x_i. The general rule allowing one to use graphs for solving the problems associated with a linear law $y = ax$ was formulated by Mason in 1955.[176] More recently, Mason and Zimmerman used this rule for the solution of Kirchhoff equations in the theory of electric networks (y is the potential difference, x is the current, and a is the resistance).[187,188] The essence of this rule can be expressed as follows:

$$\frac{x_i}{x_j} = \frac{D_i}{D_j}. \tag{2.49}$$

If applied to chemical kinetics, formula (2.49) relates concentrations x_i and x_j of intermediates expressed as vertexes of a kinetic graphs. D_i and D_j are determinants of respective graph vertexes. For deeper understanding of an analogy between linear algebraic equations and linear networks, a reader may compare formulas (2.49) and Cramer rule.

Vol'kenshtein and Gol'dshtein were first to apply this rule to chemical kinetics.[174,189–191] Yablonskii and Evstigneev proved that this rule is applicable to multiroute mechanisms (polycyclic kinetic graphs).[192]

Consider a kinetic graph of the mechanism (2.27)–(2.29). Each edge of this graph represents a reversible step (Figure 2.40a). This is a standard form of a kinetic graph. This kinetic graph can be redrawn so that an edge is replaced by two arcs having opposite directions each representing one of the unidirectional elementary reactions (see Figure 2.40b). Arcs 1, 2, 3 and −1, −2, −3 represent the forward and reverse direction of a reaction, respectively.

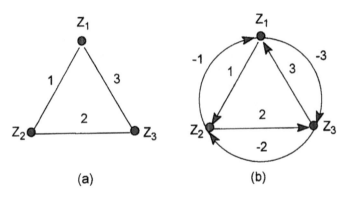

Figure 2.40. The kinetic graph of mechanism (2.27)–(2.29) in standard (a) and expanded (b) forms.

Let us introduce several terms.[193] Consider a graph $G(V(G), E(G))$ and its subgraph $H(V(H), E(H))$, where V and E are the set of vertexes and the set of edges of a graph, respectively. H is a spanning subgraph of G, if $V(G) = V(H)$ and $E(H) \subset E(G)$. If H is a *spanning subgraph* of G and a tree at a time, then H is termed a *spanning tree* of G. Removal of edge L from the graph G has the two possible results: (1) The new graph $G - \{L\}$ has two components each containing one of the ends of the edge L; that is, L is an *isthmus* (1-bridge). These components are termed *end-graphs of edge L* in a graph $G - \{L\}$; (2) $G - \{L\}$ is a connected subgraph. If all edges of a graph G are 1-bridges, then G is a *tree*.

Suppose there is a tree T having a distinguished vertex r. Let us define a digraph $\Theta(V(\Theta), W(\Theta))$, where $V(\Theta)$ is the set of edges of Θ, and $W(\Theta)$ is the set of directed edges (arcs) of a digraph Θ. T is called a *rooted tree* with root r, if there is a digraph Θ such that $V(T) = V(\Theta)$ and $W(\Theta) = E(T)$ and directions of arcs in Θ match the following rule: if the edge L in a graph T induce end-graphs K and Λ, and Λ contains a root r, then L has direction from the end in graph Λ to the end in graph K.

When T is itself a digraph, then T is termed a directed tree having root r, if direction of all arcs match the above rule.

Theorem:[193] *Let r be a vertex of a digraph G, and Φ is a spanning tree of G. Vertex r is a root of Φ if*

    1. Φ does not contains cycles (no matter regular or not);

    2. $id(r) = 0$;

    3. $id(v_i) = 1, i = 1...n, v_i \neq r$.

This theorem can be used to define a spanning tree of a digraph. Note that, according to the theorem, the tree Φ is growing from its root r. We may choose opposite direction of growth preserving the same root by saying that a tree is growing into its root. In this case indegrees should be replaced by outdegrees in the condition of the theorem. Precisely this meaning of the spanning rooted tree will be used in our further discussion. Rooted spanning trees of a digraph shown in Figure 2.40b are collected in Figure 2.41. When the graph has too many edges (arcs), one can use formal methods to generate spanning trees with assistance of computers. In the 1960s and 1970s these methods were the subject of intensive studies because of their numerous applications in different branches of science.[194-197]

The weight ω_i of the ith arc is the ratio of the rate w_i of ith step and the concentration of respective intermediate species. For example, weights of arcs of the graph shown in Figure 2.40 are as follows:

$$\omega_{+1} = \frac{w_{+1}}{[Z_1]} = \frac{k_{+1}[A][Z_1]}{[Z_1]} = k_{+1}[A], \tag{2.50}$$

$$\omega_{-1} = \frac{w_{-1}}{[Z_2]} = \frac{k_{-1}[Z_2]}{[Z_2]} = k_{-1}, \tag{2.51}$$

$$\omega_{+2} = \frac{w_{+2}}{[Z_2]} = \frac{k_{+2}[A][Z_2]}{[Z_2]} = k_{+2}[A], \tag{2.52}$$

$$\omega_{-2} = \frac{w_{-2}}{[Z_3]} = \frac{k_{-2}[B][Z_3]}{[Z_3]} = k_{-2}[B], \tag{2.53}$$

$$\omega_{+3} = \frac{w_{+3}}{[Z_3]} = \frac{k_{+3}[Z_3]}{[Z_3]} = k_{+3}, \tag{2.54}$$

$$\omega_{-3} = \frac{w_{-3}}{[Z_1]} = \frac{k_{-3}[B][Z_1]}{[Z_1]} = k_{-3}[B]. \tag{2.55}$$

The reaction weight is also termed a reaction frequency or probability and was first proposed by Christiansen.[1]

The product of weights of arcs incorporated in the kth spanning tree having a root in ith vertex T_{ki} is termed a weight of a tree D_{ik}:

$$D_{ik} = \prod_{j \in T_{ki}} \omega_j. \tag{2.56}$$

The determinant D_i of the vertex i is the sum of weights over all spanning trees having a root i:

$$D_i = \sum_k D_{ik}. \tag{2.57}$$

Several methods to derive the value of determinant D_i are known from the chemical literature.[172,175,176,198-201] The simplest among these methods is the method of Orsi.[199] This algorithm is as follows (mechanism (2.27)–(2.29) is used as an illustration):

Step 1. Write the sum of weights of all arcs that outgo from each vertex:

Z_1: $\omega_{+1} + \omega_{-3}$, the 1st sum;
Z_2: $\omega_{-1} + \omega_{+2}$, the 2nd sum;
Z_3: $\omega_{-2} + \omega_{+3}$, the 3rd sum.

Step 2. Derive $D_i{}^*$ by taking the product of sums of step 1 so that the *i*th sum is excluded from the *i*th product, while others are preserved:

$$D_1{}^* = (\omega_{-1} + \omega_{+2})(\omega_{-2} + \omega_{+3}) = \omega_{-1}\omega_{-2} + \omega_{+2}\omega_{-2} + \omega_{-1}\omega_{+3} + \omega_{+2}\omega_{+3};$$
$$D_2{}^* = (\omega_{-2} + \omega_{+3})(\omega_{+1} + \omega_{-3}) = \omega_{+1}\omega_{-2} + \omega_{-3}\omega_{-2} + \omega_{+1}\omega_{+3} + \omega_{-3}\omega_{+3};$$
$$D_3{}^* = (\omega_{-1} + \omega_{+2})(\omega_{+1} + \omega_{-3}) = \omega_{+1}\omega_{-1} + \omega_{-3}\omega_{-1} + \omega_{+1}\omega_{+2} + \omega_{-3}\omega_{+2}.$$

Step 3. In products of step 2, delete the terms that consist of weights of a circuit (in a trivial case, these are weights of forward reverse elementary steps):

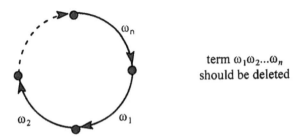

term $\omega_1\omega_2...\omega_n$
should be deleted

In the case under consideration one should delete terms $\omega_{+1}\omega_{-1}$, $\omega_{+2}\omega_{-2}$, and $\omega_{-3}\omega_{+3}$:

$$D_1 = \omega_{-1}\omega_{-2} + \omega_{-1}\omega_{+3} + \omega_{+2}\omega_{+3},$$
$$D_2 = \omega_{+1}\omega_{-2} + \omega_{-3}\omega_{-2} + \omega_{+1}\omega_{+3},$$
$$D_3 = \omega_{-3}\omega_{-1} + \omega_{+1}\omega_{+2} + \omega_{-3}\omega_{+2}.$$

Chou Kuo-Chen *et al.* proposed the method which makes it possible to exclude step 3 from the above algorithm.[201] The algorithm contains several rules to test the number of subgraphs that induce D_i.

Formula (2.49) expresses the concentration of any intermediate species through the concentration of a catalyst Z_1, the concentration of a 0-species (which is equal to unity), or the total concentration of a catalyst. Note that, in the case of a heterogeneous catalysis, the total catalyst concentration is equal to the total concentration of active sites on the catalyst surface, which also equals unity. The related expressions are given below:

$$[X_i] = [Z_1]\frac{D_i}{D_{Z_1}}, \qquad (2.58)$$

$$[X_i] = [X]_\Sigma \frac{D_i}{\sum_i D_i}, \tag{2.59}$$

where $[X]_\Sigma$ is the total concentration of all catalyst entities. If $[X_i]$ is the portion of the catalyst surface occupied by species X_i, the total catalyst concentration is $[X]_\Sigma = 1$. In the case of a noncatalytic reaction, one can derive the following expression:

$$[X_i] = D_i / D_0 \tag{2.60}$$

where D_0 is a determinant of a 0-species.

If a kinetic graph contains pendant vertexes, then edges incident with them depict equilibrium steps within steady-state or pseudo-steady-state processes. In a graph containing pendant vertexes, one can specify a function F_i associated with a vertex i, which characterize the extent to which a catalyst is bound by ligands or species. This function is defined by the formula:[202]

$$F_i = 1 + \sum_s \left(\omega_s / \omega_{-s}\right), \tag{2.61}$$

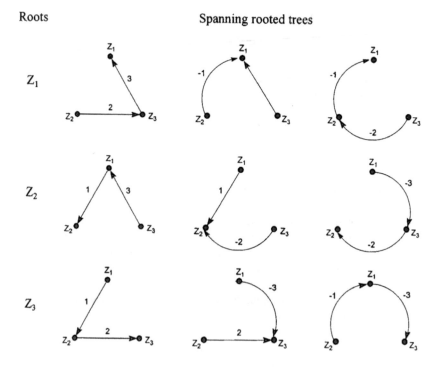

Figure 2.41. Spanning rooted trees of three vertexes of the kinetic graph shown in Figure 2.40.

where the sum is taken over all pendant vertexes s adjacent to the vertex i. The formula (2.59) can be modified in terms of the formula (2.61) as follows:

$$[X_j] = [X]_\Sigma \frac{D_i}{\sum_i F_i D_i}. \tag{2.62}$$

Thus, the kinetic graph shown in Figure 2.20 has two pendant vertexes. Functions F of vertexes Z_1 and Z_2, which are adjacent to these pendant vertexes, are $F_1 = 1 + K_1[B]$ and $F_2 = 1 + K_2[D]$, respectively. If $[Z]_\Sigma = 1$, then

$$[Z_3] = \frac{D_{Z_3}}{D_{Z_1}F_1 + D_{Z_2}F_2 + D_{Z_3}},$$

$$[Z_1] = \frac{D_{Z_1}}{D_{Z_1}F_1 + D_{Z_2}F_2 + D_{Z_3}}.$$

The rate of reaction having the mechanism depicted by the kinetic graph (Figure 2.20) can be written as

$$r = w_{+3} - w_{-3} = \omega_{+3}[Z_3] - \omega_{-3}[Z_1],$$

or

$$r = \frac{\omega_{+1}\omega_{+2}\omega_{+3} - \omega_{-1}\omega_{-2}\omega_{-3}}{D_{Z_1}F_1 + D_{Z_2}F_2 + D_{Z_3}}. \tag{2.63}$$

If $[X_j]$ in formula (2.62) is a concentration of a "free" catalyst (Z_1) the following value can be viewed as a reciprocal value of the free catalyst steady-state concentration normalized to the total catalyst concentration:

$$F_{st} = \frac{[X]_\Sigma}{[X_j]} = \frac{\sum_i F_i D_i}{D_j}. \tag{2.64}$$

Similar values are used for the description of equilibrium or pseudoequilibrium processes with a rate-limiting step. F_{st} is the ratio between the denominator of the formula for the reaction rate (2.63) and the polynomial D_j, which is the determinant of the jth vertex. Obviously, if the process approaches equilibrium (the case of an reversible process), F_{st} transforms to the reciprocal normalized concentration of a catalyst at the equilibrium point:

$$F_{eq} = \frac{[X]_\Sigma}{[\tilde{X}_j]_{eq}}. \tag{2.65}$$

Using expression (2.65), one can derive several interesting formulas, which find applications in chemical kinetics:[203-205]

$$\frac{[\tilde{X}_j]_{eq}}{[X_j]} = \frac{F_{st}}{F_{eq}} = \frac{\sum_i F_i D_i}{D_j F_{eq}}. \tag{2.66}$$

If the process approaches equilibrium, and $[\tilde{X}_j]_{eq}/[X_j] \to 1$, then:

$$\sum_i F_i \tilde{D}_i = \tilde{D}_j F_{eq}, \tag{2.67}$$

where $\tilde{D}_i$ and $\tilde{D}_j$ are determinants of ith and jth vertexes that involve equilibrium or pseudoequilibrium concentrations of reagents (and products). Equation (2.67) relates the kinetics and thermodynamics of complexation reactions and shows how coefficients of a rate law are related to the observed equilibrium constants and which of the complexes of rate constants should be associated with the equilibrium constants.

Algorithms for the Derivation of Rate Laws. Let us discuss two algorithms for the derivation of rate laws, which are, in our opinion, most useful. In the above, we already used one of them to derive equation (2.63). This algorithm is an elegant combination of graph-theoretical methods and methods of the theory of reaction routes.[206]

Proceeding from the Horiuti–Temkin equation (2.48), we can specify an elementary step (or several steps) that allow one to find the rate of route m. Suppose that this step has a serial number s and the stoichiometric number γ_s. Then,

$$r_m = w_s = w_{+s} - w_{-s}.$$

For instance, in the case of a kinetic graph shown in Figure 2.39, one can derive the rates of routes I and II that correspond to minimal cycles of this graph:

$$r_I = w_{+2} - w_{-2},$$

$$r_{II} = w_{+3} - w_{-3}.$$

For step s, we can write

$$w_s = w_{+s} - w_{-s} = \omega_{+s}[X_i] - \omega_{-s}[X_j],$$

where X_i and X_j are intermediates that take part in the forward and reverse reactions s. If the reaction is catalytic, one can use formula (2.59). Therefore,

$$r_m = [X]_\Sigma (\omega_{+s} D_i - \omega_{-s} D_j) / \sum_i D_i. \tag{2.68}$$

For a heterogeneous catalytic reaction, $[X]_\Sigma = 1$. For a noncatalytic reaction,

$$[X_i] = [X_0] D_i / D_0 \text{ and } [X_j] = [X_0] D_j / D_0.$$

Because $[X_0] = 1$,

$$r_m = [X]_\Sigma (\omega_{+s}D_i - \omega_{-s}D_j) / D_0, \qquad (2.69)$$

where D_0 is the determinant of the 0-species vertex in the kinetic graph.

The rates of reagent consumption and product formation can be expressed in terms of the rates of routes according to the following equation:

$$\mathbf{r}_N = \mathbf{B}_M^T \times \mathbf{r}_m, \qquad (2.70)$$

where $\mathbf{B}_M^T$ is the transposed stoichiometric matrix of overall equations of routes. This equation is useful when there are few reaction routes and many reaction products.

The second algorithm was proposed by Vol'kenshtein and Gol'dshtein[189–191] and modified by Yablonskii et al.[171,179,207] This algorithm uses the equation of the rate of a step and equations (2.48) and (2.70), which allow one to derive the rate laws of routes and the rates of reagent consumption and product formation. In the general case, this algorithm makes it possible to derive the rate of any step of a multiroute reaction. The rate w_s can be expressed by the following formula:

$$w_s = \frac{[X]_\Sigma}{\sum_i D_i} \sum_{n=1}^{k} \left(C_n^+ - C_n^- \right) D_n , \qquad (2.71)$$

where C_n is the weight of nth cycle, which includes the sth step. It equals the product of weights of arcs that are included in the cycle (plus and minus signs denote forward and reverse directions). D_n is the determinant of the subgraph obtained by shrinking the nth cycle into a single vertex and rooted at this vertex. The number of cycles that incorporate the sth step is k.

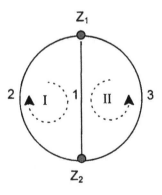

Figure 2.42. The kinetic graph of a reversible reaction: three steps, one linearly independent intermediate, and two independent reaction routes.

According to the Horiuti–Temkin rule, the steady-state rate of a single-route reaction (monocyclic kinetic graph) is equal to the rate of any step:

$$\gamma_s r = w_s = w_{+s} - w_{-s}.$$

If $\gamma_s = 1$, then for each step,

$$w_s = \frac{[X]_\Sigma}{\sum_i D_i} \sum_{n=1}^{k} \left(C^+ - C^- \right). \qquad (2.72)$$

If the graph of a catalytic mechanism has pendant vertexes, equation (2.71) should be rewritten in terms of equation (2.62):

$$w_s = \frac{[X]_\Sigma}{\sum_i F_i D_i} \sum_{n=1}^{k} \left(C_n^+ - C_n^- \right) D_n. \qquad (2.73)$$

In solutions of metal complexes, the catalyst is bound by ligands, even if reagents and products are absent. In this case, pendant vertexes, which correspond to the catalyst species bound with ligands, are adjacent to the vertex that correspond to an active site.

As an example, consider the mechanism corresponding to the kinetic graph shown in Figure 2.42. This graph is similar to that in Figure 2.39 for which a detailed mechanism was presented. Figure 2.42 shows the same graph, but in slightly more general form. Directions of routes and respective overall reactions are chosen. The mechanism contains three reversible steps and one linearly independent intermediate ($[Z_1] + [Z_2] = [Z]_\Sigma$). The number of independent routes is $M = 3 - 1 = 2$. The matrix of stoichiometric numbers can be chosen as follows (columns are numbered similarly as routes):

$$\Gamma = \begin{Vmatrix} 1 & 1 \\ 1 & 0 \\ 0 & 1 \end{Vmatrix}.$$

According to the Horiuti–Temkin rule, $w_1 = r_1 + r_{II}$, $w_2 = r_1$, $w_3 = r_{II}$. Thus, the rates of routes I and II are equal to the rates of steps 2 and 3, respectively.

According to the first algorithm,

$$r_1 = w_2 = w_{+2} - w_{-2} = \omega_{+2}[Z_2] - \omega_{-2}[Z_1] = \frac{(\omega_{+2} D_{Z_2} - \omega_{-2} D_{Z_1})[Z]_\Sigma}{\sum_i D_i}.$$

Determinants of vertexes Z_1 and Z_2 contain three spanning trees each, because the weights of edges connecting two vertexes should be added up:

$$D_{Z_1} = \omega_{+2} + \omega_{-1} + \omega_{+3},$$

$$D_{Z_2} = \omega_{-2} + \omega_{+1} + \omega_{-3}.$$

Then,

$$r_1 = [\omega_{+2}(\omega_{+2} + \omega_{-1} + \omega_{+3}) - \omega_{-2}(\omega_{-2} + \omega_{+1} + \omega_{-3})]\frac{[Z]_\Sigma}{\sum_i D_i} =$$

$$\frac{\omega_{+2}(\omega_{+1} + \omega_{-3}) - \omega_{-2}(\omega_{-1} + \omega_{+3})}{(\omega_{+2} + \omega_{-1} + \omega_{+3}) + (\omega_{-2} + \omega_{+1} + \omega_{-3})}[Z]_\Sigma.$$

(2.74)

According to the second algorithm, one should take into account that $r_1 = w_2$ and that step 2 is incorporated in two cycles: $C_1\langle 1,2\rangle$ and $C_2\langle 2,3\rangle$. It is necessary to substitute the following expressions into equation (2.75) for the rate w_2: $C_1^+ = \omega_{+1}\omega_{+2}$; $C_1^- = \omega_{-1}\omega_{-2}$; $C_2^+ = \omega_{+2}\omega_{-3}$; $C_2^- = \omega_{-2}\omega_{+3}$; and $D_1 = D_2 = 1$ to obtain

$$r_1 = w_2 = [(\omega_{+1}\omega_{+2} - \omega_{-1}\omega_{-2}) - (\omega_{+2}\omega_{-3} - \omega_{-2}\omega_{+3})]\frac{[Z]_\Sigma}{\sum_i D_i}.$$

(2.75)

Clearly, these two algorithms lead to the same result.

In the case of a single-route reaction, the cyclic characteristics of the process $C = C^+ - C^-$ corresponds to the same mass action law, which is applied to the resulting stoichiometric equation in the same manner as to a mechanism.[171,179,207] For instance, for the reaction 2A = 2B, which is discussed above, the characteristics C can be obtained by the formula

$$C = k_{+1}k_{+2}k_{-3}[A]^2 - k_{-1}k_{-2}k_{-3}[B]^2 = k_+[A]^2 - k_-[B]^2.$$

(2.76)

The denominator $\sum_i D_i$ points to the multistep nature of the process and to the mass balance of intermediates. In the case of a multiroute reaction, the mass action law, which is applied to the overall equation, corresponds to the cyclic characteristics C_1.

On the assumption that $[Z]_\Sigma = 1$, the rate law of route I for the mechanism of the coal conversion (Figure 2.39) is as follows:

$$r_1 = \frac{(k_{+1}k_{+2}[H_2O] - k_{-1}k_{-2}[H_2][CO]) - (k_{+2}k_{-3}[CO_2] - k_{-2}k_{+3}[CO]^2)}{k_{+2} + k_{-1}[H_2] + (k_{+3} + k_{-2})[CO] + k_{+1}[H_2O] + k_{-3}[CO_2]}.$$

Methods of graph theory facilitates the derivation of rate laws of processes with a rate-limiting steps. In the case of a reaction with a rate-limiting step, edges of a kinetic graph represent pseudoequilibrium steps. If the jth elementary reaction is

$$R + X_j \underset{k_{-j}}{\overset{k_{+j}}{\rightleftarrows}} P + X_{j+1}$$

where R is a reagent, P is a product, and X_j and X_{j+1} are intermediates, the weight of the jth edge in a kinetic graph should be taken as the following ratio:

$$\frac{\omega_j}{\omega_{-j}} = \frac{k_j[R]}{k_{-j}[P]} = K_j \frac{[R]}{[P]},$$

where K_j is the equilibrium constant. The product of weights of edges on the path from the ith vertex to the sth vertex is denoted D_{is}. The quantity D_{is} is termed the determinant of the sth vertex ($D_s = D_{is}$). The determinant of the 0-species vertex (noncatalytic reaction) or the vertex corresponding to the initial catalytic species (catalytic reaction) is taken to be unity: $D = 1$ ($D_{ii} = 1$).

Consider an enzymatic reaction with the rate-limiting step 3 and two inhibited catalytic species (the latter correspond to pendant vertexes in a kinetic graph:[173]

The kinetic graph of this mechanism is shown in Figure 2.43. The rate of the overall reaction

$$A + B \longrightarrow P$$

can be expressed in terms of the rate of the rate-limiting step:

$$r = \frac{D_{EAB}\omega_3[E]_0}{\sum_i D_i F_i} = \frac{D_{EAB}\omega_3[E]_0}{D_{EE} F_E + D_{EA} F_{EA} + D_{EAB}}. \tag{2.77}$$

Determinants of vertexes can be expanded into the products of appropriate edge weights:

$$D_E = D_{E \to E} = 1, \tag{2.78}$$

$$D_{EA} = D_{E \to EA} = K_1[A], \tag{2.79}$$

$$D_{EAB} = D_{E \to EA \to EAB} = K_1 K_2[A][B], \tag{2.80}$$

$$D_{EI} = D_{E \to EI} = K_4[I], \tag{2.81}$$

$$D_{EIA} = D_{E \to EA \to EIA} = K_1 K_5 [A][I].$$ (2.82)

The values of F_i are as follows:

$$F_{EA} = 1 + K_5[I],$$ (2.83)

$$F_E = 1 + K_4[I].$$ (2.84)

Here we used the formation (equilibrium) constants for complexes, while the dissociation constants can be used as well.[173]

Upon substitution of expressions (2.78)–(2.84) into formula (2.77), we arrive at the formula for the reaction rate:

$$r = \frac{k_3 K_1 K_2 [A][B][E]_0}{1 + K_4[I] + (1 + K_5[I]) K_1 [A] + K_1 K_2 [A][B]}.$$ (2.85)

As can be seen, if the mechanism contains a rate-limiting step, one can avoid the use of values of F_i. Instead, the sum $\sum_s D_{is}$ for all graph vertexes (including pendant vertexes) can be used:

$$r = \frac{D_{EAB} \omega_3 [E]_0}{\sum_s D_{is}} =$$

$$\frac{k_3 K_1 K_2 [A][B][E]_0}{1 + K_1[A] + K_1 K_2 [A][B] + K_4[I] + K_1 K_5 [A][I]}.$$ (2.86)

The presence of a rate-limiting step makes it possible to simplify the graph of a mechanism. Thus, the kinetic graph of the enzymatic reaction shown in Figure 2.43a can be reduced to the graph shown in Figure 2.43b. This graph can serve as a basis for derivation of the rate law using the expression similar to formula (2.77):

$$r = \frac{D_{EAB} k_3 [E]_0}{D_E + D_{EA} + D_{EI} + D_{EIA} + D_{EAB}} =$$

$$\frac{k_3 K_{1'} [A][B][E]_0}{1 + K_{4'}[I] + K_{1'}[A][B] + K_{2'}[A] + K_{5'}[A][I]}.$$ (2.87)

Obviously, equilibrium constants in equation (2.87) are related to those in equation (2.86): $K_{1'} = K_1 K_2$; $K_{2'} = K_1$; $K_{4'} = K_4$; and $K_{5'} = K_1 K_5$. Another variant of the algorithm for the derivation of the rate laws for mechanistic schemes of this sort was discussed in Berezin and Klesov's textbook.[173] This approach can be applied not only to enzymatic reactions, but to any catalytic reaction having a linear mechanism. Vol'kenshtein and Gol'dshtein showed also that the method

based on the formalism of kinetic graphs can be extended to the study of presta-
tionary phase of complex reactions having linear mechanisms[208] and showed that
this method can be applied using several simplifying assumptions.[209]

Evstigneev and Yablonskii applied the methods of graph theory to the derivation of characteristic polynomials of kinetic models.[171,207] The characteristic
polynomial of a kinetic model was shown to play an important role in analysis
of relaxation times. The discussion of problems related to the application of
characteristic polynomial fall outside the scope of this book.

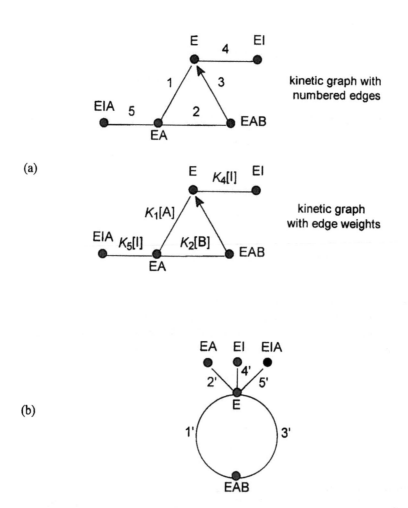

(a)

(b)

Figure 2.43. (a) The kinetic graph of an enzymatic reaction having a rate-limiting step and two inhibited catalyst entities; and the (b) simplified kinetic graph of an enzymatic reaction.

2.3. Nonlinear Reaction Networks

2.3.1. Elementary Reactions in Terms of Stoichiometry

Nonlinear mechanisms of chemical reactions are very common in current chemical studies. They are characteristic of many catalytic reactions and branched chain reactions.[210,211] Nonlinear elementary steps play a key role in understanding the critical phenomena in chemical systems (multiplicity of steady states, oscillations, relaxation behavior, etc.)[52,55,207]

Discussion in this section is based on the assumption that an elementary step of a chemical reaction consists of two reverse elementary reactions. Each of these elementary reactions is either monomolecular or bimolecular. Termolecular reactions are also possible, although they are far less probable. If the rate constant of an elementary reaction, which is one of two parts of an elementary step, is close to zero the step is said to be irreversible (or more precisely, unidirectional). Thus, if species $A_1...A_n$ are involved in the reaction mechanism, the ith elementary reaction can be written as follows:

$$a_{i1}A_1 + a_{i2}A_2 + ...a_{in}A_n \rightarrow d_{i1}A_1 + d_{i2}A_2 + ... + d_{in}A_n, \qquad (2.88)$$

or in more compact form:

$$\sum_{k=1}^{n} a_{ik}A_k \rightarrow \sum_{k=1}^{n} d_{ik}A_k. \qquad (2.89)$$

Then,

$$\sum_{k=1}^{n} a_{ik} \leq 2,$$
$$\sum_{k=1}^{n} b_{ik} \leq 2. \qquad (2.90)$$

In realistic elementary reactions, a species is usually included either on the left-hand or on the right-hand side; i.e., if $a_{ij} \neq 0$, then $d_{ij} = 0$ and *vice versa*. However, there are few reactions in which species can be involved on both sides. For instance, the following reactions taken from different fields of chemistry are known as degenerate rearrangements and contain the same species on both sides:

$$Fe^{2+}L_6 + Fe^{3+}L_6 \rightleftharpoons Fe^{3+}L_6 + Fe^{2+}L_6,$$

$$ROOH + {}^{\cdot}OOR \rightleftharpoons ROO^{\cdot} + HOOR,$$

When products are identical to the initial species, they can be distinguished by isotope labeling. A more exotic type of elementary reaction is as follows:[212]

$$Cl^- + CH_2=C=CHCH_2Cl \rightarrow CH_2=C(Cl)CH=CH_2 + Cl^-.$$

This reaction is not degenerate, but it occurs via external nucleophilic attack at C2 position, while another chlorine anion dissociates at C4 position.

Papers dedicated to stability analysis of reaction networks often deal with elementary steps in which the same species enters both sides of a chemical equation with nonzero coefficients a_{ij} and d_{ij} so that $d_{ij} > a_{ij}$. For instance,

$$A_1 + A_2 \rightarrow A_3 + 2A_2.$$

Steps of this sort were termed *autocatalytic*. Such oscillatory mechanisms as Oregonator and Brusselator contain autocatalytic steps and, therefore, are unrealistic, although these steps are possible as pseudoelementary. In further discussion, we will exclude autocatalytic steps from the consideration.

On the basis of the above assumptions, the complete list of possible stoichiometries of elementary reactions can be written as follows:[91]

(1) $A \rightarrow C$,

(2) $2A \rightarrow 2C$,

(3) $2A \rightarrow C + D$,

(4) $A + B \rightarrow 2C$,

(5) $A + B \rightarrow C + D$,

(6) $A \rightarrow C + D$,

(7) $A \rightarrow 2C$,

(8) $A + B \rightarrow C$,

(9) $2A \rightarrow C$.

Using bipartite graphs, these steps can be depicted as shown in Figure 2.44.

2.3.2. Exact Determination of Nonlinear Networks

Although bipartite graphs were already discussed in Section 2.1, we provide the reader with more precise definitions of several terms in this section.[213]

Definition 1. *A reaction network (RN) is a bipartite graph (BG). A vertex set of this bipartite graph $\mathcal{V}(BG)$ is partitioned into two proper mutually complementary subsets $\mathcal{V}(BG) = \mathcal{U}(BG) \cup \mathcal{W}(BG)$. Each species is put into correspon-*

*dence with a vertex w, w ∈ **W**(BG), and each elementary reaction is put into cor-*
*respondence with a vertex u, u ∈ **U**(BG). Vertex w_j is adjacent to u_i if the jth spe-*
*cies undergoes the ith elementary reaction, and vertex w_j is adjacent **from** u_i if the*
jth species is generated by the ith elementary reaction. Arcs of the RN can be mul-
tiple. The multiplicity of the arc e_{ij} equals to the stoichiometric coefficient of the jth
species in the ith elementary reaction.

Thus, it immediately follows from this definition that: (1) The maximum
number of arcs joining two vertexes is 2, because the molecularity of each elementary reaction is no greater than two; (2) The reaction network is loop-free; (3) No

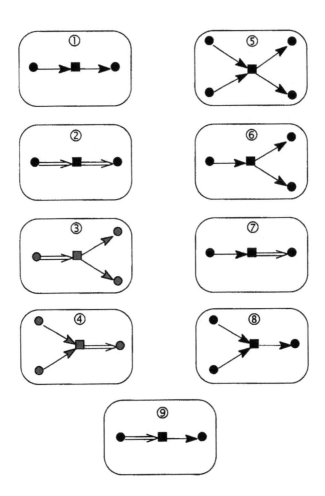

Figure 2.44. Graphical representation of elementary reactions.

arcs join vertexes from the same subset, because a graph is bipartite; (4) All cycles of the reaction network (if any) are of even length irrespective whether they are regular or not (this property follows from the König theorem); (5) If two arcs join the same pair of vertexes, they have the same direction. Reversible elementary steps are represented as two elementary reactions having opposite directions.[214]

Proposition 1. *The reaction network is a connected graph.*

Suppose u_i is a vertex belonging to the reaction network. Let the subset $\hat{W}_i'$ ($\hat{W}_i' \subset \hat{W}$) be the set of vertexes adjacent from the u_i vertex, and let the subset $\hat{W}_i''$ ($\hat{W}_i'' \subset \hat{W}$) be the set of vertexes adjacent to the u_i vertex. Then, we can write the following propositions:

Proposition 2. *In the reaction network, no two vertexes u_q and u_r can exist such that $\hat{W}_q' = \hat{W}_r'$ and $\hat{W}_q'' = \hat{W}_r''$.*

Proposition 3. *If the reaction network contains two vertexes u_q and u_r such that $\hat{W}_q' = \hat{W}_r''$ and $\hat{W}_q'' = \hat{W}_r'$, the respective reactions are reverse and represent the same reversible elementary step.*

Because the reaction network is a graph, several binary matrixes can be associated with it.

Definition 2. *Let $\mathcal{U}(RN)$ be a set of N vertexes of a reaction network: $N = n + m$, $n = |\mathcal{W}(RN)|$ and $m = |\mathcal{U}(RN)|$. A binary $N \times N$ matrix $\mathbf{A}(RN)$ is said to be an adjacency matrix of a reaction network if an element of a matrix a_{ij} is equal to the number of arcs that begin at the vertex v_i and end at vertex v_j.*

Thus, elements of the adjacency matrix can be 0, 1, or 2. Taking into account the above properties of the reaction network, we can write an adjacency matrix in the block form:[214]

$$
\mathbf{A}(RN) = \begin{array}{c|c|c} & \begin{matrix} w_1 & \cdots & w_n \end{matrix} & \begin{matrix} u_1 & \cdots & u_m \end{matrix} \\ \hline \begin{matrix} w_1 \\ \cdots \\ w_n \end{matrix} & \mathbf{0} & \mathbf{A}_{wu} \\ \hline \begin{matrix} u_1 \\ \cdots \\ u_m \end{matrix} & \mathbf{A}_{uw} & \mathbf{0} \end{array}
$$

Proposition 4. *Two reaction networks are said to be isomorphic when their adjacency matrixes can be derived from each other by the row/column permutations (in the sense that permutations of rows and columns are similar).*

Note that two different reactions may occur via mechanisms whose reaction networks are isomorphic. This makes it possible to regard reaction mechanisms and their networks as combinatorial objects.

Because the reaction network is bipartite and vertex bipartition is fixed, one can assign another matrix with the reaction network, which is derived from $\mathbf{A}(RN)$.

Definition 3. *Let* $n = |\mathcal{W}(RN)|$ *and* $m = |\mathcal{U}(RN)|$. *A binary* $m \times n$ *matrix* $\mathbf{B}(RN) = \mathbf{A}_{uw} - \mathbf{A}_{wu}^{\mathrm{T}}$ *is said to be a reduced adjacency matrix of a reaction network.*[91]

The reduced adjacency matrix, thus defined, is also called a stoichiometric matrix of the reaction mechanism.[213] An element b_{ij} of the stoichiometric matrix $\mathbf{B}(RN)$ is equal to the stoichiometric coefficient of the jth species in the ith elementary reaction. Thus, a reaction

$$a_{i1}\mathrm{A}_1 + a_{i2}\mathrm{A}_2 + \ldots a_{in}\mathrm{A}_n \rightarrow d_{i1}\mathrm{A}_1 + d_{i2}\mathrm{A}_2 + \ldots + d_{in}\mathrm{A}_n$$

can be rewritten as follows:

$$0 \rightarrow (d_{i1} - a_{i1})\mathrm{A}_1 + (d_{i2} - a_{i2})\mathrm{A}_2 + \ldots + (d_{in} - a_{in})\mathrm{A}_n.$$

In this expression, all terms are placed on the right-hand side. Then, the stoichiometric coefficient of the kth term is $b_{ik} = d_{ik} - a_{ik}$:

$$0 \rightarrow b_{i1}\mathrm{A}_1 + b_{i2}\mathrm{A}_2 + \ldots + b_{in}\mathrm{A}_n.$$

Proposition 5. *Two reaction networks are said to be isomorphic when their reduced adjacency matrixes can be derived from each other by the row and column permutations (not necessarily similar).*

A controlling role in the dynamic behavior of a chemical reaction belongs to intermediates.[52] Therefore, it is often important to derive a bipartite graph of the reaction mechanism from which vertexes corresponding to nonintermediate species are removed. If the overall reaction equation is known in advance, all species involved in this equation are either reactants or products. They were termed *terminal species*.[210,211] In Clarke's terms, these species are *external*. Other species that are involved in elementary reactions but not in the overall equation are termed *intermediate species* (*internal* species[52]). The overall reaction equation is generally unknown. Moreover, several overall equations can exist for a given reaction mechanism, because a mechanism may contain several routes, each producing a different overall equation. No mathematically rigorous definition for the term *intermediate* exists. So, to develop the theory of reaction networks, we should first unambiguously define this term and thus answer the question, *Which species are intermediates and which are not?*.

In principle, several definitions of this term may be extracted from the literature, but there is no complete agreement between them. Below is a list of some typical examples, which illustrates that these definitions often may be partially or completely incorrect.

1. Suppose that a steady-state reaction occurs in the open continuous well-stirred tank reactor. The species is said to be terminal if its concentration remains constant at the expense of entering this species to (or expelling from) the reactor. The intermediate species is that for which the constant concentration is maintained by its consumption in some elementary reactions that compensate for the generation of this species by some other elementary reactions.[184]

2. The *terminal species* is an observed species, and *intermediate* is species that is not observed.[171]

3. An *intermediate* is a species not involved in each possible overall equation corresponding to a reaction route.[216,217]

4. *Initial species* are supplied to the chemical system and undergo further chemical transformations. *Reaction products* are those generated by chemical transformations and undergoing no further transformations. *Intermediate species* are produced by chemical transformations in the course of the chemical reaction and undergo further transformations.[218]

All of these definitions are either completely or partially incorrect. First, it is unreasonable to define *intermediate* through the *steady-state* reaction mode or any other kinetic modes of a reaction. A particular species may be intermediate or terminal no matter whether the reaction is steady-state or not and regardless of what kinetic behavior is observed. Second, the question, *Which species are observed?* is answered depending on what stage the reaction is stopped and how one performs observations. Third, the *intermediate* cannot be defined using such concepts as the overall equation, the stoichiometric number, the reaction route, etc., because these are themselves defined through the term *intermediate*. Finally, when defining the term *intermediate*, it should be taken into account that among the reactants that are supplied to the reactor or observed among the products, some can belong to intermediates. Examples of catalytic reactions given in Chapter 3 are very illustrative in this case: a catalyst (as well as all terminal species) is supplied to the reaction system and undergoes several transformations to give different intermediate catalytic species. After each turnover it returns to its initial state. Thus, the catalyst is among those reactants which are supplied to the reactor and plays the role of an intermediate at a time. We propose the definition based on the concept of a reaction network.

Proposition 6. *If a w-vertex in the reaction network has either zero out degree (od(w) = 0) or indegree (id(w) = 0), the corresponding species is terminal.*

If one uses this proposition to define the terminal species, all species involved in reversible elementary steps will appear to be intermediates. However, it is chemically incorrect. To avoid this disagreement, we assume that if a species is to be termed an intermediate only because it is involved in a reversible step, it should be regarded as terminal. That is, if a participant of the two reverse reactions is not involved in any other reactions, it is also should be regarded as terminal.[211]

Definition 4. *An ith species is terminal if one of the following conditions is met:*

1. *$od(w_i) = 0$.*
2. *$id(w_i) = 0$.*
3. *There are two vertexes that correspond to reverse reactions such that one is adjacent to w_i while the other is adjacent from w_i, and the deletion of one of these two vertexes makes one of the above conditions valid.*

All species other than terminal are intermediate.

Definition 5. *A bipartite graph that is obtained from the reaction network by the removal of w-vertexes corresponding to terminal species is said to be a subnetwork of intermediates (SI).*

Proposition 7. *The subnetwork of intermediates of the reaction network is a connected graph. Otherwise, the set of elementary reactions represent several disjoint chemical processes occurring in the same system.*

Definition 6. *The reduced adjacency matrix of the subnetwork of intermediates $\mathbf{B}(SI)$ is a stoichiometric matrix of intermediates.*

Strictly speaking, an arbitrary set of reactions can be represented by the reaction network, which in turn can be reduced to the subnetwork of intermediates. This subnetwork of intermediates is not necessarily a connected graph. Hereafter, this kind of sets of reactions will not be regarded as mechanisms.

Intermediate summary. *Thus, the reaction network is a connected graph whose vertexes are partitioned into two nonoverlapping subsets. Arcs of this graph are directed and, in the general case, multiple. No greater than two arcs can join a pair of vertexes. Two arcs that join a pair of vertexes should have the same direction. The reaction network does not contain loops. The subnetwork of intermediates of the reaction network is a connected graph. Each u-vertex is unique in the sense that no two u-vertexes have similar adjacencies.*

2.3.3. Stoichiometric Relationships in Nonlinear Reaction Networks

Suppose there is a set of reactions written as

$$0 \rightarrow b_{i1}A_1 + b_{i2}A_2 + \ldots + b_{in}A_m, \, i = 1\ldots m. \tag{2.91}$$

Let us replace the arrow in equation (2.91) with an equality sign and, then, arrange all intermediate species X_j, $j = 1...x$, before terminal species P_j, $j = (x + 1)...n$. All symbols A_j will be substituted by either X_j or P_j:

$$b_{i1}X_1 + b_{i2}A_2 + ... + b_{ix}X_x + b_{i(x+1)}P_{x+1} + b_{i(x+2)}A_{(x+2)} + ... + b_{in}P_n = 0, \quad (2.92)$$
$$i = 1...m.$$

The system (2.92) of linear homogeneous algebraic equations can be rewritten in the matrix form:

$$\mathbf{B}(RN) \times \mathbf{a} = 0 \tag{2.93}$$
$$(m \times n) \times (n \times 1) = (m \times 1),$$

where $\mathbf{a}(n \times 1)$, is a species column-vector. The stoichiometric matrix $\mathbf{B}(RN)$ and each vector $\mathbf{a}$ can be spliced into two blocks, one of which corresponds to intermediate species, while the other corresponds to terminal species:

$$\mathbf{B}(RN)\mathbf{a} = \left\| \mathbf{B}_X(RN) \mid \mathbf{B}_P(RN) \right\| \times \left\| \begin{matrix} \mathbf{x} \\ \mathbf{p} \end{matrix} \right\| = 0. \tag{2.94}$$

Then, we can write the following expression (recall that $\mathbf{B}_X(RN) \equiv \mathbf{B}(SI)$):

$$\mathbf{B}(SI)\mathbf{x} + \mathbf{B}_P(RN)\mathbf{p} = 0. \tag{2.95}$$

Definition 7. *The vector of stoichiometric numbers γ is a set of integers (not necessarily distinct) each of which can be assigned to a mechanistic step so that the sum of products of each equation of step and the respective integer gives the overall equation in which all intermediate species vanish.*

The vector of stoichiometric numbers is also called a *reaction route*. To use Definition 7, let us take the matrix product of γ^T and the above expression:

$$\gamma^T\mathbf{B}(SI)\mathbf{x} + \gamma^T\mathbf{B}_P(RN)\mathbf{p} = 0. \tag{2.96}$$

According to the definition,

$$\gamma^T\mathbf{B}(SI)\mathbf{x} = 0. \tag{2.97}$$

Then, we can simplify this expression by writing:

$$\gamma^T\mathbf{B}(SI) = 0 \quad \text{or} \quad \mathbf{B}(SI)^T\gamma = 0. \tag{2.98}$$

Stoichiometric numbers for reversible steps can be positive, negative, or zero integers. If a step is irreversible, the respective stoichiometric numbers must be nonnegative integers. Because in our notation all elementary reactions are written explicitly (i.e., not in the form of elementary steps), all stoichiometric numbers must be nonnegative integers. Equation (2.98) reflects the balance of intermediate species in the mechanism. This means that all intermediates formed in the course of the reaction should be consumed.

Definition 8. *Any subset of the set of mechanistic steps is called a submechanism of the reaction mechanism.*

The number of submechanisms is finite because submechanisms can be viewed as subgraphs of the reaction network in which u-vertexes preserve all their adjacencies. Among all possible submechanisms, there are those for which respective subsets of stoichiometric numbers can be found such that they contain only positive (nonzero) integers. Hereinafter, unless otherwise stated, we will discuss only submechanisms of this sort. Thus, the *subnetwork of intermediates of a submechanism (SIS)* is a graph such that there exists a vector γ that is orthogonal to the transposed matrix $\mathbf{B}(SIS)$

$$\mathbf{B}(SIS)^{\mathsf{T}}\gamma = 0 \qquad\qquad (2.99)$$

and all components of γ are positive integers.

Suppose we selected several elementary reactions from the mechanism and ensured that the above condition is met. Let us now forget about the rest of the reactions and analyze this submechanism as a mechanism in its own right. Two results of this analysis are possible: (1) This submechanism is further inseparable into simpler submechanisms; (2) The submechanism can be subdivided into several simpler ones. In the first case, we say that the *submechanism is simple.* Each simple submechanism has its own unique overall equation, which does not contain intermediate species. A mechanism that contains several submechanisms has several (similar or distinct) overall equations. Then, the following statement is critical for the further discussion.

Proposition 8. *The set of elementary reactions of the mechanism induce a finite set of simple submechanisms.*

This fact is important from the standpoint of mechanistic classification and coding. In the case of linear mechanisms, it was sufficient to use the finite set of linearly independent reaction routes. Indeed, the choice of routes for the basis can be carried out in a unified manner because of two facts: (1) The routes for the basis always can be chosen so that all stoichiometric numbers will be either zero or unity; (2) All submechanisms of linear mechanisms are cycles, i.e., they are identical graphical substructures of the kinetic graph. This serves as the basis for the classification.[89-93] In the case of nonlinear mechanisms, submechanisms correspond to the subgraphs of the reaction network having different structures and the choice of routes for the basis has many alternatives. Then, classification and coding techniques require the special procedure maintaining the canonical choice of routes for the basis. When solving this problem, it is reasonable to use the unique and finite set of simple submechanisms rather than an arbitrary, though also finite, set of independent (basis) routes.

The concept of a simple submechanism can be also defined through its properties:

Definition 9. *Suppose that a set of reactions contains s steps, and* $\mathbf{B}(SIS)$ *is the reduced adjacency (stoichiometric) matrix of intermediates for this set of reactions. Then, this set of reactions is a simple mechanism if the following conditions are met:*

(1) $s - \text{rank}[\mathbf{B}(SIS)] = 1.$

(2) *There exists a vector* γ *that is orthogonal to the transposed matrix* $\mathbf{B}(SIS)$: $\mathbf{B}(SIS)^{\mathrm{T}}\gamma = 0$ *and all components of the vector are nonzero integers.*

The idea of simple submechanisms resembles the idea of *direct routes* proposed by Happel and Sellers,[216,217,219] but they are different. Thus, for example, a direct route is found proceeding from the given overall equation; that is, intermediate and terminal species are known in advance. This overall equation can be multiple in the sense that it can match a combination of simple submechanisms. Another distinguishing point is that *cycles*[216,217,219] or *empty routes*,[184] which correspond to simple submechanisms with the overall equation $0 = 0$, are not treated in Happel and Sellers' approach as normal routes. Thus, if an overall equation is chosen so that all terminal species included in this equation match Definition 4 and if this equation is not multiple, the direct route is a simple submechanism.[220]

To illustrate the idea of a simple mechanism consider the following example of the platinum-catalyzed heterogeneous reaction of the nitrogen dioxide reduction by carbon monoxide. Two reaction routes were proposed for this reaction, which are the dissociative and bimolecular mechanisms.[221] The product of NO_2 reduction is nitrogen monoxide, which, in turn, can undergo the reduction to produce N_2:[222]

$$NO_2 + CO \rightarrow NO + CO_2, \qquad (2.100)$$

$$2NO + 2CO \rightarrow N_2 + 2CO_2. \qquad (2.101)$$

Platinum catalyzes reactions (2.100) and (2.101) under the same conditions and, for the second reaction, the dissociative and bimolecular mechanisms were also proposed.[222] Let us discuss the mechanism composed of the elementary reactions proposed in Refs. 220 and 221 (Z is an active site on the metal surface):

(1) $NO_2 + 2Z \rightarrow ZNO + ZO,$

(2) $ZNO \rightarrow NO + Z,$

(3) $CO + Z \rightarrow ZCO,$

(4) $ZCO + ZO \rightarrow 2Z + CO_2,$ \qquad (2.102)

(5) $NO_2 + Z \rightarrow ZNO_2,$

(6) $ZCO + ZNO_2 \rightarrow 2Z + NO + CO_2$,

(7) $ZCO + ZNO \rightarrow Z + ZN + CO_2$,

(8) $2ZN \rightarrow 2Z + N_2$,

(9) $ZNO + Z \rightarrow ZN + ZO$,

(10) $NO + Z \rightarrow ZNO$.

According to the Definition 4, NO_2, CO, CO_2, and N_2 are terminal species because the respective w-vertexes in the reaction network have either zero indegree or zero outdegree. NO is also a terminal species, but $id(w_{NO}) = 2$ and $od(w_{NO}) = 1$. Note, however that reactions (2) and (10) are reverse. If one deletes reaction (2), w_{NO} will become terminal and its outdegree turns zero. All other species are intermediates (Z, ZCO, ZO, ZNO_2, ZN, and ZNO). The reaction network and the subnetwork of intermediates for this mechanism are too complicated to be illustrative. Therefore, no figures are given for them. The number of independent routes M can be derived from the Horiuti–Temkin rule:

$$M = s - \text{rank}\mathbf{B}(SI) = 9 - 5 = 4,$$

where s is the number of elementary steps (not elementary reactions!). The number of simple submechanisms is greater than M. Thus, one can identify six simple submechanisms, which can be expressed by ordinal numbers of steps in the mechanism: $\langle 1,2,3,4 \rangle$, $\langle 3,5,6 \rangle$, $\langle 3,4,8,9,10 \rangle$, $\langle 3,7,8,10 \rangle$, $\langle 1,3,4,8,9 \rangle$, and $\langle 1,3,4,7, 8 \rangle$. The first two simple submechanisms were discussed as dissociative and bimolecular routes for the reaction $NO_2 \rightarrow NO$. The two next were discussed as dissociative and bimolecular routes for the reaction $NO \rightarrow N_2$, and the two mentioned last in this series were found using the formal procedure for the derivation of complete set of simple submechanisms (*vide infra*).

To illustrate that some submechanisms are not simple, consider the set of reactions $\langle 1,2,3,4,5,6 \rangle$, which can be separated into two simple submechanisms (the first two submechanisms in the above series). Subnetworks of intermediates of simple submechanisms (*SISS*) for the above set of reactions are shown in Figure 2.45. Note that equation $\mathbf{B}(SI)^{\mathsf{T}}\gamma = 0$ can be interpreted in terms of bipartite graphs which are derivative from subnetworks of intermediates. Consider the subnetwork of intermediates for the submechanism $\langle 3,7,8,10 \rangle$ (see Figure 2.45). The presence of a submechanism implies that there exists the vector of stoichiometric numbers that is orthogonal to $\mathbf{B}(SIS)^{\mathsf{T}}$. For instance, stoichiometric numbers for steps 3, 7, 8, and 10 in this submechanism are 2, 2, 1, and 2, respectively. Each of these numbers should be assigned to arcs incident with respective u-vertexes. Thus, the number of arcs incident with u_3, u_7, u_{10}, should be doubled, while the number of arcs incident with u_8 should remain the same. We arrive at the graph shown in Figure 2.46. Graphs of this sort, which are derived from SI of submechanisms, were termed *current diagrams*.[52] Note that after assigning the stoichiometric numbers to the arcs incident with u-vertexes, the indegree and outdegree of each w-vertex become equal:

$$\text{id}(w_i') = \text{od}(w_i'), \; \forall \, i, \; i = 1...k, \; k = |\boldsymbol{\mathcal{W}}(SI)|. \tag{2.103}$$

This expression is true of an arbitrary subnetwork of intermediates; however, in some instances, one or all stoichiometric numbers of steps must be zero. For simple submechanisms, these numbers should be positive integers.

Lemma. *If stoichiometric numbers are assigned to the arcs of the SISS incident with u-vertexes, which increase their multiplicity, indegrees and outdegrees of u-vertexes become related by the expression:*

$$\sum_{i=1}^{s} \text{od}(u_i') = \sum_{i=1}^{s} \text{id}(u_i'), \quad s = |\boldsymbol{\mathcal{U}}(SISS)|.$$

Proof. Using expression (2.103), we have

$$\text{id}(w_j') = \text{od}(w_j'), \; \forall \, j, \; j = 1...k, \; k = |\boldsymbol{\mathcal{W}}(SISS)|.$$

Then,

$$\sum_{j=1}^{k} \text{od}(w_j') = \sum_{j=1}^{k} \text{id}(w_j').$$

Because *SISS* is bipartite,

$$\sum_{i=1}^{s} \text{od}(u_i') = \sum_{j=1}^{k} \text{id}(w_j')$$

and

$$\sum_{j=1}^{k} \text{od}(w_j') = \sum_{i=1}^{s} \text{id}(u_i').$$

Thus we have

$$\sum_{j=1}^{k} \text{od}(w_j') = \sum_{i=1}^{s} \text{id}(u_i') = \sum_{i=1}^{s} \text{od}(u_i') = \sum_{j=1}^{k} \text{id}(w_j'),$$

and among other things, we proved

$$\sum_{i=1}^{l} \text{od}(u_i') = \sum_{i=1}^{l} \text{id}(u_i'). \; \square \tag{2.104}$$

Theorem. *If the simple submechanism involves at least one elementary reaction whose equation contains more intermediates in the left member than in the right, this submechanism must also contain the elementary reaction whose equation has more intermediates in the right than in the left and vice versa.*

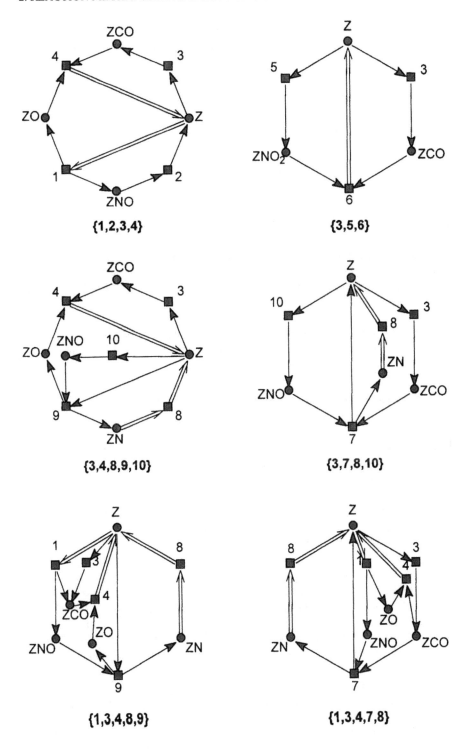

Figure 2.45. *SISS*s of the mechanism (2.102).

Proof. Suppose that *SISS* contains a u-vertex u_t for which $od(u_t) < id(u_t)$ and $od(u) \neq id(u)$. Then, let us apply formula (2.104) and place all terms on the left-hand side:

$$\sum_{i=1}^{s}[od(u_i') - id(u_i')] = 0.$$

Then, rewrite it as follows:

$$\gamma_t[od(u_t) - id(u_t)] + \sum_{\substack{i=1 \\ i \neq t}}^{l}\gamma_i[od(u_i) - id(u_i)] = 0. \qquad (2.105)$$

Because γ_i is a nonzero positive integer and because the sum in the above expression is zero, indegree and outdegree of u_t must be equal: $od(u_t) = id(u_t)$, which is contradictory to the above statement. If there are several vertexes like u_t, the sum in expression (2.105) still must be zero, and the first term is negative for all positive stoichiometric numbers γ_t. Therefore, expression (2.105) cannot be truth. The theorem is proved. $\square$

Note that if a set of reactions contains elementary reactions of both types (see the condition of the theorem), then it is not necessarily a simple submechanism.

2.3.4. The Algorithm and Computer Program GERM for the Search for Simple Submechanisms

The mechanism of a complex reaction can be described by the $m \times n$ $\mathbf{B}(RN)$ matrix in which rows stand for m elementary reactions and the $m \times k$ block corresponding to the subnetwork of intermediates $\mathbf{B}(SI)$. The algorithm can be subdivided into three blocks: (1) the generation of the t-subsets of an m-set,[223] (2) the calculation of the matrix rank, and (3) the selection of sets that do not contain given subsets. At the first stage, 2-subsets of the m-set are generated, which identify the numbered rows of the $\mathbf{B}(SI)$ matrix (elementary reactions). This is followed by the calculation of ranks for the submatrices of $\mathbf{B}(SI)$ that correspond to each combination. Combinations corresponding to the submatrices whose rank is less by one than the number of reactions are stored as solutions and in the "filter". At the second stage, 3-subsets of the m-set are generated. A 3-subset should be deleted if it contains a 2-subset stored in a filter. Then, the ranks of submatrices that remain after checking the content of the filter should be calculated. This is followed by finding the new selection of submatrices the rank of which is less by one than the number of steps. New solutions are stored and the filter is replenished. These operations are consecutively repeated until t becomes m. At this point, the generation of the t-subsets is completed and results of searches can be read. Because the search is conducted so that the number of steps is increased from iteration to iteration, unnecessary steps are not included in submechanisms. After 2-subsets are stored as solutions (or when all solutions are found) it is necessary to check whether all of them match simple submechanisms or part of them are pairs of reverse reactions. Note that part of "false" solutions, which consist of two reverse reactions, is known in advance. The number of reversible steps is $(m - s)$, where m is the number of elementary reactions and s is the number of elementary steps.

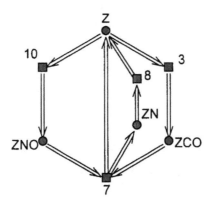

Figure 2.46. The current diagram for the submechanism $\langle 3,7,8,10 \rangle$.

This algorithm underlies the computer program GERM (Generic Ensembles of Reaction Mechanism) and can be used for a systematic search for simple submechanisms in the space of elementary reactions.

The total number of t-subsets generated increases with m very rapidly (in proportion to $2^m - m - 2$). Because of this, it is sometimes useful to modify the algorithm by introducing a special procedure facilitating the decrease in the number of elementary reactions. Sometimes it is possible to choose a pair of elementary reactions that involve at least one common intermediate not involved in any third reaction. In this pair of reactions, the intermediate is the product of one reaction and the reactant of the other reaction. Stoichiometric numbers for this pair should be chosen so that, in the sum of reaction equations, the intermediate vanishes. An ordinal number is assigned to the resulting reaction and the initial pair of reactions is further excluded from the consideration. Obviously, reactions from this pair enter any simple submechanism only together.

For instance, reactions 5 and 6 of the mechanism (2.102) have a common intermediate, ZNO_2, that does not participate in any other reaction. A new overall reaction can be written as

$$(5\text{-}6) \quad NO_2 + ZCO \rightarrow Z + NO + CO_2.$$

As can be seen, reactions 5 and 6 are both involved in the simple submechanism $\langle 3,5,6 \rangle$ and do not enter all other routes.

The representation of the mechanism by the subnetwork of intermediates facilitates the procedure of simplifying the reaction network. In the subnetwork of intermediates of the overall mechanism, one should find two adjacent u- and w-vertexes, each having indegree and outdegree equal to unity. Then, one should delete both vertexes and add an arc instead. This procedure is shown in Figure 2.47.

A more general method to simplify the reaction network was proposed by Clarke.[224] Clarke's algorithm is very simple and can be easily adapted to match the purpose of the search for simple submechanisms. Let us illustrate it by example of mechanism (2.102). Suppose that in addition to intermediate ZNO_2 we wish to exclude intermediate ZNO:

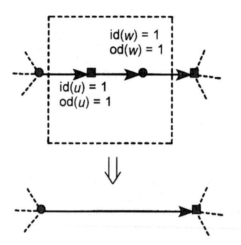

Figure 2.47. The procedure of simplifying the reaction network while preserving the number and structure of simple submechanisms. Part of the subnetwork of intermediates to be removed is put in the area with a dashed boundary.

Step 1.   Construct the subnetwork of intermediates containing only intermediates (w-vertexes) to be removed. The subnetwork should contain
only reactions (u-vertexes) that involve these intermediates. In the
case under consideration the subnetwork contains vertex w_{ZNO} and
vertexes u_1, u_2, u_7, u_9, and u_{10}:

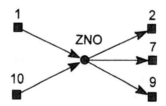

Step 2.   Find all combinations of elementary reactions in which all intermediates to be removed vanish after appropriate stoichiometric numbers are assigned to the steps. That is, find all simple submechanisms
in this more simple subnetwork. In the case under consideration
these are $\langle 1, 2 \rangle$, $\langle 1, 7 \rangle$, $\langle 1, 9 \rangle$, $\langle 7, 10 \rangle$, and $\langle 9, 10 \rangle$. A combination $\langle 2,
10 \rangle$ is simply a pair of reverse reactions.

Step 3.   Replace elementary reactions of the mechanism included in the subnetwork of Step 1 with overall reactions of submechanisms of Step
2. The new (simplified) mechanism will be as follows:

$$
\begin{align}
&(3) \quad CO + Z \rightarrow ZCO, \\
&(4) \quad ZCO + ZO \rightarrow 2Z + CO_2, \\
&(5\text{-}6) \quad NO_2 + ZCO \rightarrow Z + NO + CO_2, \\
&(8) \quad 2ZN \rightarrow 2Z + N_2, \\
&(1\text{-}2) \quad NO_2 + Z \rightarrow ZO + NO, \\
&(1\text{-}7) \quad NO_2 + Z + ZCO \rightarrow ZO + ZN + CO_2, \\
&(1\text{-}9) \quad NO_2 + 3Z \rightarrow ZN + 2ZO, \\
&(9\text{-}10) \quad NO + 2Z \rightarrow ZN + ZO, \\
&(7\text{-}10) \quad ZCO + NO \rightarrow ZN + CO_2.
\end{align} \tag{2.106}
$$

Simple submechanisms of mechanism (2.106) are completely identical to
those of (2.102). Clarke's algorithm is useful when the mechanism is too large.
Then, the exponential algorithm of the search for simple submechanisms can be
conducted as several stages. Each stage is also exponential but the dimensionality
of the search space is lower, which allows one to save the time for computation.

Stricly speaking, when one wishes to exclude intermediate X, while $id(v_X) = \mu$
and $od(v_X) = \eta$, the algorithm replaces $(\mu + \eta)$ reactions with $(\mu \times \eta)$ reactions.
However, among these $(\mu \times \eta)$ reactions, may be those containing no intermediate
species, having degenerate overall equations $0 = 0$, or combining pairs of reverse
elementary reactions. "New" reactions of this sort should be further excluded from
the consideration, and the number of new reactions thus can become less than the
number of old ones. This is a driving force of reducing the number of reactions in
the network. Clearly, if the number of new reactions is greater than the number of
old ones, the method does not work and should not be applied. Otherwise, the
search space will be even larger.

2.3.5. The Structure of the Solution to the Equation $B(SI)^T\gamma = 0$ in which γ Is Unknown

Consider equation $B(SI)^T\gamma = 0$ in which $B(SI)$ is the $m \times k$ stoichiometric matrix of
intermediates; k is the number of intermediates; γ is the unknown vector of stoichiometric numbers. Apparently, matrix equation (2.98), viewed as a system of

linear homogeneous algebraic equations, always has solutions to it.[225] Let us discuss different variants of solutions. In what follows, we use the following designation: $\text{rank}[\mathbf{B}(SI)] \equiv r$.

Variant 1. Let $r = s = m$. Equation (2.98) has a unique solution, which is trivial: $\gamma = 0$. In this case, all stoichiometric numbers of elementary reactions are zeros and no submechanisms can be found in the reaction network. The condition $r = m$ implies that $\mathbf{B}(SI)$ contains a column (or several columns) for the intermediate species that in no case vanishes in the overall equation.

Simple example. Consider the mechanism of alkyne oxidative chlorination, which involves three steps. Steps 1 and 2 are reversible, but we neglect the reverse elementary processes in this example. X_1 and X_2 are intermediate species, whose structure is unessential for the discussion at this point.[226]

$$(1) \ HC \equiv CR + CuCl \longrightarrow X_1 + HCl,$$
$$(2) \ X_1 + 2CuCl_2 \longrightarrow X_2 + CuCl, \qquad (2.107)$$
$$(3) \ X_2 \longrightarrow RC \equiv CCl + 2CuCl.$$

The subnetwork of intermediates for the mechanism (2.107) is shown in Figure 2.48. Species $HC \equiv CR$, $CuCl_2$, $RC \equiv CCl$, and HCl are terminal. Other species (X_1, X_2, and $CuCl$) are intermediate. Thus, we have the stoichiometric matrix:

$$\mathbf{B}(SI) =$$

	CuCl	X_1	X_2
(1)	−1	1	0
(2)	1	−2	1
(3)	2	0	−1

As can be seen, $r = m = s$ and the solution is unique and trivial: $\gamma = 0$. That is, each component of the γ vector must be zero. CuCl plays the role of a catalyst (i.e., intermediate species) and the role of a product at a time. The above mechanism exemplifies the case of an autocatalysis in which one of the products catalyzes the intermediate transformations of other species.

This example is taken from real chemical practice. The overall equation of this mechanism (as it was observed in experiments) is as follows:

$$HC \equiv CR + 2CuCl_2 \longrightarrow RC \equiv CCl + HCl + 2CuCl.$$

However, if we deal with the mechanistic scheme in the abstract form (see the next example) it is impossible to avoid ambiguity in determining the overall equation.[210] It can be shown that autocatalysis is not the necessary condition for all components of γ to be zeros. The following example, although not from real chemical practice, serves well as an illustration.

Another example. Consider the schematic mechanism[227,228]

$$(1) \ A + B \rightarrow X + Y,$$
$$(2) \ B + X \rightarrow 2Z, \qquad\qquad (2.108)$$
$$(3) \ Y + Z \rightarrow T,$$

where X, Y, and Z are intermediates. It can be easily checked that the only solution to equation (2.108) is all zeros for γ_1, γ_2, and γ_3, which are components of the vector γ. That is, "intermediates cannot have zero net formation, unless no reaction occurs at all."[227,228]

Variant 2. Let $r = s < m$, and the system has nontrivial solutions. The example given below shows that, even if $r < m$, simple submechanisms can be absent.

Example. Suppose that steps 1 and 2 in mechanism (2.107) are reversible; that is, two more equations are added:

$$(4) \ X_1 + HCl \longrightarrow HC \equiv CR + CuCl,$$
$$(5) \ X_2 + CuCl \longrightarrow X_1 + 2CuCl_2.$$

The sets of intermediate and terminal species remain as before. The stoichiometric matrix is as follows:

		CuCl	X_1	X_2
	(1)	−1	1	0
	(2)	1	−2	1
$\mathbf{B}(SI) =$	(3)	2	0	−1
	(4)	1	−1	0
	(5)	−1	2	−1

In this case, r remains the same, because two linearly dependent rows were added to the matrix. The number of elementary reactions increased: $m - r = 2$. The number of elementary steps remained as before. The GERM program found only two solutions: $\langle 1,4 \rangle$ and $\langle 2,5 \rangle$. Both of these are simply pairs of reverse reactions and should be discarded as submechanisms. Thus, the situation holds. No simple submechanisms can be found in the system. After checking the condition $r < m$, one should verify whether any of solutions correspond to simple submechanisms, or they all correspond to pairs of reverse reactions. If $r = s$, there are no simple submechanisms and the number of reversible steps is $(m - s)$.

Variant 3. Let $r < s \leq m$, and the system has nontrivial solutions. It may appear that all solutions to equation (2.98) will necessarily contain zero components. Interpretation of this situation is simple: some reactions will not be involved in any simple submechanisms. Let us discuss this case.

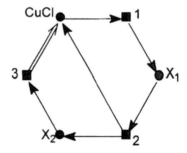

Figure 2.48. The subnetwork of intermediates for the mechanism of oxidative chlorination (2.107). Note that there are two reactions in which the number of intermediates on the right is greater than on the left.

A simple criterion exists to identify whether all elementary reactions are involved in simple submechanisms or not.[229]

Suppose that, among s elementary steps, s_0 steps are linearly independent; that is, $\mathbf{B}(SI)^{\mathsf{T}}$ has s_0 independent columns. These columns can be also placed on the first s_0 places in the matrix by column permutations. The result will be a modified $\mathbf{B}(SI)^{\mathsf{T}}$ matrix designated as $\mathbf{G}$. Components of vector γ can be rearranged as appropriate (the resulting vector will be γ^*). Thus, we arrive at the system of equations, which is equivalent to the system (2.98):

$$\mathbf{G}\gamma^* = \left\| \begin{array}{c|c} \mathbf{G}_0 & \mathbf{G}_1 \end{array} \right\| \times \left\| \begin{array}{c} \gamma_0^* \\ \gamma_1^* \end{array} \right\| = \mathbf{0}. \tag{2.109}$$

Equation (2.109) can be rewritten as follows:

$$\mathbf{G}_0\gamma_0^* + \mathbf{G}_1\gamma_1^* = \mathbf{0}. \tag{2.110}$$

Taking into account that $\mathbf{G}_0$ is not generally a square matrix, we place the second term on the right-hand side and consecutively multiply from left both sides of equation by $\mathbf{G}_0^{\mathsf{T}}$ and $(\mathbf{G}_0^{\mathsf{T}}\mathbf{G}_0)^{-1}$. Then we arrive at:

$$\gamma_0^* = -(\mathbf{G}_0^{\mathsf{T}}\mathbf{G}_0)^{-1}\mathbf{G}_0^{\mathsf{T}}\mathbf{G}_1\gamma_1^*. \tag{2.111}$$

Using the designation

$$\mathbf{L} \equiv -(\mathbf{G}_0^{\mathsf{T}}\mathbf{G}_0)^{-1}\mathbf{G}_0^{\mathsf{T}}\mathbf{G}_1, \tag{2.112}$$

we have

$$\gamma^* = \left\| \begin{array}{c} \gamma_0^* \\ \gamma_1^* \end{array} \right\| = \left\| \begin{array}{c} \mathbf{L}\gamma_1^* \\ \gamma_1^* \end{array} \right\| = \left\| \begin{array}{c} \mathbf{L} \\ \mathbf{E} \end{array} \right\| \gamma_1^*. \tag{2.113}$$

In formula (2.113), $\mathbf{E}$ is the $m_1 \times m_1$ unity matrix; $m_1 = m - s_0$.

Theorem. *Each solution to equation (2.98) must contain zero components iff the $\mathbf{L}$ matrix, which is derived from $\mathbf{B}(SI)$ by formula (2.112) using (though not uniquely defined) transformation to $\mathbf{G}$, contains completely zero rows.*

Proof. Let columns of the $\mathbf{L}$ matrix induce the subspace $\mathcal{L}$ of the linear space $\mathcal{E}$ having the orthonormal basis $(e_1, e_2, ..., e_{s_0})$.

1. *Sufficiency.* Let there be m_1 independent vectors of $\mathcal{L}$ that are orthogonal to one of the vectors e_i of the basis of $\mathcal{E}$. Then, vector e_i is also orthogonal to the entire subspace $\mathcal{L}$. In this case, each vector from the subspace $\mathcal{L}$ contains a zero ith component, while the entire matrix $\mathbf{L}$ contains a zero row.

2. *Necessity.* Let for each e_i, $i = 1 ... s_0$, of the basis of $\mathcal{E}$, none of the m_1 independent vectors of $\mathcal{L}$ be orthogonal to the vector e_i. Then, none of the vectors of the orthonormal basis of $\mathcal{E}$ is orthogonal to $\mathcal{L}$. Therefore, for each i, $i = 1 ... s_0$, there exists a vector z ($z \in \mathcal{L}$) such that the ith component of this vector is not zero. Let us show that these s_0 vectors can be used to obtain the vector that does not contain zero components and thus prove the existence of the this vector.

Let us construct the $(s_0 \times s_0)$ matrix $\mathbf{Z}$ such that columns of the matrix are vectors, and each jth vector contains nonzero jth component z_{jj}. In the first column of the matrix $\mathbf{Z}$, find the kth zero component. Add to the first column the kth column multiplied by the quantity ξ_k, $\xi_k \neq (-z_{1j}/z_{kj})$, $j = 1 ... s_0$. Clearly, this quantity always can be found. The result of these operations will be the linear combination of the first and kth columns such that this combination of columns is the vector l ($l \in \mathcal{L}$) and the ith component is nonzero, $l_i \neq 0$. It is not improbable that some other zero components of the first column-vector will also disappear. If the first column-vector still has zero components, it should be replaced with vector l and the procedure should be repeated. Thus, all zero components will be eventually reduced. The resulting vector l, $l \in \mathcal{L}$, will be completely nonzero, which is the required result. $\square$

As can be seen from the above, in some cases $\mathbf{L}$ contains a row that entirely consists of zeros. This means that the reaction corresponding to this row is not involved in simple submechanisms.

Example. Consider the simplified mechanism of 1-butene to 2-butene isomerization in the solution of palladium complexes:[230]

(1) $C_4H_8 + 2PdCl_4^{2-} + H_2O \rightarrow Pd_2Cl_4^{2-} + C_4H_8O + 2H^+ + 4Cl^-,$

(2) $Pd_2Cl_4^{2-} \rightarrow PdCl_4^{2-} + Pd^0,$

(3) $Pd_2Cl_4^{2-} + 1\text{-}C_4H_8 \rightarrow Pd_2Cl_4(1\text{-}C_4H_8)^{2-},$

(4) $Pd_2Cl_4(1\text{-}C_4H_8)^{2-} \rightarrow Pd_2Cl_4(2\text{-}C_4H_8)^{2-},$

(5) $Pd_2Cl_4(2\text{-}C_4H_8)^{2-} \rightarrow Pd_2Cl_4^{2-} + 2\text{-}C_4H_8.$

$$(2.114)$$

Steps of this mechanism are pseudoelementary. Using the designations $X_1 = PdCl_4^{2-}$, $X_2 = Pd_2Cl_4^{2-}$, $X_3 = Pd_2Cl_4(1\text{-}C_4H_8)^{2-}$, $X_4 = Pd_2Cl_4(2\text{-}C_4H_8)^{2-}$, $P_1 = C_4H_8O$, $P_2 = Pd^0_{solv}$, $P_3 = 2\text{-}C_4H_8$, and $P_4 = 1\text{-}C_4H_8$, we can write the stoichiometric matrix:

$$
B = \begin{Vmatrix}
 & X_1 & X_2 & X_3 & X_4 \\
\hline
(1) & -2 & 1 & 0 & 0 \\
(2) & 1 & -1 & 0 & 0 \\
(3) & 0 & -1 & 1 & 0 \\
(4) & 0 & 0 & -1 & 1 \\
\hline
(5) & 0 & 1 & 0 & -1
\end{Vmatrix}
\begin{matrix} \\ \\ B_0 \\ \\ \\ \hline B_1 \end{matrix}
$$

There is only one simple submechanism, which involves reactions 3, 4, and 5. Reactions 1 and 2 are not included in the submechanism. Elementary reactions of the mechanism are numbered so as to avoid the transformation of $B(SI)^T$ into G; that is, $G = B(SI)^T$ and $\text{rank}G = \text{rank}B(SI)^T = 4$. Then, the matrix L defined by the formula

$$L = -(G_0^T G_0)^{-1} G_0^T G_1 = -(B_0 B_0^T)^{-1} B_0 B_1^T$$

contains a zero row for the first two elementary reactions: $L = [0\ 0\ 1\ 1]^T$. All possible routes can be obtained by the choice of the free term γ_5:

$$\gamma = [0\ 0\ 1\ 1\ 1]^T \gamma_5.$$

If $s < m$, part of the solutions to equation (2.98) are pairs of reverse elementary reactions. If $s = m$, all solutions found by the computer program GERM are simple submechanisms.

Definition 10. *The subgraph of the reaction network that contains all u-vertexes corresponding to elementary reactions not involved in any simple submechanisms and w-vertexes adjacent to or from these u-vertexes is termed an unbalanced component of the reaction network.*

Definition 11. *The subgraph of SI that contains all u-vertexes corresponding to elementary reactions not involved in any simple submechanisms and w-vertexes adjacent to or from these u-vertexes is termed an unbalanced component of the subnetwork of intermediates.*

Definition 12. *Each connected component of the unbalanced component of the subnetwork of intermediates is termed a defect.*

Variant 4. Let $r < s = m$, and the system has nontrivial solutions. All elementary reactions are involved in any of simple submechanisms. This variant is most common. Numerous examples are discussed in the next chapter.

It is interesting to note that, if $m - r = 1$, equation (2.98) also has an infinity of solutions. All of them are reducible to the simplest one by finding a maximum common divisor for the components of each particular solution. In other words, in this case there is one independent route and one simple submechanism. Mechanisms of this sort were termed *single-route mechanisms*.[210]

Thus, equation (2.98) can have either a unique solution, which is trivial, $(r = m)$ or an infinity of solutions $(r < m)$. In the latter case, there exists a criterion, which makes it possible to check whether all steps are involved in simple submechanisms or not. Note, however, that both the program GERM and this criterion fail to distinguish between simple submechanisms and pairs of reverse reactions. Because a pair of reverse reactions resembles a simple submechanism, it is always necessary to exclude them from the list of simple submechanisms.

Generally, three possible cases can exist: (1) Each step is involved in any simple submechanism. (2) Several (but not all) steps are involved in simple submechanisms. (3) Simple submechanisms do not exist.

In Ref. 211 mechanisms corresponding to the above cases were termed *balanced*, *partly balanced*, and *unbalanced*, respectively, because each simple submechanism corresponds to the balance of intermediate species such that in the overall of the simple submechanism all intermediates vanish. The analysis of the mechanism can be summarized by the block scheme shown in Figure 2.49.

2.3.6. The Material Balance in Reaction Networks

The transformation of the reaction network into subnetwork of intermediates usually results in the loss of material (weight) balance of elementary reactions. This transformation is tantamount to crossing out terminal species from equations of elementary reactions. Let the elementary reaction be written as

$$A + X \rightarrow B + Y, \tag{2.115}$$

where X and Y are intermediates, while A and B are terminal species. Equation (2.115) implies that weights of species fulfill the following equation:

$$m_A + m_X = m_B + m_Y. \tag{2.116}$$

In the space of intermediates equation (2.115) should be rewritten as

$$X \rightarrow Y. \tag{2.117}$$

Equation (2.117) does NOT imply that

$$m_X = m_Y.$$

Residual part of an elementary step, such as

$$A \rightarrow B, \tag{2.118}$$

also violates the material balance ($m_A \neq m_B$). The overall equation of a simple submechanism is constructed from "residual parts" of elementary reactions similar to equation (2.118). Because all intermediates vanish in the overall equation of a simple submechanism, the material balance is violated in residual parts of elementary steps, but not in the overall equation.

Suppose we have the set of balanced reactions:

$$
\begin{align}
&(1)\ A + X \rightarrow B + Y, \\
&(2)\ Y + C \rightarrow Z, \\
&(3)\ Z \rightarrow D + X.
\end{align}
\tag{2.119}
$$

The overall equation can be derived by adding up equations (2.119):

$$A + C \rightarrow B + D. \tag{2.120}$$

Let us cross out intermediates from equations (2.119):

$$
\begin{align}
&(1)\ A \Rightarrow B, \\
&(2)\ C \Rightarrow nothing, \\
&(3)\ nothing \Rightarrow D\,.
\end{align}
\tag{2.121}
$$

The resulting equations violate the material balance, but their sum produces a balanced equation (2.120) and

$$m_A + m_C \rightarrow m_B + m_D. \tag{2.122}$$

Reaction networks constructed on the formal basis using combinatorial algorithms and applying constraints that act on graphs may contain inconsistent elementary reactions, which violate the material balance without crossing out any species. For instance, elementary reactions

$$
\begin{align}
&A + B \rightarrow D, \\
&A \rightarrow D
\end{align}
\tag{2.123}
$$

are inconsistent because equations

$$
\begin{align}
&m_A + m_B - m_D = 0, \\
&m_A - m_D = 0
\end{align}
\tag{2.124}
$$

cannot be solved in positive integers. If m_D is chosen as independent variable, $m_A = m_D$ and $m_B = 0$; that is, species B has zero weight, which is impossible.

To test whether the set of reactions can constitute a reaction network, the following algorithm should be applied:

Step 1. Construct the $m \times n$ matrix $\mathbf{B}(RN)$.
Step 2. Calculate rank$\mathbf{B}(RN)$.
Step 3. Calculate the number of species weights that can serve as independent variables using the formula

$$n - \text{rank}\mathbf{B}(RN).$$

Step 4. Find species, the weights of which can serve as independent variables.
Step 5. Set their weights to unity.
Step 6. Solve the system of equations

$$\mathbf{B}(RN)\mathbf{m} = \mathbf{0},$$

with respect to unknown $\mathbf{m}$ (the vector of species weights).
Step 7. Check whether components of $\mathbf{m}$ are all positive. If some of the components must be negative or zero, the reaction network cannot exist.

Let us illustrate the above algorithm. Suppose we have a set of equations which is a candidate for the mechanism:

$$
\begin{aligned}
&(1)\ A + B \rightarrow C, \\
&(2)\ 2C \rightarrow E, \\
&(3)\ E \rightarrow A + B.
\end{aligned}
\qquad (2.125)
$$

The 3×4 matrix $\mathbf{B}(RN)$ is as follows:

$$
\mathbf{B}(RN) =
\begin{array}{c|cccc}
 & A & B & C & E \\
\hline
(1) & -1 & -1 & 1 & 0 \\
(2) & 0 & 0 & -2 & 1 \\
(3) & 1 & 1 & 0 & -1 \\
\end{array}
$$

and rank$\mathbf{B}(RN) = 3$. The number of independent species is $m - \text{rank}\mathbf{B}(RN) = 4 - 3 = 1$. Let us choose A as an independent species. Indeed, the removal of column for species A does not change the rank of the matrix. By setting $m_A = 1$, we arrive at the following system of equations:

$$
\begin{aligned}
-m_B + m_C &= 1, \\
-2m_C + m_E &= 0, \\
m_B - m_E &= -1.
\end{aligned}
$$

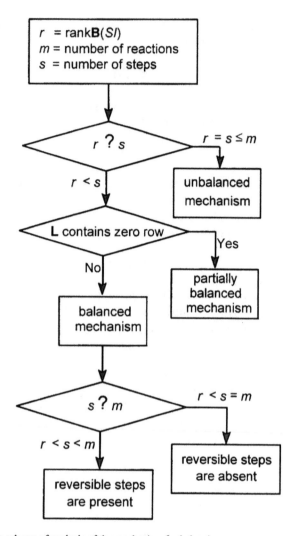

Figure 2.49. The scheme of analysis of the mechanism for balancing.

The solution to this system of equations is as follows:

$$m_A = 1,$$
$$m_B = -m_A = -1,$$
$$m_C = 0,$$
$$m_E = 0.$$

Because zero weights of B, C, and E are impossible, the mechanism should be discarded. More generally, this solution can be interpreted so that weights of A and B has opposite signs. Therefore, one of the species must have a negative weight, which is impossible.

Thus, the set of reactions written in terms of A, B, C, etc., symbols can violate the condition of material balance even if compositions of A, B, C, etc., are unknown. On generating a list of mechanism graphs (reaction networks), one should check whether the material balance of elementary steps can exist or not.

2.3.7. Applications of Simple Submechanisms and Nonlinear Networks

1°. When constructing kinetic models of complex reactions, the pseudo-steady-state approximation is often used to describe the kinetics. The essence of this method is that concentrations of intermediate species are taken to be constant in time:

$$\frac{d\mathbf{C_X}}{dt} = 0. \tag{2.126}$$

In this expression, $\mathbf{C_X}$ is the vector of concentrations of intermediate species. Sometimes the true steady state is also considered when this equality is fulfilled exactly rather than approximately. In Ref. 218, it is argued that expression $\mathbf{B}(SI)^T\gamma = 0$ is the necessary condition for the fulfillment of equation (2.98). The expression $\mathbf{B}(SI)^T\gamma = 0$ was termed the *condition for the steady state*. Let us discuss in which instances this is true.

Clearly, the true steady state is feasible only in open systems. Then, dC_i/dt depends on three factors: (1) the inflow of the ith species into the system, (2) the outflow of the ith species, and (3) the consumption (and/or generation) of the ith species at the expense of elementary reactions occurring within the system. Often, it is assumed that the first two factors are absent: intermediate species are not inputted to and outputted from the system. In this instance, expression $\mathbf{B}(SI)^T\gamma = 0$ actually is the necessary condition for the fulfillment of (2.98). In the general case, intermediates can enter and exit the system. Then, the condition of the steady state becomes more complicated, while equation (2.98) requires the overall rates of transformations of intermediate species to be zeros. Apparently, if $s = r$ or if part of the elementary reactions are not involved in simple submechanisms, the condition (2.126) cannot be valid. If all intermediates are located in the system (for instance, when all intermediates are located on the surface of a heterogeneous catalyst), condition $s = r$ and the presence of zero rows in the $\mathbf{L}$ matrix play the role of stoichiometric constraints on the feasibility of a steady state.

2°. There is a direct analogy between simple submechanisms and Clarke's current diagrams. Suppose we have the SI of Oregonator (Figure 2.50). The reaction mechanisms is as follows:

$$(1) \ Y \rightarrow X,$$

$$(2) \ X + Y \rightarrow product(s),$$

$$(3) \ X \rightarrow 2X + 2Z, \tag{2.127}$$

$$(4)\ 2X \rightarrow \text{product(s)},$$

$$(5)\ 2Z \rightarrow fY.$$

X, Y, and Z are intermediate species; reaction 3 is autocatalytic (and is far from being realistic as an elementary one, but this is of no concern). Suppose the stoichiometric coefficient f is unity. Then, we can find all submechanisms (simple and not simple). Note that stoichiometric numbers of elementary reactions involved in the submechanism must be nonzero positive integers. In the case of Oregonator, there are three submechanisms, $\langle 1,2,3,4,5\rangle$, $\langle 2,3,5\rangle$, and $\langle 1,3,4,5\rangle$. The first one among them is not simple.

To obtain current diagrams, one should assign minimum-integer stoichiometric numbers to arcs incident with u-vertexes. The stoichiometric number of reaction 4 in the submechanism $\langle 1,3,4,5\rangle$ is 2. Other stoichiometric numbers in these submechanisms are equal to 1. Then, subnetworks of intermediates for submechanisms $\langle 1,2,3,4,5\rangle$ and $\langle 2,3,5\rangle$ coincide with current diagrams, while SI of submechanism $\langle 1,3,4,5\rangle$ should be slightly rearranged (Figure 2.51). The arc $w_X \rightarrow u_4$ should be doubled.

Thus, we arrive at current diagrams of submechanisms, which match the rule: $\mathrm{id}(w_i') = \mathrm{od}(w_i')$, $\forall\ i$, $i = 1 \ldots k$, $k = |\mathcal{W}(SI)|$. Clarke also introduced the notion of a *current cycle* in the current diagram. A current cycle is the irreversible cycle on a current diagram. Current cycles of the Oregonator network are shown in Figure 2.52.

Clarke classified current cycles and showed that networks that do not have any current cycles are stable and provide the method for the instability analysis based on the classification of current cycles. Similar results were obtained by Vol'pert.

3°. The idea of simple submechanisms was shown to be applicable to the construction of classification principles for reaction networks. This problem will be discussed in the next chapter.

4°. Finally, simple submechanisms serve as the basis for the formulation of the complexity index of reaction mechanisms (see Chapter 4).

Thus, there are at least four areas in which the idea of simple submechanisms can be used: (1) chemical kinetics; (2) stability analysis of chemical dynamic systems; (3) classification of reaction mechanisms; and (4) formulation of the complexity measures.

2.3.8. An Alternative Approach to the Choice of Intermediates

In Section 2.3.2, we discussed several incorrect definitions of the term *intermediate* and gave our own definition. In this section, we discuss alternative definitions of this term, which are also correct.

According to the approach proposed by Happel and Sellers,[216,217,219] the overall equation should be predetermined. The species involved in the overall equation are thus defined as terminal. However, the overall equation is not always known. Therefore, this approach leads to the loss of generality.

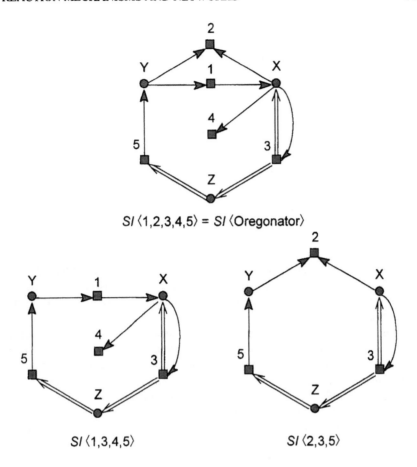

Figure 2.50. The subnetwork of intermediates of Oregonator and its simple submechanisms.

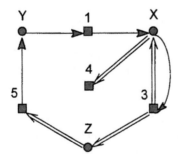

Figure 2.51. The current diagram of the submechanism ⟨1,3,4,5⟩ of the Oregonator.

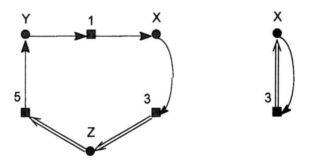

Figure 2.52. The current cycles of the Oregonator network.

Valdés-Pérez proposed the formalism in terms of linear programming.[227] This formalism allows one, among other things, to distinguish between intermediate and terminal species. In the case of a balanced mechanism, there is no need to canonically choose intermediates.

Suppose we have the set of reactions that constitute an unbalanced mechanism and no overall equation is predetermined. Different sets of stoichiometric numbers can be assigned to elementary reactions. Each variant of assignment corresponds to a particular overall equation. Thus, for the set of reactions discussed above (variant 1, equation (2.108)), the vector of stoichiometric numbers can be chosen as [1 1 1], [2 1 1], or [1 1 2] (other assignments of stoichiometric numbers are also possible). These vectors give overall equations (2.128), (2.129), and (2.130), respectively:

$$A + 2B \rightarrow T + Z, \tag{2.128}$$

$$2A + 3B \rightarrow 2T + X, \tag{2.129}$$

$$A + 2B + Y \rightarrow 2T. \tag{2.130}$$

Each of these overall equations contains X, Y, or Z. According to our approach each of these species should be marked as intermediate. In Ref. 210 they were termed dualistic, because they exhibit the features of both terminal and intermediate species. According to the linear programming approach proposed by Valdés-Pérez, one or several of these species can be marked as products and only one overall equation should be chosen. To do this, one can fix the ideal concentration of the target product T after the reaction completion at 1 mol/l; all other quantities (stoichiometric numbers and concentrations of other species after the reaction completion) are set to be nonnegative but receive no definite values. In this case, the task of linear programming is to minimize expenses of starting products at a fixed value of the concentration of T after the reaction completion. In the context of this approach, equation (2.130) is incorrect, because Y is not contained in the initial system, while equation (2.129) improves on equation (2.128). Thus, X rather than Z is the "product".

Although this method is elegant and mathematically rigorous, it implicitly contains several weak points. First, one should also fix initial concentrations of all species. In the above example, the initial concentration of Y was fixed at 0 mol/l. That is, a somewhat ambiguous choice of initial concentrations affects the result. Indeed, intermediate (dualistic) species can be present in the initial system. For example, in mechanism (2.107), CuCl, which is an intermediate, was included in the first step to initiate the reaction. Then, a minimization criterion is also not a uniquely defined function; that is, the choice of a minimization criterion also may affect the result.

Several methods to define the term intermediate can be found, which are equally correct and can serve as basis for the construction of mechanistic graph theory or classification and nomenclature of reaction mechanisms.[210,211] According to which method is used, the general theory may be slightly changed. This is the reason to either adopt a unified description of the term intermediate or to specify each time which of the meanings of this term is used.

References and Notes

1. Christiansen, J. A. The Elucidation of Reaction Mechanisms by the Method of Intermediates in Quasi-Stationary Concentrations. In *Advances in Catalysis*; Frankenburg, W. G.; Komarewsky, V. I.; Rideal, E. K., Eds.1953, Academic: New York, Vol. 5, pp 311–353. See also a more recent paper of Helfferich in which Christiansen's approached was also used: Helfferich, F. G. Systematic Approach to Elucidation of Multistep Reaction Networks. *J. Phys. Chem.* **1989**, *93*, 6676–6681.

2. King, E. L.; Altman, C. A Schematic Method of Deriving the Rate Laws for Enzyme-Catalyzed Reactions. *J. Phys. Chem.* **1956**, *60*, 1375–1378.

3. Temkin, M. I. Graphical Method for the Derivation of the Rate Laws of Complex Reactions. *Dokl. Akad. Nauk SSSR* **1965**, *165*, 615–618.

4. Balaban, A. T.; Fărcaşiu, D.; Bănică, R. Graphs of Multiple 1,2-Shifts in Carbonium Ions and Related Systems. *Rev. Roum. Chim.* **1966**, *11*, 1205–1227.

5. Whitlock, H. W., Jr.; Siefkon, M. W. The Tricyclo[4.4.0.0^{3,8}]decane to Adamantane Rearrangement. *J. Am. Chem. Soc.* **1968**, *90*, 4929–4939.

6. Muetterties, E. L. Topological Representation of Stereoisomerism. I. Polytopal Rearrangements. *J. Am. Chem. Soc.* **1969**, *91*, 1636–1643.

7. Johnson, C. K.; Collins, C. J. An Algebraic Model for the Rearrangements of 2-Bicyclo[2.2.1]hepthyl Cations. *J. Am. Chem. Soc.* **1974**, *96*, 2514–2523.

8. Collins, C. J.; Johnson, C. K.; Raaen V. F. Molecular Rearrangements. XXX. Applications of an Algebraic–Graphical Model for Analyzing Rearrangements of Bicyclo[2.2.1]hepthyl Cations. *J. Am. Chem. Soc.* **1974**, *96*, 2524–2531.

9. Gielen, M. Applications of Graph Theory to Organometallic Chemistry. In *Chemical Applications of Graph Theory*; Balaban, A. T., Ed.; Academic: London, 1976, pp 211–298.

10. Gund, T. M.; von R. Schleyer, P.; Gund, P. H.; Wipke, W. T. Computer Assisted Graph Theoretical Analysis of Complex Mechanistic Problems in Polycyclic Hydrocarbons. The Mechanism of Diamantane Formation from Various Pentacyclotetradecanes. *J. Am. Chem. Soc.* **1975**, *97*, 743–751.

11. Balasubramanian, K. Enumeration of Internal Rotation Reactions and Their Reaction Graphs. *Theor. Chim. Acta* **1979**, *53*, 129–146.

12. Ôsawa, E.; Tahara, Y.; Togashi, A.; Iizuka, T.; Tanaka, N.; Kan, T.; Fărcaşiu, D., Kent, G. J., Engler, E. M.; von R. Schleyer, P. Fused and Bridged Tetracyclic C_{13} and C_{14} Adamantanes. Synthesis of Methyl-2,4-ethano-, 1,2-Trimethylene, 2,4-Trimethylene, and 1,2-Tetramethyleneadamantane. *J. Org. Chem.* **1982**, 47, 1923–1932.

13. For a review on isomerization graphs see: Balaban, A. T. Reaction Graphs. In *Graph Theoretical Approaches to Chemical Reactivity*; Bonchev, D., Mekenyan, O., Eds.; Kluwer: Dordrecht, 1994, pp 137–180; see also: Balaban, A. T. Applications of Graph Theory in Chemistry. *J. Chem. Inf. Comput. Sci.* **1985**, 25, 334–343.

14. Dugundji, J.; Gillespie, P.; Marquarding, D.; Ugi, I.; Ramirez, F. Metric Spaces and Graphs Representing the Logical Structure of Chemistry. In *Chemical Applications of Graph Theory*; Balaban, A. T., Ed.; Academic: London, 1976, pp 107–174.

15. Dugundji, J; Ugi, I. An Algebraic Model of Constitutional Chemistry as a Basis for Chemical Computer Programs. *Top. Curr. Chem.* **1973**, 39, 19–64.

16. The method for the linearization of networks based on graphs of FIEM is not universal. Sometimes it fails when the reaction network describes a nonlinear multiroute mechanism for which several overall reactions exist and when some stoichiometric numbers of steps of this mechanism are not zero or unity.

17. Koča, J.; Kratochvíl, M.; Kvasnička, V. Reaction Mechanism Graphs. *Collect. Czech. Chem. Commun.* **1985**, 50, 1433–1449.

18. Koča, J. A Graph Model of the Synthon. *Collect. Czech. Chem. Commun.* **1988**, 53, 3108–3118.

19. Koča, J. The Reaction Distance. *Collect. Czech. Chem. Commun.* **1988**, 53, 3119–3130.

20. Koča, J.; Kratochvíl, M.; Kvasnička, V.; Matyska, L.; Pospíchal, J. *Synthon Model of Organic Chemistry and Synthesis Design*; Springer: Berlin, 1989 (Lecture Notes in Chemistry, Vol. 51).

21. Kvasnička, V.; Pospíchal, J. Two Metrics for a Graph-Theoretical Model of Organic Chemistry. *J. Math. Chem.* **1989**, 3, 161–191.

22. Kvasnička, V.; Pospíchal, J. Graph-Theoretical Interpretation of Ugi's Concept of the Reaction Network. *J. Math. Chem.* **1990**, 5, 309–322.

23. Pospíchal J.; Kvasnička, V. Graph Theory of Synthons. *Int. J. Quantum Chem.* **1990**, 38, 253–278.

24. Pospíchal J.; Kvasnička, V. Reaction Graphs and a Construction of Reaction Networks. *Theor. Chim. Acta* **1990**, 76, 423–435.

25. Kvasnička, V.; Pospíchal J.; Baláž, V. Reaction and Chemical Distances and Reaction Graphs. *Theor. Chim. Acta* **1991**, 79, 65–79.

26. Johnson, M. Graph Transforms: A Formalism for Modeling Chemical Reaction Pathways. In *Graph Theory, Combinatorics, and Applications*; Alavi, Y. et al., Eds.; Wiley: New York, 1991, pp 725–738.

27. Johnson, M. Relating Metrics, Lines and Variables Defined on Graphs to Problems of Medicinal Chemistry. In *Graph Theory and Its Applications to Algorithms and Computer Science*; Alavi, Y. et al; Wiley: New York, 1985, pp 457–470.

28. Johnson, M. Structure-Activity Maps for Visualizing the Graph Variables Arising in Drug Design. *J. Biopharm. Stat.* **1993**, 3, 203–236.

29. Willamowski, K.-D.; Rössler, O. E. Contributions to the Theory of Mass Action Kinetics. I. Enumeration of Second Order Mass Action Kinetics. *Z. Naturforsch., A: Phys. Sci.* **1978**, 33, 827–833.

30. Horn, F.; Jackson, R. General Mass Action Kinetics. *Arch. Ration. Mech. Anal.* **1972**, 47, 81–116.

31. Feinberg, M.; Horn, F. G. M. Dynamics of Open Chemical Systems and the Algebraic Structure of the Underlying Reaction Network. *Chem. Eng. Sci.* **1974**, 29, 775–787.

32. To our knowledge, the idea of complexes was first used by Balandin in derivation of the rate laws: Balandin, A. A. Calculation of Intermediate Equilibria in Kinetics of Catalytic Reactions and Analysis Situs. *Dokl. Akad. Nauk SSSR* **1939**, *24*, 741–747.

33. Henrici-Olivé, G.; Olivé, S. *The Chemistry of the Catalyzed Hydrogenation of Carbon Monoxide*; Springer: Berlin, 1984.

34. Othmer, H. G. The Global Dynamics of a Class of Reaction Networks. In *Chemical Applications of Topology and Graph Theory*; King, R. B., Ed.; Elsevier: Amsterdam, 1983, pp 285–306.

35. Veitsman, E. V. Application of Graph Theory (Ford–Fulkerson Theorem) to the Study of Processes Combining Diffusion and Kinetic Units. *Zh. Fiz. Khim.* **1986**, *60*, 1140–1144.

36. Ford, L. R., Jr.; Fulkerson, D. R. *Flows in Networks*; Princeton Univ.: Princeton, NJ, 1962.

37. Busacker, R. G.; Saaty, T. L. *Finite Graphs and Networks*; McGraw-Hill: New York, 1973.

38. Nemes, I.; Vidóczy, T.; Botár, L.; Gál, D. A Possible Construction of a Complex Chemical Reaction Network. I. Definitions and Procedure for Construction. *Theor. Chim. Acta* **1977**, *45*, 215–223.

39. Nemes, I.; Vidóczy, T.; Botár, L.; Gál, D. A Possible Construction of a Complex Chemical Reaction Network. II. Applications. *Theor. Chim. Acta* **1977**, *45*, 225–233.

40. Nemes, I.; Vidóczy, T.; Gál, D. A Possible Construction of a Complex Chemical Reaction Network. III. The Systematization of Elementary Processes. *Theor. Chim. Acta* **1977**, *46*, 243–250.

41. Sokolov, V. I.; Nikonorov, V. A. Genetic Relationships among Neutral Molecules, Ions, Radicals, Carbenes, and Other Species of the Same Series: An Operator Representation. *Zh. Strukt. Khim.* **1975**, *16*, 1068–1070.

42. Nikonorov, V. A.; Sokolov, V. I. Use of Operator Networks To Interpret Evolutionary Interrelations between Chemical Entities. In *Chemical Graph Theory: Reactivity and Kinetics*; Bonchev, D., Rouvray, D. H., Eds.; Abacus Press/Gordon & Breach: Philadelphia, 1992, pp 155–198.

43. Balandin, A. A. *Multiplet Theory of Catalysis: Theory of Hydrogenation. Classification of Organic Catalytic Reactions. Algebra Applied to Structural Chemistry*; Moscow State Univ.: Moscow, **1970**; Vol. 3, (in Russian). The book containing application of bipartite graphs was published in 1970. However, these results were obtained much earlier. In 1967 Balandin died. This material was carefully collected and published by his disciples, who used his earliest papers and notes. The major portion of the book is a compilation of articles in Russian journals, but some material (including the idea of bipartite graphs) was not published elsewhere. We have reasons to suggest that his studies on bipartite graphs and classification of multistep reactions belong to the period of 1936–1939. During this period Balandin was exiled to Orenburg. In one of his letters from exile, he wrote: "...A new branch of science has become open to me. I don't know how to call it yet. Maybe, structural algebra or geometry with reach applications in chemistry..." (Citation by Solov'ev, Yu. I. Academician Aleksei Aleksandrovich Balandin (Biographical Essay). In *Aleksei Aleksandrovich Balandin*; Nauka: Moscow, 1995 (in Russian)).

44. In numerous chemical papers, these graphs are termed *bipartite*. However, this is not fully correct. A vertex bipartition of graph *BG* is an ordered pair (*U, W*) of the mutually complementary subsets of the vertex set *V(BG)* such that one of the ends of each edge belongs to the subset *U*, and the other, to the subset *W*. A graph that possesses a bipartition is bipartite. If one *can* assign two colors to vertexes of a graph so that no vertexes of the same color are adjacent, the graph is said to be *bicolorable*. The term *bicolorable* is a synonym of *bipartite*. If vertexes of a bipartite graph are *colored* so

that no vertexes of the same color are adjacent, the graph is *bicolored*. Many authors neglect this difference between bicolorable and bicolored graphs. We also will follow this, though not very good, tradition.

45. Shestakov, G. K.; Temkin, O. N.; Vsesvyatskaya, N. Yu.; Stepanov, A. M. A New Method to Control the Selectivity of the Hydration of a Triple Bond. *Zh. Org. Khim.* **1979**, *15*, 248–251.
46. Clarke, B. L. Graph-Theoretic Approach to the Stability Analysis of Steady State Chemical Reaction Networks. *J. Chem. Phys.* **1974**, *60*, 1481–1492.
47. Clarke, B. L. Stability Analysis of a Model Reaction Network Using Graph Theory. *J. Chem. Phys.* **1974**, *60*, 1493–1501.
48. Clarke, B. L. Theorems on Chemical Network Stability. *J. Chem. Phys.* **1975**, *62*, 773–775.
49. Clarke, B. L. Stability of Topologically Similar Chemical Networks. *J. Chem. Phys.* **1975**, *62*, 3726–3738.
50. Clarke, B. L. Stability of the Bromate–Cerium–Malonic Acid Network: I. Theoretical Formulation. *J. Chem. Phys.* **1976**, *64*, 4165–4178.
51. Clarke, B. L. Stability of the Bromate–Cerium–Malonic Acid Network: II. Steady-State Formic Acid Case. *J. Chem. Phys.* **1976**, *64*, 4179–4191.
52. Clarke, B. L. Stability of Complex Reaction Networks. In *Advances in Chemical Physics*; Prigogine, I., Rice, S. A., Eds.; Wiley: New York, 1980, Vol. 43, pp 7–215.
53. Clarke, B. L. Qualitative Dynamics and Stability of Complex Reaction Networks. In *Chemical Applications of Topology and Graph Theory*; King, R. B., Ed.; Elsevier: Amsterdam, 1983, pp 322–357 (Stud. Phys. Theor. Chem., Vol. 28).
54. Vol'pert, A. I. Differential Equation on Graphs. *Mat. Sb.* **1972**, *88(130)*, 578–588.
55. Vol'pert, A. I.; Khudyaev, S. I. *Analysis in Classes of Discontinuous Functions and Equations of Mathematical Physics*; Nauka: Moscow, 1975 (in Russian).
56. Vol'pert, A. I. *Qualitative Methods for the Study of Equations in Chemical Kinetics*, Preprint of the Inst. of Chem. Physics, Chernogolovka, 1975 (in Russian).
57. Vol'pert, A. I.; Gel'man, E. A.; Ivanova, A. N. *Some Problems of Qualitative Theory of Differential Equations on Graphs*, Preprint of the Inst. of Chem. Physics, Chernogolovka, 1975 (in Russian).
58. Ivanova, A. I. The Condition for the Uniqueness of the Steady State of Kinetic Systems Related to the Structure of Reaction Scheme. Part I. *Kinet. Katal.* **1979**, *20*, 1019–1023.
59. Ivanova, A. I. The Condition for the Uniqueness of the Steady State of Kinetic Systems Related to the Structure of Reaction Scheme. Part II. Open Systems. *Kinet. Katal.* **1979**, *20*, 1024–1028.
60. Ivanova, A. I.; Tarnopol'skii, B. L. On the Method for the Solution of a Number of Qualitative Problems on Kinetic Systems and Its Computer Implementation (Critical Conditions and Autooscillations). *Kinet. Katal.* **1979**, *20*, 1541–1548.
61. Ivanova, A. I.; Gol'dshtein, B. N. A Kinetic Model of the Bifunctional Enzyme: Phosphorylation of 6-Phosphofructo-2-kinase (Fructoso-2,6-bisphosphatase). *Mol. Biol.* (Moscow), **1986**, *20*, 1522–1529.
62. Gol'dshtein, B. N.; Ivanova, A. I. Simple Kinetic Models Able To Account for Critical Phenomena in Enzymatic Reactions that Involve Isomerizations of Enzyme and Substrate. *Mol. Biol.* (Moscow), **1988**, *22*, 1381–1392.
63. Balandin, A. A. Theory of Complex Reactions. 1. Fundamentals of Theory. Kinetic Simplexes. *Zh. Fiz. Khim.* **1941**, *15*, 615–628.
64. Balandin, A. A. Theory of Complex Reactions. 2. Kinetic Complexes. Diagrams of Properties. Hyperreactions. *Zh. Fiz. Khim.* **1941**, *15*, 629–644.
65. Solimano, F.; Beretta, E. Tree Graphs Theory for Enzymatic Reactions, *J. Theor. Biol.* **1976**, *59*, 159–166.

66. Solimano, F.; Beretta, E.; Piatti, E. Tree Graphs Theory for Enzymatic Reactions: A Theorem for the Reactions among the Non-Enzymatic Species, *J. Theor. Biol.* **1977**, *64*, 401–412.

67. Beretta, E.; Vertrano, F.; Solimano, F.; Lazzari, C. Some Results about Chemical Systems Represented by Trees and Cycles. *Bull. Math. Biol.* **1979**, *41*, 641–664.

68. Horn, F. J. M. On a Connection between Stability and Graphs in Chemical Kinetics: I. Stability and the Reaction Diagram. *Proc. Royal Soc. London, A.* **1973**, *334*, 299–312.

69. Horn, F. J. M. On a Connection between Stability and Graphs in Chemical Kinetics: II. Stability and the Complex Graph. *Proc. Royal Soc. London, A.* **1973**, *334*, 313–330.

70. Delattre, P. *L'evolution des Systèmes Moléculaires. Bases Théoretiques. Applications à la Chimie et à la Biologie*; Malone: Paris, 1971.

71. Hyver, C. Valeurs Propres des Systèms de Transformation Répresentables par des Graphes en Arbres. *J. Theor. Biol.* **1973**, *42*, 397–409.

72. Perelson, A. S.; Oster, G. F. Chemical Reaction Dynamics. Part II. Reaction Networks. *Arch. Ration. Mech. Anal.* **1974**, *57*, 31–98.

73. Sinanoğlu, O. Theory for Generating All Synthetic Pathways or Reaction Mechanisms with Specified Number of Steps – Three-Step Mechanisms. *Chim. Acta Turcica* **1975**, *3*, 155–161.

74. Sinanoğlu, O. Theory of Chemical Reaction Networks. All Possible Mechanisms or Synthetic Pathways with Given Number of Reaction Steps or Species. *J. Am. Chem. Soc.* **1975**, *37*, 2307–2320.

75. Sinanoğlu, O.; Lee, L. S. Finding All Possible a priori Mechanisms for a Given Type of Overall Reaction. The Cases of (1) Molecular Rearrangements (A → B); and (2) Molecular Associations (A + B → C) Reaction Types, *Theor. Chim. Acta* **1978**, *48*, 287–299; Finding the Possible Mechanisms for a Given Type of Overall Reaction. The Case of the (A + B → C + D) Reaction Types, *Theor. Chim. Acta* **1979**, *51*, 1–9.

76. Lee, L. S.; Sinanoğlu, O. Reaction Mechanisms and Chemical Networks – Types of Elementary steps and Generation of Laminar Mechanisms. *Z. Phys. Chem.* **1981**, *125*, 129–160.

77. Sinanoğlu, O. 1- and 2-Topology of Reaction Networks. *J. Math. Phys.* **1981**, 22, 1504–1512.

78. Sinanoğlu, O. Autocatalytic and Other General Networks for Chemical Mechanisms, Pathways, and Cycles: Their Systematic and Topological Generation. *J. Math. Chem.* **1993**, *12*, 319–363.

79. In part of Sinanoğlu's papers, species-vertexes of degree two are contracted to form a species-line (■———●———■ ⇒ ■———■).

80. Sellers, P. Algebraic Complexes which Characterize Chemical Networks. *SIAM J. Appl. Math.* **1967**, *15*, 13–68.

81. Sellers, P. H. Combinatorial Analysis of a Chemical Network. *J. Franklin Institute*, **1970**, *290*, 113–130.

82. Sellers, P. H. An Introduction to a Mathematical Theory of Chemical Reaction Network. I. *Arch. Ration. Mech. Anal.* **1971**, *44*, 23–40.

83. Sellers, P. H. An Introduction to a Mathematical Theory of Chemical Reaction Network. II. *Arch. Ration. Mech. Anal.* **1971**, *44*, 376–386.

84. The material of this section is original and was not published elsewhere.

85. Zykov, A. A. Hypergraphs. *Usp. Mat. Nauk* **1974**, *29*, 89–154.

86. Emelichev, V. A.; Mel'nikov, O. I.; Sarvanov, V. I.; Tyshkevich, R. I. *Lecture Course on Graph Theory*; Nauka: Moscow, 1990 (in Russian).

87. Harary, F.; Prince, G.; Tutte, W. T. The Number of Plane Trees. *Indagationes Math.* **1964**, *26*, 319–329.

88. Khomenko, A. A.; Apel'baum, L. O.; Shub, F. S.; Snagovskii, Yu. S.; Temkin, M. I. Kinetics of the Reaction of Methane with Water Vapor and the Reverse Reaction of

Hydrogenation of Carbon Monoxide on the Surface of Nickel. *Kinet. Catal. Engl. Transl.* **1971**, *12*, 367–373.

89. Bonchev, D.; Temkin, O. N.; Kamenski, D. Graph-Theoretical Classification and Coding of Chemical Reactions with a Linear Mechanism. *J. Comput. Chem.* **1982**, *3*, 95-110.
90. Temkin, O. N.; Bonchev, D. Classification and Coding of Chemical Reaction Mechanisms. In *Mathematical Chemistry. Chemical Graph Theory. Reactivity and Kinetics;* Bonchev, D.; Rouvray, D. H., Eds.; Gordon and Breach: Chichester, 1992, Vol. 2, Chapter 2.
91. Temkin, O. N.; Zeigarnik, V. A; Bonchev, D. Graph-Theoretical Models of Complex Reaction Mechanisms and Their Elementary Steps. In *Graph Theoretical Approaches to Chemical Reactivity;* Bonchev, D. and Mekenyan, O., Eds.; Kluwer Academic: Dordrecht, 1994, pp 241–275.
92. Bonchev, D.; Temkin, O. N.; Kamenski, D. On the Classification and Coding of Linear Reaction Mechanisms. *React. Kinet. Catal. Lett.* **1980**, *15*, 113-118.
93. Gordeeva, E.; Bonchev, D.; Kamenski, D.; Temkin, O. N. Enumeration, Coding, and Complexity of Linear Reaction Mechanisms. *J. Chem. Inf. Comput. Sci.* **1994**, *34*, 436–445.
94. Zeigarnik, A. V. A Graph-Theoretical Model of Complex Reaction Mechanisms: Special Graphs for Characterization of the Linkage between the Routes in Complex Reactions Having Linear Mechanisms. *Kinet. Katal.* **1994**, *35*, 711-713; *Kinet. Catal. Engl. Transl.* **1994**, *35*, 656–658.
95. Harary, F. *Graph Theory;* Addison-Wesley: Reading, MA, 1969.
96. Balaban, A. T., Ed. *The Chemical Applications of Graph Theory;* Academic: New York, 1976.
97. Bonchev, D.; Balaban, A. T.; Mekenyan, O. Generalization of the Graph Center Concept, and Derived Topological Indexes. *J. Chem. Inf. Comput. Sci.* **1980**, *20*, 106-113.
98. Bonchev, D.; Balaban, A. T.; Randić, M. The Graph Center Concept for Polycyclic Graphs. *Int. J. Quantum Chem.* **1981**, *19*, 61-82.
99. Bonchev, D. The Concept for the Center of a Chemical Structure and Its Applications. *Theochem* **1989**, *185*, 155-168.
100. Bonchev, D.; Mekenyan, O.; Balaban, A. T. An Iterative Procedure for the Generalized Graph Center in Polycyclic Graphs. *J. Chem. Inf. Comput. Sci.* **1989**, *29*, 91–97.
101. Temkin, O. N.; Bruk, L. G.; Zeigarnik, A. V. Some Aspects of the Methodology of Mechanistic Studies and Kinetic Modeling of Catalytic Reactions. *Kinet. Katal.* **1993**, *34*, 445–462; *Kinet. Catal. Engl. Transl.* **1993**, *34*, 389–405.
102. Zeigarnik, A. V.; Bruk, L. G.; Temkin, O. N.; Likholobov, V. A.; Maier, L. I. Computer-Assisted Mechanistic Studies. *Usp. Khim.* **1996**, *65*, 125–139.
103. Temkin, O. N.; Bruk, L. G.; Bonchev, D. G. Topological Structure of Complex Reaction Mechanisms. *Teor. Eksp. Khim.* **1988**, *23*, 282-291.
104. Gordeeva, E. V.; Molchanova, M. S.; Zefirov, N. S. General Methodology and Computer Program for the Exhausted Restoring of Chemical Structures by Molecular Connectivity Index. Solution of the Inverse Problem in QSAR/QSPR. *Tetrahedron Comp. Method.* **1990**, *3*, 389-415.
105. Faradzhev, I. A. Generation of Nonisomorphic Graphs with a Given Distribution of Vertex Types. In *Algorithmic Investigations in Combinatorics;* Nauka: Moscow, 1978, pp 11–19 (in Russian).
106. Balaban, A. T.; Harary, F. Chemical Graphs. V. Enumeration and Proposed Nomenclature of Benzenoid Cata-Condensed Polycyclic Aromatic Hydrocarbons. *Tetrahedron* **1968**, *24*, 2505-2516.
107. Bonchev, D.; Temkin, O. N.; Kamenski, D. On the Complexity of Linear Reaction Mechanisms. *React. Kinet. Catal. Lett.* **1980**, *15*, 119–124.

108. Bonchev, D.; Kamenski, D.; Temkin, O. N. Complexity Index for the Linear Mechanisms of Chemical Reactions. *J. Math. Chem.* **1987**, *1*, 345–388.

109. Morgan, H. L. Generation of Unique Machine Description of Chemical Structures. *J. Chem. Doc.* **1965**, *5*, 107–112.

110. Wipke, W. T.; Dyott, T. M. Stereochemically Unique Naming Algorithm. *J. Am. Chem. Soc.* **1974**, *96*, 4834–4842.

111. Jochum, C.; Gasteiger, J. Canonical Numbering and Constitutional Symmetry. *J. Chem. Inf. Comput. Sci.* **1977**, *17*, 113–117.

112. Randić, M. On Unique Numbering of Atoms and Unique Codes for Molecular Graphs. *J. Chem. Inf. Comput. Sci.* **1975**, *15*, 105–108; On Canonical Numbering of Atoms and Graph Isomerism. *J. Chem. Inf. Comput. Sci.* **1977**, *17*, 171–180; Randić, M.; Brissey, G. M.; Wilkins, C. L. Computer Perception of Topological Symmetry via Canonical Numbering. *J. Chem. Inf. Comput. Sci.* **1981**, *21*, 52–59.

113. *Algorithmic Investigations in Combinatorics*; Faradzhev, I. A., Ed.; Nauka: Moscow, 1978, pp 11–19 (in Russian).

114. Shelley, C. A.; Munk, M. E. An Approach to the Assignment of Canonical Connection Tables and Topological Symmetry Perception. *J. Chem. Inf. Comput. Sci.* **1979**, *19*, 247–250;

115. Brandt, J.; von Scholley, A. An Efficient Algorithm for the Computation of the Canonical Numbering of Reaction Matrices. *Comput. Chem.* **1983**, *7*, 51–59.

116. Balaban, A. T.; Mekenyan, O.; Bonchev, D. Unique Description of Chemical Structures Based on Hierarchically Ordered Extended Connectivities. I. Algorithms for Finding Graph Orbits and Canonical Numbering of Atoms, Unique Description of Chemical Structures Based on Hierarchically Ordered Extended Connectivities (HOC Procedures). *J. Chem. Inf. Comput. Sci.* **1985**, *6*, 538-551.

117. Stankevich, M. I.; Tratch, S. S.; Zefirov, N. S. Combinatorial Models and Algorithms in Chemistry. Search for Isomorphism and Automorphisms of Molecular Graphs. *J. Comput. Chem.* **1988**, *9*, 303–314.

118. Kvasnička, V.; Pospíchal, J. Canonical Indexing and Constructive Enumeration of Molecular Graphs. *J. Chem. Inf. Comput. Sci.* **1990**, *30*, 99–105 and references therein.

119. Razinger, M.; Balasubramanian, K.; Munk, M. E. Graph Automorphism Perception Algorithms in Computer-Enhanced Structure-Elucidation. *J. Chem. Inf. Comput. Sci.* **1993**, *33*, 197–201.

120. Hu, C.-Y.; Xu, L. A New Scheme for Assignment of a Canonical Connection Table. *J. Chem. Inf. Comput Sci.* **1994**, *34*, 840–844.

121. Bonchev, D.; Balaban, A. T. Topological Centric Coding and Nomenclature of Polycyclic Compounds. I. Condensed Benzenoid Systems (Polyhexes, Fusenes). *J. Chem. Inf. Comput. Sci.* **1981**, *21*, 223-229.

122. Bonchev, D. Principles of a Novel Nomenclature of Organic Compounds. *Pure Appl. Chem.* **1983**, *55*, 221-228.

123. Rozovskii, A. Y.; Lin, G. I. *Theoretical Foundations of the Methanol Synthesis Process*; Khimiya: Moscow, 1990 (in Russian).

124. Nagishkina, I. S.; Kiperman, S. L. Kinetics of Dehydrogenation of Formic Acid over a Nickel Catalyst. *Kinet. Katal.* **1965**, *6*, 1010–1017.

125. Kliger, G. A.; Bashkirov, A. N.; Glebov, L. S.; Lessik, O. A.; Marchevskaya, E. V.; Fridman, R. A. The Mechanism of Amination of Alcohols and Carbonyl-Containing Compounds in the Presence of Melted Iron Catalysts. In *Reports of the All-Union Conference on Catalytic Reaction Mechanisms*, Vol. 1. Nauka: Moscow, 1978, pp 251–258 (in Russian).

126. Yatsimirskii, K. B. *Application of the Graph Method in Chemistry*. Khimiya: Moscow, 1971 (in Russian).

127. Brailovskii, S. M.; Temkin, O. N.; Flid, R. M. Kinetics and Mechanism of the Oxidative Chlorination of Acetylene. *Kinet. Catal. Engl. Transl.* **1971**, *12*, 1025–1029.

128. Shestakova, V. C.; Kuperman, A. F.; Brailovskii, S. M.; Temkin, O. N. Kinetics and Mechanism of Carbon Monoxide Oxidation by Ferric Chloride in Aqueous Palladium Chloride Solutions. *Kinet. Catal. Engl. Transl.* **1981**, *22*, 279–283.

129. Brailovskii, S. M.; Temkin, O. N.; Kostyushin, A. S.; Odintsov, K. Y. Kinetics of Catalytic Synthesis of 1,1- and *trans*-1,2-Dichloroethylenes from Acetylene in the CuCl–CuCl₂–NH₄Cl-H₂O System. *Kinet. Katal.* **1990**, *31*, 1371–1376.

130. Kiperman, S. L. From the Kinetic Model to the Reaction Mechanism. In *Mechanism of Catalysis: The Nature of the Catalytic Action*; Boreskov, G. K.; Andruskevich, T. V., Eds.; Nauka: Moscow, 1984, Vol. 1.

131. Kiperman, S. L. Some Problems of Kinetic Modeling. In *Kinetics of Heterogeneous Catalytic Reactions*; Kiperman, S. L., Ed.; Inst. Chem. Phys.: Chernogolovka, 1983, pp 5–16.

132. Schmid, R.; Sapunov, V. N. *Non-Formal Kinetics*; Chemie: Weinheim, 1984, p. 264.

133. Howe, G. R.; Hiatt, R. R. Metal-Catalyzed Hydroperoxide Reactions. II. Molybdenum-Catalyzed Epoxidation of Styrene and Some Substituted Styrenes. *J. Org. Chem.* **1971**, *36*, 2493–2497.

134. Sobczak, J.; Ziolkowski, J. J. The Catalytic Epoxidation of Olefins with Organic Hydroperoxides. *J. Mol. Catal.* **1981**, *13*, 11–42.

135. Su, C. C.; Reed, J. W.; Gould, E. C. Metal Ion Catalysis of Oxygen-Transfer Reactions. II. Vanadium and Molybdenum Chelates as Catalysts in the Epoxidation of Cycloalkenes. *Inorg. Chem.* **1973**, *12*, 337–342.

136. Sapunov, V. N.; Margitfal'vi, I.; Lebedev, N. N. Epoxidation of Cyclohexene by Ethylbenzene Hydroperoxide in the Presence of Molybdenum Catalyst. I, II. *Kinet. Catal. Engl. Transl.* **1974**, *15*, 1046–1050, 1051–1054.

137. Sheldon, R. A. Synthetic and Mechanistic Aspects of Metal-Catalyzed Epoxidations with Hydroperoxides. *J. Mol. Catal.* **1980**, *7*, 107–126.

138. Metelitsa, D. I. Reaction Mechanism of the Direct Epoxidation of Alkenes in the Liquid Phase. *Russ. Chem. Rev.* **1972**, *41*, 807–821.

139. Skibda, I. P. Kinetics and Mechanism of the Decomposition of Organic Hydroperoxides in the Presence of Transition Metal Complexes. *Russ. Chem. Rev.* **1975**, *44*, 789–800.

140. Moiseev, I. I.; Vargavtic, M. N. Metal Complex Catalysis of Oxidation Reactions; Principles and Problems. *Russ. Chem. Rev.* **1990**, *59*, 1133–1149.

141. Tolstikov, G. A. *Oxidation Reactions by Hydroperoxides*; Nauka: Moscow, 1976 (in Russian).

142. Jørgensen, K. A. Transition-Metal-Catalyzed Epoxidations. *Chem. Rev.* **1989**, *89*, 431–458.

143. Fontain, E.; Bauer, J.; Ugi, I. Computer Assisted Bilateral Generation of Reaction Networks from Educts and Products. *Chem. Lett.* **1987**, 37–40.

144. Ugi, I.; Bauer, J.; Baumgartner, R.; Fontain, E., Forstmeyer, D.; Lohberger, S. Computer Assistance in the Design of Syntheses and a New Generation of Computer Programs for the Solution of Chemical Problems by Molecular Logic. *Pure Appl. Chem.* **1988**, *60*, 1573–1586.

145. Bauer, J.; Fontain, E.; Ugi, I. Computer-Assisted Bilateral Solution of Chemical Problems and Generation of Reaction Networks. *Anal. Chim. Acta* **1988**, *210*, 123–134.

146. Ugi, I.; Fontain, E.; Bauer, J. Transparent Formal Methods for Reducing the Combinatorial Abundance of Conceivable Solutions to a Chemical Problem: Computer-Assisted Elucidation of Complex Reaction Mechanisms. *Anal. Chim. Acta* **1990**, *235*, 155–161.

147. Fontain, E.; Bauer, J.; Ugi, I. Reaction Pathways on a PC. In *PCs for Chemists*; Zupan, J., Ed.; Elsevier: Amsterdam, 1990, pp 135–154.

148. Fontain, E.; Reitsam, K. The Generation of Reaction Networks with Rain. 1. The Reaction Generator. *J. Chem. Inf. Comput. Sci.* **1991**, *31*, 96–101.

149. Fontain, E. The Generation of Reaction Networks with Rain. 2. Resonance Structures and Tautomerism. *Tetrahedron Comput. Methodol.* **1990**, *3*, 469–477.

150. Yoneda, Y. A Computer Program Package for the Analysis, Creation, and Estimation of Generalized Reactions – GRACE. 1. Generation of Elementary Reaction Network in Radical Reactions – A/GRACE. *Bull. Chem. Soc. Jpn.* **1979**, *52*, 8–14.

151. Shtokolo, L. I.; Shteingauer, L. G.; Likholobov, V. A.; Fedotov, A. V. Generation of Mechanisms of Chemical Reactions. Gas-Phase $CO + H_2$ Reaction in the Absence of Catalysts. *React. Kinet. Catal. Lett.* **1984**, *26*, 227–233.

152. Likholobov, V. A.; Mayer, L. I.; Bulgakov, N. N.; Fedotov, A. V.; Shteingauer, L. G. Computer-Aided Prognosis of the Catalytic Properties of Metal Atoms in the Gas-Phase Hydrogenation of Carbon Monoxide; In *Homogeneous and Heterogeneous Catalysis*; Yermakov, Yu.; Likholobov, V., Eds.; Inst. of Catalysis: Novosibirsk, 1986, pp 229–244.

153. Shteingauer, L. G.; Mayer, L. I.; Bulgakov, N. N.; Fedotov, A. V.; Likholobov, V. A. Generation of Chemical Reaction Mechanisms. CO Hydrogenation Catalyzed by Metal Atoms in the Gas Phase. *React. Kinet. Catal. Lett.* **1988**, *36*, 139–143.

154. Zabolotnaya, L. G. Algorithms and Programs To Predict the Catalytic Activity of Metal Atoms in Reactions Catalyzed with Metal Complexes. *Ph.D. Dissertation.* Inst. of Catalysis: Novosibirsk, 1989(in Russian).

155. Maier, L. I. Prediction of Catalytic Properties of Metal Atoms Using the Methods of Computer Synthesis by the Example of CO Hydrogenation. *Ph.D. Dissertation.* Inst. of Catalysis: Novosibirsk, 1990 (in Russian).

156. Barone, R.; Chanon, M.; Green, M. L. H. Possible Mechanisms for Ethylene Dimerization Derived by Use of a New Computer Program. *J. Organometal. Chem.* **1980**, *185*, 85–93.

157. Theodosiou, I.; Barone, R.; Chanon, M. New Computer Program Able To Propose Mechanistic Schemes and Secondary Products for Transformations Catalyzed by Transition Metal Complexes: Possible Mechanisms for Hydroformylation of Ethylene. *J. Mol. Catal.* **1985**, *32*, 27–50.

158. Theodosiou, I.; Barone, R.; Chanon, M. Computer Aids for Organometallic Chemistry and Catalysis. *Adv. Organometal. Chem.* **1986**, *26*, 165–216.

159. Ostrovskii, G. M.; Zyskin, A G.; Snagovskii, Yu. S. Computer-Assisted Kinetic Modeling of complex heterogeneous catalytic reactions. In *Physical Chemistry: Current Problems*; Kolotyrkin, Ya. M., Ed.; Khimiya: Moscow, 1986, pp 84–155.

160. Snagovskii, Yu. S.; Ostrovskii, G. M. Kinetic Modeling of Heterogeneous Catalytic Processes; Khimiya: Moscow, 1976 (in Russian).

161. Valdés-Pérez, R. E. Machine Discovery of Chemical Reaction Pathways. *Ph.D. Thesis*, Carnegie Mellon University, 1990 (School of Computer Science, CMU-CS-90-191).

162. Valdés-Pérez, R. E. Algorithm to Generate Reaction Pathways for Computer-Assisted Elucidation, *J. Comput. Chem.* **1992**, *13*, 1079–1088.

163. Valdés-Pérez, R. E. Symbolic Computing on Reaction Pathways. *Tetrahedron Comput. Methodol.* **1990**, *3*, 277–285.

164. Valdés-Pérez, R. E. Human/Computer Interactive Elucidation of Reaction Mechanisms: Application of Catalyzed Hydrogenolysis of Ethane. *Catal. Lett.* **1994**, *28*, 79–87.

165. Valdés-Pérez, R. E. Heuristics for Systematic Elucidation of Reaction Pathways. *J. Chem. Inf. Comput. Sci.* **1994**, *34*, 976–983.

166. Valdés-Pérez, R. E. Algorithm to Test the Structural Plausibility of a Proposed Elementary Reaction. *J. Comput. Chem.* **1993**, *14*, 1454–1459.

167. Valdés-Pérez, R. E. Algorithm to Infer the Structures of Molecular Formulas within a Reaction Pathway, *J. Comput. Chem.* **1994**, *15*, 1266–1277.

168. Valdés-Pérez, R. E. Conjecturing Hidden Entities by Means of Simplicity and Conservation Laws: Machine Discovery in Chemistry. *Artificial Intelligence*, **1994**, *65*, 247–280.

169. Kamenski, D.; Temkin, O. N.; Bonchev, D. Reaction Network for the Epoxidation Reaction of Alkenes with Organic Hydroperoxides. *Appl. Catal. A* **1992**, *88*, 1–22.

170. Jacimirskij, K. B. Anwendung der Graphenmethode in der Chemie. *Z. Chem.* **1973**, *13*, 201–213.

171. Yablonskii, G. S.; Bykov, V. I.; Gorban', A. N. *Kinetic Models of Catalytic Reactions*; Nauka: Novosibirsk, 1983 (in Russian). This book contains a comprehensive review of graph-theory applications in chemical kinetics.

172. Cornish-Bowden, A. *Fundamentals of Enzyme Kinetics*; Butterworth: London, 1979.

173. Berezin, I. V.; Klesov, A. A. *Practical Course of Chemical and Enzymatic Kinetics*; Moscow State Univ.: Moscow, 1976 (in Russian).

174. Vol'kenshtein, M. V. *Physics of Enzymes*; Nauka: Moscow, 1967 (in Russian).

175. Petrov, L. A. Application of Graph Theory to Study of the Kinetics of Heterogeneous Catalytic Reactions. In *Chemical Graph Theory: Reactivity and Kinetics*; Bonchev, D., Rouvray, D. H., Eds.; Abacus Press/Gordon & Breach: Philadelphia, 1992, pp 1–52.

176. Petrov, L. A. Application of Graph Theory in Chemical Kinetics. In *Kinetics of Heterogeneous Catalytic Reactions*; Kiperman, S. L., Ed.; Inst. of Chem. Phys.: Chernogolovka, 1983, pp 39–47 (in Russian).

177. Kiperman, S. L. *Fundamentals of Chemical Kinetics of Heterogeneous Catalysis*; Khimiya: Moscow, 1979 (in Russian).

178. Bezdenezhnykh, A. A. *Engineering Methods for the Derivation of Rate Laws and Estimation of Kinetic Constants*; Khimiya: Leningrad, 1973 (in Russian).

179. Temkin, O. N.; Bonchev, D. G. Application of Graph Theory to Chemical Kinetics. Part 1. Kinetics of Complex Reactions. *J. Chem. Educ.* **1992**, *69*, 544–550.

180. Horiuti, J. Stoichiometrische Zahlen und die Kinetik der chemischen Reactionen. *J. Res. Inst. Catal., Hokkaido Univ.* **1957**, *5*, 1–26.

181. Horiuti, J. How To Find Chemical Equation of a Reverse Reaction? In *Problems of Chemical Kinetics*; Goskhimizdat: Moscow, 1953, pp 39–55 (in Russian).

182. Horiuti, J.; Nakamura, T Stoichiometric Number and the Theory of Steady Reaction. *Z. Phys. Chem., Neue Folge* **1957**, *11*, 358–3655.

183. Temkin, M. I. Kinetics of Steady-State Reactions. *Dokl. Akad. Nauk SSSR* **1963**, *152*, 156–159.

184. Temkin, M. I. Kinetics of Complex Steady-State Reactions. In *Mechanism and Kinetics of Complex Catalytic Reactions*; Isagulyants, G. V.; Tret'yakov, I. I., Ed.; Nauka: Moscow, 1970, pp 57–71 (in Russian).

185. Temkin, M. I. Kinetics of Heterogeneous Catalytic Reactions. *Zh. Vses. Khim. Ob-va im. D.I. Mendeleeva* **1975**, *20*, 7–14; Theory of Steady-State Reactions. In *Fundamentals of the Catalyst Selection and Production*; USSR Academy of Sciences: Novosibirsk, 1964, pp 46–100; The Kinetics of Some Industrial Heterogeneous Catalytic Reactions. In *Advances in Catalysis*; Academic: New York 1979, Vol. 28, pp 173–291.

186. Zyskin, A G.; Snagovskii, Yu. S.; Ostrovskii, G. M. Some Problems of the Theory of Steady-State Complex Reactions. In *Chemical Kinetics in Catalysis*; Kiperman, S. L., Ed.; Inst. of Organic Chem.: Chernogolovka, 1985, pp 56–62.

187. Mason, S. J.; Zimmerman, H. J., *Electronic Circuits, Signals, and Systems*; Wiley: New York, 1963.

188. An analogy between graphs and electric networks as well as their relevance to linear algebraic equations was used even by Kirchhoff: Kirchhoff, G. Uber die Auflösung der Gleichungen, auf welche man bei der Untersuchung der linearen Verteilung galva-

nischer Ströme geführt wird. *Ann. Phys Chem.* **1847**, *72*, 497–508). One of the graph theoretical generalization of Kirchhoff laws can be found in the textbook written by Tutte: Tutte W. T. *Graph Theory*; Addison-Wesley: Reading, 1984.

189. Vol'kenshtein, M.V.; Gol'dshtein, B.N. Application of Graph Theory to Modeling of Complex Reactions. *Dokl. Akad. Nauk SSSR* **1966**, *170*, 963–965.

190. Volkenstein, M.V.; Goldstein, B.N. A New Method for Solving the Problems of Stationary Kinetics of Enzymological Reactions. *Biochim. Bipophys. Acta* **1966**, *115*, 471–477. For applications of this method see also: Volkenstein, M.V.; Goldstein, B.N. Allosteric Enzyme Models and Their Analysis by the Theory of Graphs. *Biochim. Bipophys. Acta* **1966**, *115*, 478–485. (Hereinafter, the transliteration of authors' names may vary from paper to paper.)

191. Vol'kenshtein, M.V.; Gol'dshtein, B.N. A New Method for Solving the Problems of Steady-State Kinetics of Enzymatic Reactions. *Biokhimiya* **1966**, *31*, 541–547.

192. Evstigneev, V. A.; Yablonskii, G.S. Proof of One Formula of the Steady-State Kinetics of Catalytic Reactions. *Kinet. Katal.* **1979**, *20*, 1549–1555.

193. Tutte W. T. *Graph Theory*; Addison-Wesley: Reading, 1984.

194. Chen, W.-K. On the Directed Trees and Directed *k*-Trees of a Digraph and Their Generation. *J. SIAM Appl. Math.* **1966**, *14*, 550–560.

195. Chen, W.-K. *Applied Graph Theory*; North Holland: Amsterdam, 1971 and references therein.

196. Mayeda, W. *Graph Theory*; Wiley: New York, 1972.

197. Christofides, N. *Graph Theory: An Algorithmic Approach*; Academic: New York, 1975 and references therein.

198. From, H. J. A Simplified Schematic Method for Deriving Steady-State Rate Equations Using a Modification of the "Theory of Graphs" Procedure. *Biochem. Biophys. Res. Commun.* **1970**, *40*, 692–697.

199. Orsi, B. A. A Simple Method for the Derivation of the Steady-State Rate Equations for an Enzymatic Mechanism. *Biochim. Biophys. Acta* **1972**, *258*, 4-8.

200. Petrov, L. A.; Shopov, D. M., A Method for the Construction of Kinetic Graph Determinants. *React. Kinet. Catal. Lett.* **1977**, *7*, 273-278.

201. Kuo-Chen Chou; Carter, R. E.; Forsén, S. A New Graphical Method for Deriving Rate Equations for Complicated Mechanisms. *Chem. Scripta* **1981**, *18*, 82–86. More recent papers of Kuo-Chen Chou contain some extensions of the algorithm onto the non-steady-state kinetics: Kuo-Chen Chou. Applications of Graph Theory to Enzyme Kinetics and Protein Folding Kinetics: Steady and Non-Steady-State Systems. *Biophys. Chem.* **1990**, *35*, 1–24; Graphic Rules in Steady and Non-Steady State Enzyme Kinetics. *J. Biol. Chem.* **1989**, *264*, 12074–12079.

202. Temkin, O. N.; Bonchev, D. Classification and Coding of Chemical Reaction Mechanisms. In *Graph Theory and Its Applications to Chemistry*; Tyutyulkov, N.; Bonchev, D., Eds.; Nauka Izkustvo: Sofia, 1987, pp166–200 (in Bulgarian).

203. Pryakhin, A. N. On Multiple Proportions in Steady-State Kinetics, *Vestn. Mosk. Univ.*, *Ser 2: Khim.* **1987**, *28*, 127–131.

204. Pryakhin, A. N. The Method to Calculate Equilibrium Constants in Reactions of Binding Reagents and Effectors with Enzyme in Steady-State Kinetics. *Zh. Fiz. Khim.* **1986**, *60*, 2288–2292.

205. Pryakhin, A. N. *Kinetic and Thermodynamic Analysis of Steady-State Rate Laws of Enzymatic Reactions*; Moscow State Univ.: Moscow, 1988 (in Russian).

206. Klibanov, M. V.; Slin'ko, M. G.; Spivak, S. I.; Timoshenko, V. I. Applications of Graph Theory to the Formulation of a Mechanism and Derivation of Rate Laws of a Complex Chemical Reaction. In *Controlled Systems (Upravlyaemye Sistemy)*; Institute of Mathematics: Novosibirsk, 1970, Vol. 7, pp 64–69 (in Russian).

207. Yablonskii, G. S.; Evstigneev, V. A.; Bykov, V. I. Graphs in Chemical Kinetics. In *Applications of Graph Theory in Chemistry*; Zefirov, N. S., Kuchanov, S. I., Eds.; Nauka: Novosibirsk, 1988, pp 70–143 (in Russian).

208. Vol'kenshtein, M. V.; Gol'dshtein, B. N.; Stefanov, V. E. The Study of Non-Steady-State Enzymatic Reactions by the Method of Graphs. *Mol. Biol.* (Moscow) **1967**, *1*, 52–58.

209. Gol'dshtein, B. N.; Magarshak, Yu. B.; Vol'kenshtein, M. V. Analysis of Multisubstrate Enzymatic Reactions Using the Method of Graphs. *Dokl. Akad. Nauk SSSR* **1970**, *191*, 1172–1174.

210. Temkin, O. N.; Zeigarnik, A. V.; Bonchev, D. G. Application of Graph Theory to Chemical Kinetics. Part 2. Topological Specificity of Single-Route Reaction Mechanisms. *J. Chem. Inf. Comput. Sci.* **1995**, *35*, 729–737.

211. Zeigarnik, A. V.; Temkin, O. N.; Bonchev, D. G. Application of Graph Theory to Chemical Kinetics. Part 3. Topological Specificity of Multiroute Reaction Mechanisms. *J. Chem. Inf. Comput. Sci.* (in press)

212. Temkin, O. N.; Shestakov G. K.; Treger Yu. A., *Acetylene: Chemistry, Reaction Mechanisms, and Technology*; Khimiya: Moscow, 1991 (in Russian).

213. Zeigarnik, A. V.; Temkin, O. N. A Graph-Theoretical Model of Complex Reaction Mechanisms: Bipartite Graphs and the Stoichiometry of Complex Reactions. *Kinet. Katal.* **1994**, *35*, 702–710; *Kinet. Catal. Engl. Transl.* **1994**, *35*, 691–701.

214. To describe an autocatalytic reaction or a reaction containing the same species on both sides of a chemical equation, the reaction network should include a cycle of length 2 in which two vertexes are joined by arcs having opposite directions.

215. Zeigarnik, A. V.; Temkin, O. N. Graph-Theoretical Model of Complex Reaction Mechanisms: A New Complexity Index for Reaction Mechanisms. *Kinet. Katal.* **1996**, *37*, 372–385; *Kinet. Catal. Engl. Transl.* **1996**, *37*, 347–360.

216. Happel, J.; Sellers, P. H. Multiple Reaction Mechanisms in Catalysis. *Ind. Eng. Chem. Fundam.* **1982**, *21*, 67–76.

217. Happel, J.; Sellers, P. H. The Characterization of Complex Systems of Chemical Reactions. *Chem. Eng. Commun.* **1982**, *83*, 221–240.

218. Gorskii, V. G. *Kinetic Experiment Design*; Nauka: Moscow, 1984 (in Russian).

219. Sellers, P. H. The Classification of Chemical Reaction Mechanisms from a Geometric Viewpoint. In *Chemical Applications of Topology and Graph Theory*; King, R. B., Ed.; Elsevier: Amsterdam, 1983, pp 420–429.

220. Recently, M. Mavrovouniotis proposed a generalization of the Sellers and Happel's idea of a direct route. He argued that, given a set of species and experimental evidence about their relative reactivities, one can regard part of them as intermediates, and the others as terminal ones. With this knowledge the overall equation is not needed. He also proposed a series of algorithms for the construction of the set of direct mechanisms and extended the method to the mechanisms in which steps are both reversible and irreversible reactions. Mavrovouniotis, M. L.; Stephanopoulos, G. Synthesis of Reaction Mechanisms Consisting of Reversible and Irreversible Steps. 1. A Synthesis Approach in the Context of Simple Examples. *Ind. Eng. Chem. Res.* **1992**, *31*, 1625–1637; Mavrovouniotis, M. L. Synthesis of Reaction Mechanisms Consisting of Reversible and Irreversible Steps. 2. Formalization and Analysis of the Synthesis Algorithm. *Ind. Eng. Chem. Res.* **1992**, *31*, 1637–1653. S. Schults and C. Hilgetag proposed the concept of an elementary flux mode, which is similar to the generalized concept of a direct route: Schults, S.; Hilgetag, C. On Elemntary Flux Modes in Biochemical Reaction Systems at Steady State. *J. Biol. Systems* **1994**, *2*, 165–182.

221. Wickham, D. T.; Koel, B. E. Steady-State Kinetics of The Catalytic Reduction of Nitrogen Dioxide by Carbon Monoxide on Palladium. *J. Catal.* **1988**, *114*, 207–216.

222. Banse, B. A.; Wickham, D. T.; Koel, B. E. Transient Kinetic Studies of the Catalytic Reduction of NO by CO o Platinum. *J. Catal.* **1989**, *119*, 238–248.
223. Lipski, W. *Combinatorics for Programmers*; Naukowo-Techniczne Wydawnictwa: Warsaw, 1982 (in Polish).
224. Clarke, B. L. General Method for Simplifying Chemical Networks while Preserving Overall Stoichiometry in Reduced Mechanisms. *J. Chem. Phys.* **1992**, *97*, 4066–4071.
225. Kurosh, A. *Higher Algebra*; Mir: Moscow, 1975.
226. Shchel'tsyn, L. V., Brailovskii, S. M.; Temkin, O. N. Kinetics and Mechanism of Transformations of Chloroalkynes in Solutions of Copper(I) and (II) Chlorides. *Kinet. Katal.* **1990**, *31*, 1316–1370; *Kinet. Catal.* **1990**, *31*, 1191–1200.
227. Valdés-Pérez, R. E. On the Concept of Stoichiometry of Reaction Mechanisms. *J. Phys. Chem.* **1991**, *95*, 4918–4921.
228. Valdés-Pérez, R. E. A Necessary Condition for Catalysis in Reaction Pathways the Concept of Stoichiometry of Reaction Mechanisms. *J. Phys. Chem.* **1992**, *96*, 2394–2396.
229. Zeigarnik, A. V. A Graph-Theoretical Model of Complex Reaction Mechanisms. Basic Principles of Classifying Complex Reactions. *Kinet. Katal.* **1995**, *36*, 653–657.
230. Moiseev, I. I. Catalysis by Palladium Clusters. In *The Mechanism of Catalysis: The Nature of the Catalytic Action*; Boreskov, G. K., Andrushkevich, T. V., Eds.; Nauka: Novosibirsk, 1984, Vol. 1, pp 72–86.

Chapter 3

Classification of Reaction Mechanisms Based on Bipartite Graphs

3.1. Classification of Simple Submechanisms

In Chapter 2, we discussed several basic concepts such as the simple submechanism, the subnetwork of intermediates, etc. In this chapter, the discussion is completely based on these concepts.

Simple submechanisms are minimum balanced structural units underlying the structure of reaction networks. Therefore, the classification of reaction mechanisms can be based on the classification of simple submechanisms. Simple submechanisms can be of two types: catalytic (**C**) and noncatalytic (**N**).[1]

A catalyst is a species that is introduced to the reaction system, undergoes consecutive transformations, and returns to its initial state within one turnover of the reaction. The catalysis implies the presence of several (intermediate) species, each being the catalytic entity. If one of these entities is introduced into the system instead of the catalyst, the reaction scheme is not changed. The catalyst is a non-stoichiometric species in the sense that it is not involved in the overall equation; that is, a catalytic submechanism should be balanced. Two features of the catalytic mechanism are of importance for our analysis:

Condition 1. Each intermediate species is produced from another intermediate species (or several species).

Condition 2. Each intermediate species produce another intermediate species (or several species).

Note that these conditions are not equivalent.

Consider the mechanism of ammonia synthesis proposed by M.I. Temkin.[2] This is an example of the mechanisms that consists of one simple submechanism:

$$(1) \ Z + N_2 \rightarrow ZN_2,$$

$$(2) \ ZN_2 + H_2 \rightarrow ZN_2H_2,$$

$$(3) \ ZN_2H_2 + Z \rightarrow 2ZNH,$$

$$(4) \ ZNH + H_2 \rightarrow Z + NH_3.$$

$$(3.1)$$

The overall equation is as follows:

$$N_2 + 3H_2 \rightarrow 2NH_3.$$

The reaction network is shown in Figure 3.1a. The removal of terminal w-vertexes results in the subnetwork of intermediates Figure 3.1b, which is the graph called *SISS* (see Chapter 2).

This example illustrates the simple idea that the subnetwork of intermediates of simple C-submechanism does not contain terminal u-vertexes (that is, vertexes having zero indegree or outdegree), while terminal w-vertexes were deleted when deriving the subnetwork of intermediates from the reaction network. In Christiansen's terms, catalytic submechanisms are *closed* sequences of steps.[3]

A noncatalytic simple submechanism is *open* in the sense that terminal u-vertexes can be involved in the subnetwork of intermediates. This means that Condition 1 or 2 or both are violated. A typical example of the noncatalytic network is the mechanism of the Kharash reaction:[4]

$$(1) \ PhMgBr + CoCl_2 \rightarrow PhCoCl + MgBrCl,$$

$$(2) \ 2PhCoCl \rightarrow Ph_2 + 2^{\cdot}CoCl,$$

$$(3) \ ^{\cdot}CoCl + PhBr \rightarrow CoClBr + ^{\cdot}Ph,$$

$$(4) \ 2^{\cdot}Ph \rightarrow Ph_2.$$

$$(3.2)$$

The reaction network and the subnetwork of intermediates of this reaction are shown in Figure 3.2. The overall equation corresponding to the minimum integer stoichiometric numbers is as follows:

$$2PhMgBr + 2CoCl_2 + 2PhBr \rightarrow 2Ph_2 + 2MgBrCl + 2 \ CoClBr.$$

The example of a simple submechanism in Figure 3.2b shows that the subnetwork of intermediates of a noncatalytic submechanism may contain two terminal u-vertexes. One of these vertexes begins the transformations of species (u_1), while the other completes transformations (u_4). These two vertexes are termed *source* and *sink* of *SISS*, respectively. Thus, a C-submechanism contains neither source nor sink. An N-submechanism must contain a terminal u-vertex or several u-vertexes, among which can be sources and sinks. That is, a noncatalytic simple submechanism contains an intermediate that produces (or is generated by) terminal species. Table 3.1 illustrates different variants of N-submechanisms. As can be seen from this table, an N-submechanism can involve cycles, including those with regular orientation. The presence of cycles with regular orientation (also termed *circuits*) points to the possible manifestation of exotic dynamic behavior, such as oscillations, multiplicity of steady states, etc. Both catalytic and noncatalytic simple submechanisms may contain circuits and display these phenomena.

Table 3.1 also shows that noncatalytic simple submechanisms may be sink-free or source-free. In submechanisms without a source, multiplication of intermediates in branching steps is compensated by sinks. If a submechanism has no sink, the source of intermediates compensates for their coupling in elementary reactions. N-Submechanisms may also be classified according to the presence or absence of cyclic subgraphs in *SISS*. Informally, these cyclic subgraphs induce a reaction chain like those in catalytic reactions, but as differentiated from catalytic chains, noncatalytic chains are always unbalanced. Only the presence of a sink (or a source) makes the entire submechanism balanced. For example record no. 5 in Table 3.1 presents the subnetwork of intermediates for the simple N-submechanism. This subnetwork obtains a circuit $Z \rightarrow \blacksquare \rightarrow X \rightarrow \blacksquare \rightarrow Y \rightarrow \blacksquare \rightarrow Z$, which involves first, second, and third steps. These steps are unbalanced because intermediate W cannot vanish in the overall equation unless the fourth step is added.

To distinguish between submechanisms that contain and do not contain chains, one should apply the following algorithm:[5]

Step 1. Construct the *SISS* of the N-submechanism.
Step 2. Delete all terminal u-vertexes.
Step 3. Delete all terminal w-vertexes.
Step 4. If any of terminal u-vertexes remain after the deletion of terminal w-vertexes repeat steps 2 and 3. If there are no terminal u-vertexes go to step 5.
Step 5. Check whether *SISS* is transformed to the "residual" subnetwork or the entire *SISS* is deleted. In the former case the N-submechanism is categorized as N_1; in the latter case the N-submechanism is categorized as N_2.

For instance, consider the mechanism of a noncatalytic reaction placed as record no. 4 in Table 3.1. The *SISS* for this mechanism is placed in the second column. It contains four terminal u-vertexes, which should be deleted (step 2). Step 2 produces a graph with four terminal w-vertexes (w_X, w_Y, w_Z, and w_W). They should be deleted (step 3). Then, step 2 should be repeated because a terminal u-vertex remains after step 3. As a result, the entire *SISS* is deleted and does not contain circuits. Consider, next, record no. 5. The *SISS* contain only one terminal u-vertex, which should be deleted (step 2). Then, the terminal vertex w_W should be deleted (step 3). After steps 2 and 3, we arrive at a graph, which is termed the residual subnetwork. It does not contain further terminal vertexes and, therefore, contains circuits. These examples show that the fourth and fifth records in the table contain mechanisms of N_1 and N_2 type, respectively.

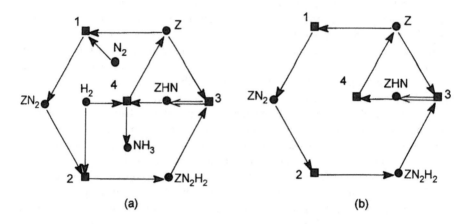

Figure 3.1. (a) The reaction network and (b) the subnetwork of intermediates for the mechanism of ammonia synthesis (3.1).

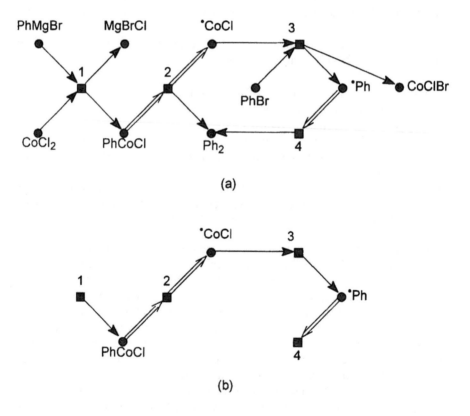

Figure 3.2. (a) The reaction network and (b) the subnetwork of intermediates for the mechanism of the Kharash reaction (3.2).

Table 3.1. Different variants of N-submechanisms

No.	Scheme	SISS	Description
1	$A + B \rightarrow X$ $X \rightarrow Y + P_1$ $2Y \rightarrow P_2 + P_3$		One source and one sink
2	$A \rightarrow X$ $B \rightarrow Y$ $X + Y \rightarrow Z$ $2Z \rightarrow P$		Several (two) sources and one sink
3	$A \rightarrow Z$ $2Z \rightarrow X + Y$ $X \rightarrow P_1$ $Y \rightarrow P_2$		One source and several (two) sinks
4	$A \rightarrow X$ $B \rightarrow Y$ $X + Y \rightarrow Z + W$ $Z \rightarrow P_1$ $W \rightarrow P_2$		Several (two) sources and several (two) sinks
5	$A + Z \rightarrow W + X$ $B + X \rightarrow Y + P_1$ $C + Y \rightarrow Z + W$ $2W \rightarrow 2P_2$		No source and one sink
6	$2A \rightarrow 2W$ $Z + W \rightarrow P_1 + Y$ $Y + B \rightarrow X + P_2$ $W + X \rightarrow Z + P_2$		No sink and one source
7	$A + Z \rightarrow W + X$ $B + X \rightarrow Y + P_1$ $C + Y \rightarrow Z + W$ $2W \rightarrow V + Q$ $V \rightarrow P_2$ $Q \rightarrow P_2$		No source and several (two) sinks
8	$A \rightarrow Q$ $A \rightarrow V$ $Q + V \rightarrow 2W$ $Z + W \rightarrow P_1 + Y$ $Y + B \rightarrow X + P_2$ $W + X \rightarrow Z + P_2$		No sink and several (two) sources

	SISS contains regular cycles	*SISS* does not contain regular cycles
SISS contains terminal *u*-vertexes	N_2	N_1
SISS does not contain terminal *u*-vertexes	C	impossible situation

Figure 3.3. The classificational scheme for simple submechanisms.

Apparently, if the *SISS* contains circuits, the submechanisms belongs to N_2-type; if not, the submechanisms is of N_1-type. Thus, simple submechanisms can be **C**, N_1, or N_2, and are defined by the classificational scheme shown in Figure 3.3.

The classificational scheme of simple submechanisms poses an intriguing issue: whether a circuit-free digraph[6] must contain vertexes with zero outdegree (or indegree). The two theorems that follows are extracted from Harary's textbook and provides the solution to this problem.[7]

Theorem 3.1. *A circuit-free digraph contains at least one vertex having zero outdegree.*

Theorem 3.2. *A circuit-free digraph contains at least one vertex having zero indegree.*

Because *SISS* does not contain terminal *w*-vertexes, one of the *u*-vertexes must have zero indegree (outdegree). A *u*-vertex that has zero indegree (outdegree) is terminal.

3.2. Submechanism Graphs

Suppose the mechanism contains several simple submechanisms. Subnetworks of intermediates of simple submechanisms may have common subgraphs. If two simple submechanisms have a common subgraph, we say that these submechanisms are *conjugated*. In this case, overall reactions that correspond to these submechanisms are also said to be conjugated. This term is widely used in mechanistic chemistry. Note, however, that for the conjugation of two overall reactions, the overlap of *SISSs* rather than overlap of *RNs* of simple submechanisms is needed. The types of simple submechanisms and their conjugation define the class of the overall mechanism and the submechanism graph

Definition 3.1. *The submechanism graph is a triple* $SG(\vartheta, \Sigma, \Lambda)$ *in which*
    1. ϑ *is the set of vertexes;*

2. Σ is the set of edges; and

3. Λ is the set of vertex labels.

Each vertex corresponds to the simple submechanism or a defect of the mechanism. An edge joins two vertexes if the corresponding submechanisms are conjugated. An edge is multiple if the maximum common subgraph of the corresponding SISSs is disconnected. The set of vertex labels is $\Lambda = \{C, N_1, N_2, D\}$ (letter D denotes a defect).

To illustrate the concept of the submechanism graph (SG), consider the mechanism of butane hydrogenation (for the sake of simplicity, some mechanistic steps are treated as irreversible):[8]

$$(1)\ Z + C_4H_{10} \rightarrow ZC_4H_{10},$$

$$(2)\ ZC_4H_{10} \rightarrow ZC_4H_8 + H_2,$$

$$(3)\ ZC_4H_8 \rightarrow ZC_4H_6 + H_2,$$

$$(4)\ ZC_4H_8 \rightarrow Z + C_4H_8, \qquad\qquad (3.3)$$

$$(5)\ ZC_4H_6 \rightarrow Z + C_4H_6,$$

$$(6)\ ZC_4H_6 + ZC_4H_{10} \rightarrow 2ZC_4H_8,$$

$$(7)\ Z + C_4H_6 \rightarrow ZC_4H_6.$$

The subnetwork of intermediates of this mechanism is shown in Figure 3.4.

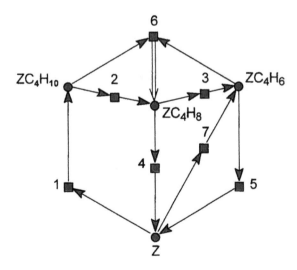

Figure 3.4. The subnetwork of intermediates of mechanism (3.3).

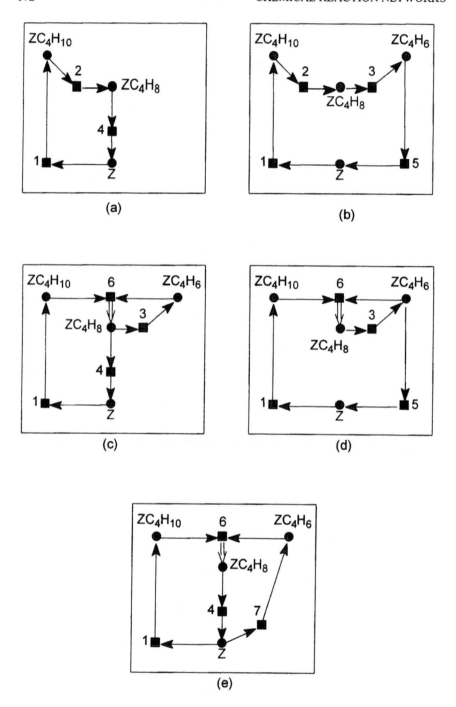

Figure 3.5. *SISS*s for mechanism (3.3) and their overall equations: (a) $C\langle 1,2,4\rangle$, $C_4H_{10} \rightarrow C_4H_8 + H_2$; (b) $C\langle 1,2,3,5\rangle$, $C_4H_{10} \rightarrow C_4H_6 + 2H_2$; (c) $C\langle 1,3,4,6\rangle$, $C_4H_{10} \rightarrow C_4H_8 + H_2$; (d) $C\langle 1,3,5,6\rangle$, $C_4H_{10} \rightarrow C_4H_6 + 2H_2$; (e) $C\langle 1,4,6,7\rangle$, $C_4H_{10} + C_4H_6 \rightarrow 2C_4H_8$.

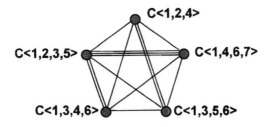

Figure 3.6. The submechanism graph for mechanism (3.3).

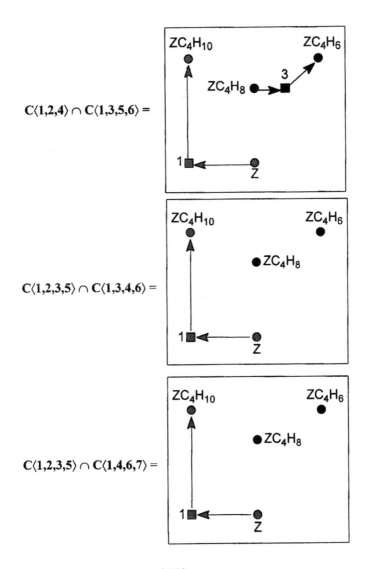

Figure 3.7. Maximum common subgraphs of *SISS*s.

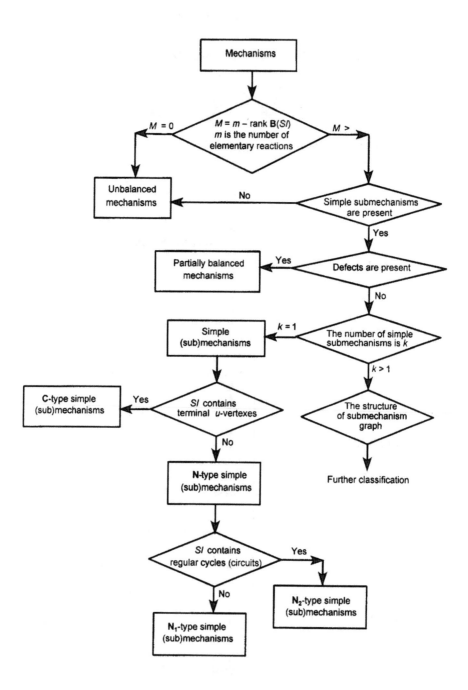

Figure 3.8. Classificational scheme of reaction mechanisms.

As was shown in Chapter 2, mechanisms can be balanced, partially balanced, or unbalanced. The submechanism graph of a partially balanced mechanism contains vertexes labeled with **D**. All unbalanced mechanisms are described by the same submechanism graph because they consist of the single defect. Balanced and partially balanced mechanisms can be classified in terms of the submechanism graphs. Sometimes the structure of the submechanism graph correlates with commonly accepted concepts, which are widely accepted by chemists. Typical examples are catalytic, noncatalytic conjugated, and branched and nonbranched chain reactions. Before proceeding to the analysis of mechanisms that consist of several simple submechanisms, let us summarize the above discussion by introducing the general classification of complex mechanisms. This classification can be represented in the form of a block-scheme in which diamonds represent classification criteria, and rectangles represent classes and subclasses (Figure 3.8).

3.3. Balanced Mechanisms

3.3.1. Purely Catalytic Reactions

Purely catalytic reactions occur via mechanisms from which only **C**-type submechanisms can be extracted. The submechanism graph of a purely catalytic reaction contains only vertexes that are labeled with letter **C**. This simplifies both analytical and constructive enumeration of submechanism graphs. Thus, a method for analytical enumeration was reported by Cadogan.[9,10] To conduct the constructive enumeration one can adopt one of the procedures applied previously to the generation of carbon skeletons.[11] Even deletion of symbols of carbon atoms will not be needed!

An example of purely catalytic reaction is given by the simplified mechanism of CO to CO_2 oxidation on the heterogeneous catalyst surface:[12]

$$(1)\ O_2 + 2Z \rightarrow 2ZO,$$
$$(2)\ CO + Z \rightarrow ZCO,$$
$$(3)\ ZO + ZCO \rightarrow 2Z + CO_2, \qquad (3.4)$$
$$(4)\ ZO + CO \rightarrow Z + CO_2.$$

The reaction network, subnetworks of intermediates for the overall mechanism and simple submechanisms, and the submechanism graph are shown in Figure 3.9. Another example of purely catalytic reaction is represented by the submechanism graph shown in Figure 3.6 and reactions (3.3).

3.3.2. Noncatalytic Conjugated Reactions

This class of reactions describes mechanisms involving at least two simple **N**-submechanisms and not involving **C**-submechanisms. Although a variety of noncatalytic simple submechanisms can exist (see Table 3.1 and Figure 3.3), non-catalytic conjugated reactions are traditionally thought of as those occurring via N_1-submechanisms only. Others are usually regarded as chain reactions. The sub-mechanisms must have a common subgraph to make the submechanism graph connected. The conjugation reveals itself in the fact that if two *SISS*s have a common subgraph, one of the overall reactions cannot be observed in the absence of another. An example of the noncatalytic reaction was discussed in Ref. 5. Thus, the addition of acetylene to the system containing the Fenton reagent (H_2O_2–Fe^{2+}) produces two simple submechanisms whose overall equations are

$$H_2O_2 + 2Fe^{2+} \rightarrow 2OH^- + 2Fe^{3+}, \tag{3.5}$$

$$H_2O_2 + 2Fe^{2+} + C_2H_2 + H_2O \rightarrow CH_3CHO + 2OH^- + 2Fe^{3+}. \tag{3.6}$$

Although equation (3.5) seems to be part of equation (3.6), these are quite different reactions. It is important that reaction (3.6) cannot occur without reaction (3.5). At the same time, the reaction

$$C_2H_2 + H_2O \rightarrow CH_3CHO$$

does not occur at all in this system. It is noteworthy that the latter reaction and re-action (3.5) are not stoichiometrically independent in this case.

Another example of noncatalytic conjugated reaction mechanism is the mechanism of the reaction between the Grignard reagent and dibromoethane:[13]

$$(1) \; RMgCl + BrCH_2CH_2Br \rightarrow R^{\cdot} + BrCH_2CH_2^{\cdot} + MgClBr,$$

$$(2) \; R^{\cdot} + BrCH_2CH_2^{\cdot} \rightarrow RCH_2CH_2Br, \tag{3.7}$$

$$(3) \; RMgCl + BrCH_2CH_2^{\cdot} \rightarrow R^{\cdot} + CH_2{=}CH_2 + MgClBr,$$

$$(4) \; 2R^{\cdot} \rightarrow R{-}R.$$

This mechanism contains two simple N_1-submechanisms, $N_1\langle 1,2 \rangle$ and $N_1\langle 1,3,4 \rangle$, with overall equations (3.8) and (3.9), respectively:

$$RMgCl + BrCH_2CH_2Br \rightarrow RCH_2CH_2Br + MgClBr, \tag{3.8}$$

$$2RMgCl + BrCH_2CH_2Br \rightarrow + 2MgClBr + CH_2{=}CH_2. \tag{3.9}$$

Subnetworks of intermediates of the overall mechanism and simple submecha-nisms and the submechanism graph for the set of reactions (3.7) are shown in Figure 3.10.

It is noteworthy that, theoretically, the simple submechanisms involved in the mechanism of conjugated reactions may correspond to the same overall equation. In other words, two similar processes, which can be expressed by the same overall equation, may be conjugated. In this case, a reaction goes by different routes, each starting from similar reactants and ending at similar products. A situation when overall equations of conjugated reactions have opposite directions is impossible because in this case all reagents and products become intermediates and the simple submechanism is of **C**-type.

3.3.3. Chain Reactions

Although numerous examples of chain reactions are discussed in the literature, there is little agreement between different authors in defining the chain process. Many aspects of chain mechanisms are usually under consideration except for the structure of the mechanism. Thus, in many classical textbooks on chemical kinetics, chain mechanisms are associated with free radicals as carriers of kinetic chains. In several papers, authors categorize the reaction as a chain process relying on the form of the rate law.[14,15] The growth of the polymer chain is often misinterpreted as chain propagation. For instance, "living" ionic polymerization having the overall equation

$$A + nB \rightarrow AB_n$$

occurs via the mechanism involving consecutive monomer (B) addition to the initiator (A):

$$A + B \xrightarrow{k_0} AB,$$

$$AB + B \xrightarrow{k_1} AB_2,$$

$$AB_2 + B \xrightarrow{k_2} AB_3,$$

$$\ldots,$$

$$AB_{n-1} + B \xrightarrow{k_{n-1}} AB_n.$$

Although rate constants of elementary steps are very close to each other ($k_1 \approx k_2 \approx k_3 \approx \ldots \approx k_{n-1}$), these steps are different. Therefore, the word "chain" refers to a polymer chain rather than to a kinetic chain. It will become evident from the succeeding discussion that this is a single route. For the reaction to be a chain one, at least the presence of several routes (simple submechanisms) is necessary. Thus, it is beyond reason to categorize this reaction as a chain one.

In our approach, we proceeded from the common fact that a chain mechanism involves several typical steps: initiation, propagation, and termination steps (and, in the case of branched chain reactions, branching steps), which are defined intuitively. We examined a rich variety of chain reactions and found that graph-theoretical standpoint lends a clearer understanding of the concept of the chain *mechanism*.[5] The reaction mechanism structure directly answers the question of whether the reaction is a chain one or not. Surprisingly, the graph-theoretical analysis revealed that nonbranched and branched chain mechanisms are structurally distinct cases. Thus, it was shown that the necessary condition for the nonbranched chain mechanism is the presence of conjugated **C**- and N_1-submechanisms.

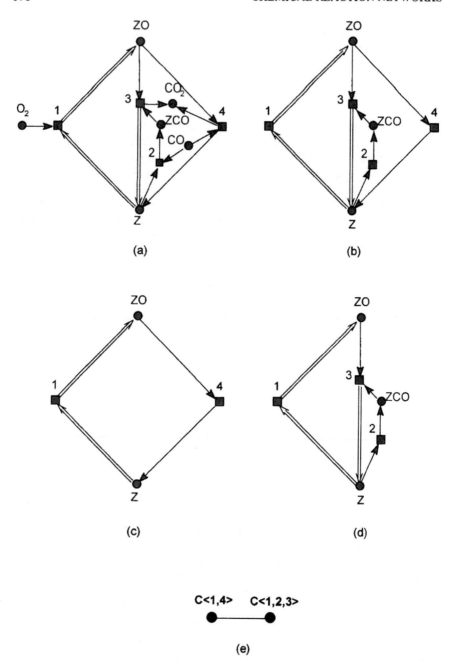

Figure 3.9. (a) The reaction network; (b) the subnetwork of intermediates of the overall mechanism; (c) and (d) subnetworks of intermediates of simple submechanisms $C\langle 1,4 \rangle$ and $C\langle 1,2,3 \rangle$, respectively; (e) the submechanism graph for mechanism (3.4). The overall equation is the same for both submechanisms: $2CO + O_2 \rightarrow 2CO_2$.

The presence of one submechanism of each type is the obligatory minimum. The mechanism of a branched chain reaction involves N_1- and N_2-submechanisms. The presence of C-submechanisms is possible but not necessary.

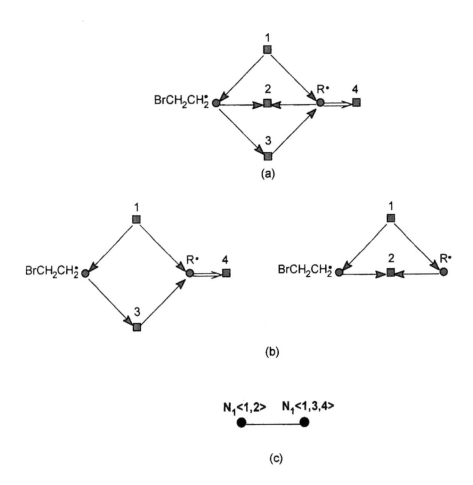

Figure 3.10. (a) The subnetwork of intermediates of the overall mechanism, (b) subnetworks of intermediates of simple submechanisms $N_1\langle1,3,4\rangle$ and $N_1\langle1,2\rangle$, and (c) the submechanism graph for the mechanism (3.7).

Nonbranched Chain Reactions. Many typical examples of nonbranched chain reactions are discussed in the textbook of Emanuel' and Knorre.[16] Mechanisms discussed in this textbook are somewhat simplified, but they serve well as an illustration. For instance, the photochemical chlorinaion of formic acid can be described by the following set of reactions:

(1) $Cl_2 \rightarrow 2Cl^{\cdot}$,   initiation

(2) $Cl^{\cdot} + HCOOH \rightarrow HCl + {}^{\cdot}COOH$,   propagation

(3) $Cl_2 + {}^{\cdot}COOH \rightarrow HCl + CO_2 + Cl^{\cdot}$,   propagation   (3.10)

(4) $Cl^{\cdot} + wall \rightarrow Cl \cdot wall$,   termination

Overall equations of submechanisms are as follows:

$$Cl_2 + 2wall \rightarrow 2Cl \cdot wall,$$

$$Cl_2 + HCOOH \rightarrow 2HCl + CO_2.$$

Examples of nonradical chain reactions can be found in Refs. 5, 17, and 18. Thus, isotope-labeled CO exchange according to the reaction

$$Co_2(CO)_8 + {}^{14}CO \rightarrow Co_2(CO)_7{}^{14}CO + CO$$

occurs via the following nonbranched chain mechanism:

(1) $Co_2(CO)_8 \rightarrow Co_2(CO)_7 + CO$,

(2) $Co_2(CO)_7 + {}^{14}CO \rightarrow Co_2(CO)_6{}^{14}CO + CO$,   (3.11)

(3) $Co_2(CO)_8 + Co_2(CO)_6{}^{14}CO \rightarrow Co_2(CO)_7 + Co_2(CO)_7{}^{14}CO$,

(4) $Co_2(CO)_7 + {}^{14}CO \rightarrow Co_2(CO)_7{}^{14}CO$.

Both submechanisms have the same overall equation:

$$Co_2(CO)_8 + {}^{14}CO \rightarrow Co_2(CO)_7{}^{14}CO + CO.$$

Subnetworks of intermediates for mechanisms (3.10) and (3.11) are shown in Figure 3.11. As can be seen, these subnetworks are almost identical except for vertex labels and multiplicity of the edge incident to vertex w_1 (stoichiometry of the first step). Reaction mechanisms contain two simple submechanisms each [$N_1\langle 1,4 \rangle$ and $C\langle 2,3 \rangle$], which have a common subgraph consisting of only one vertex. This example shows that radicals are not necessary for the mechanism to account for the chain process.

In some instances, the maximum common subgraph of two *SISS*s in chain mechanisms contains more than one vertex.[5] In this case, N_1-submechanism contains one or several steps of the catalytic submechanism and *vice versa*.

Branched Chain Reactions. Branched chain mechanisms are characterized by the presence of conjugated N_1- and N_2-submechanisms. An N_2-submechanism involved in a branched chain reaction contains a circuit (chain) and a terminal *u*-vertex (chain termination) that compensates for the chain branching. Terminal *u*-vertexes of this sort are often involved in N_1-submechanisms as well. This is necessary to compensate for the chain branching. Without the chain termination step, the N_2-submechanism cannot be balanced. The difference between the branched and nonbranched chain mechanisms is as follows. Circuits in the *SI* of nonbranched chain are parts of a C-submechanism. In the *SI* of a branched chain mechanism, circuits are parts of an N_2-submechanism. Examples of the branched chain reactions are the mechanism of reaction between oxygen and hydrogen,[5] which involves free radicals as active species, and the mechanism of reaction between H_2 and F_2, which involves free radicals and molecules excited at the expense of reaction heat.[5,19]

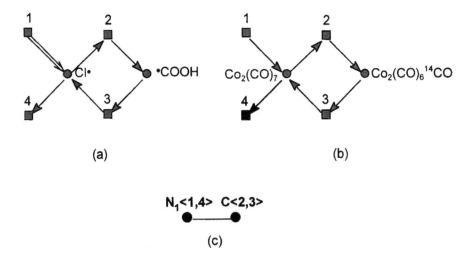

Figure 3.11. (a) The subnetwork of intermediates for mechanism (3.9); (b) the subnetwork of intermediates for mechanism (3.10); and (c) the submechanism graph, common to both mechanisms.

An example of the branched chain reaction is the mechanism of interaction between F_2 and CH_3I.[16,19,20] The asterisk denotes an excited molecule:

(1) $F_2 + CH_3I \rightarrow HF + \cdot F + \cdot CH_2I$,	initiation	
(2) $\cdot F + CH_3I \rightarrow HF + \cdot CH_2I$,	propagation	
(3) $\cdot CH_2I + F_2 \rightarrow CH_2FI^* + \cdot F$,	branching	
(4) $CH_2FI^* \rightarrow \cdot CH_2F + \cdot I$,	branching	
(5) $\cdot CH_2F + F_2 \rightarrow CH_2F_2 + \cdot F$	propagation	(3.12)
(6) $\cdot F + CH_3I \rightarrow CH_3F + \cdot I$,	propagation	
(7) $CH_2FI^* + M \rightarrow CH_2FI + M^*$,	termination	
(8) $2\cdot I \rightarrow I_2$	termination	

The simple submechanisms and respective overall equations for this set of reactions are given below:

Submechanism	Overall equation
$N_2\langle 2,3,7\rangle$	$CH_3I + F_2 + M \rightarrow CH_2FI + HF + M^*$
$N_1\langle 1,3,4,5,6,8\rangle$	$3F_2 + 4CH_3I + \rightarrow HF + CH_2F_2 + 3CH_3F + 2I_2$
$N_1\langle 1,3,6,7,8\rangle$	$2F_2 + 3CH_3I + M \rightarrow I_2 + CH_2FI + 2CH_3F + HF + M^*$
$N_2\langle 2,3,4,5,6,8\rangle$	$2CH_3I + 2F_2 \rightarrow HF + CH_2F_2 + CH_3F + I_2$

Each step is involved in one (or several) submechanisms and, therefore, the mechanism is balanced.

Figure 3.12 shows the *SISS*s and the submechanism graph for mechanism (3.12). Because both N_1- and N_2-submechanisms are present, mechanism (3.12) should be categorized as a branched chain mechanism.

Other Chain Reactions. We discussed examples of chain reactions, which are in agreement with classical theory of chain processes.[21,22] However, there exists a rich variety of reactions that are also categorized as chain processes but the structures of their mechanisms may be quite different. First, mention should be made of mechanisms containing defects. They will be discussed in Section 3.4. Others may contain combinations of several N_2-submechanisms or N_2- and C-submechanisms. The two examples that follow are discussed in the literature as branched chain processes. The first example refers to the mechanism of methane chlorination in the presence of salt catalysts.[23] This mechanism can be described by the following set of elementary reactions (Z is an active site of the catalyst):

$$(1)\ Cl_2 + 2Z \rightarrow 2ZCl,$$
$$(2)\ ZCl \rightarrow Z + {}^\bullet Cl,$$
$$(3)\ ZCl + CH_4 \rightarrow {}^\bullet CH_3 + HCl + Z,$$
$$(4)\ {}^\bullet Cl + CH_4 \rightarrow {}^\bullet CH_3 + HCl,$$
$$(5)\ Cl_2 + {}^\bullet CH_3 \rightarrow CH_3Cl + {}^\bullet Cl,$$
$$(6)\ {}^\bullet CH_3 + {}^\bullet Cl \rightarrow CH_3Cl.$$

(3.13)

This mechanism contains four simple submechanisms ($C\langle 4,5\rangle$, $N_2\langle 1,2,3,6\rangle$, $N_2\langle 1,2,4,6\rangle$, $N_2\langle 1,3,5,6\rangle$). The fourth submechanism corresponds to the overall equation:

$$2Cl_2 + 2CH_4 \rightarrow 2HCl + 2CH_3Cl.$$

Others correspond to the equation

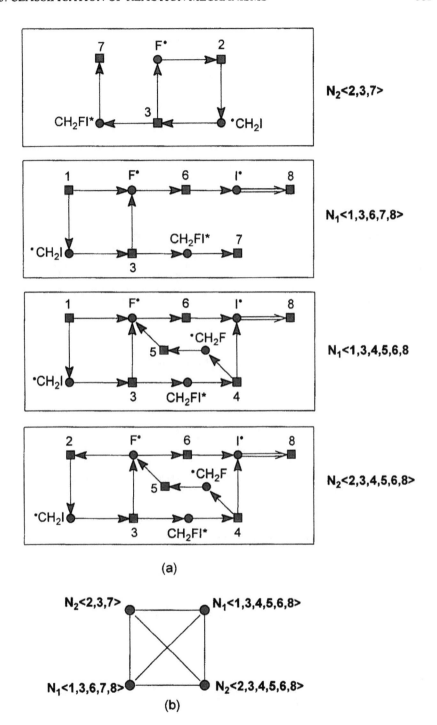

(a)

(b)

Figure 3.12. (a) Subnetworks of intermediates for the simple submechanisms and (b) the submechanism graph for mechanism (3.12).

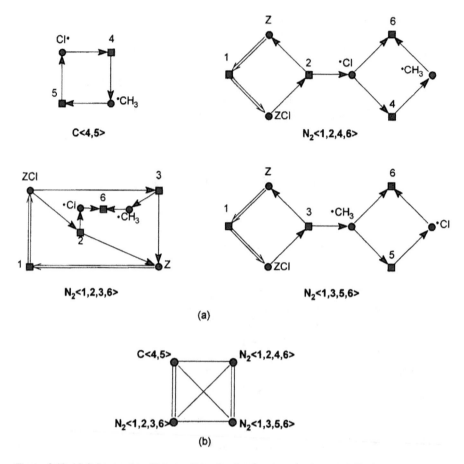

Figure 3.13. (a) Subnetworks of intermediates for simple submechanisms and (b) the submechanism graph for mechanism (3.13).

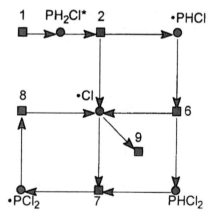

Figure 3.14. The subnetwork of intermediates for the simple submechanism $N_2\langle 1,2,6,7,8,9\rangle$ of mechanism (3.14).

$$Cl_2 + CH_4 \rightarrow HCl + CH_3Cl.$$

*SISS*s and the submechanism graph for the set of reactions (3.13) are shown in Figure 3.13.

Another example is the mechanism for the reaction between phosphine and Cl_2.[24]

$$(1)\ Cl_2 + PH_3 \rightarrow HCl + PH_2Cl^*,$$

$$(2)\ PH_2Cl^* + Cl_2 \rightarrow {}^{\bullet}PHCl + HCl + {}^{\bullet}Cl,$$

$$(3)\ PH_3 + {}^{\bullet}Cl \rightarrow {}^{\bullet}PH_2 + HCl,$$

$$(4)\ {}^{\bullet}PH_2 + Cl_2 \rightarrow PH_2Cl + {}^{\bullet}Cl,$$

$$(5)\ {}^{\bullet}Cl + PH_2Cl \rightarrow {}^{\bullet}PHCl + HCl,$$

$$(6)\ {}^{\bullet}PHCl + Cl_2 \rightarrow PHCl_2 + {}^{\bullet}Cl,$$

$$(7)\ PHCl_2 + {}^{\bullet}Cl \rightarrow {}^{\bullet}PCl_2 + HCl,$$

$$(8)\ {}^{\bullet}PCl_2 + Cl_2 \rightarrow PCl_3 + {}^{\bullet}Cl,$$

$$(9)\ {}^{\bullet}Cl + wall \rightarrow Cl{\cdot}wall,$$

$$(10)\ {}^{\bullet}PH_2 + Cl_2 \rightarrow PH_2Cl^* + {}^{\bullet}Cl.$$

(3.14)

The computer program detected three simple submechanisms: $C\langle 3,4,5,6,7,8\rangle$, $N_2\langle 1,2,6,7,8,9\rangle$, and $N_2\langle 2,3,6,7,8,9,10\rangle$. One of these simple submechanisms ($N_2\langle 1,2,6,7, 8,9\rangle$) is of special sort. It contains two terminal vertexes: w_1 and w_9. Vertex w_1 corresponds to the initiation step, while vertex w_9 corresponds to the termination step. Nevertheless, the submechanism is of N_2-type because it contains a circuit (regular cycle). The subnetwork of intermediates for this submechanism is shown in Figure 3.14.

As can be seen, the concept of chain reaction is based on the structure of the mechanism, which can be viewed in terms of the submechanism classes and their conjugation. This concept is independent of the type of active species involved in the reaction. At the same time, an exact definition of a chain process is not yet given. On one hand, this concept implies the presence of circuits, which are also characteristic of catalytic mechanisms. On the other hand, a chain mechanism also contains noncatalytic routes. Defects may be also present.

Interrelations between noncatalytic and catalytic components of the chain reaction may be very complex. The submechanism graph exactly defines the structure of the mechanism and can be used instead of ill-defined notions of branched and nonbranched chain reactions.

3.3.4. "Supercatalysis"

It can be shown that the same mechanism can be categorized differently depending on the extent to which various mechanistic details are included in the scheme of the mechanism. For instance, the mechanistic scheme may be written as a set of pseudoelementary reactions, each being a multistep process. These pseudoelementary reactions can be catalyzed by species not included in the stoichiometry of pseudoelementary reactions. A typical example is a branched chain reaction in which initiation, termination, and branching steps are catalyzed by hidden species.[25,26] Consider the mechanistic scheme of the branched chain reaction:

$$(1)\ A \rightarrow X, \qquad \text{initiation}$$
$$(2)\ X \rightarrow T, \qquad \text{termination}$$
$$(3)\ X \rightarrow Y + P, \quad \text{propagation}$$
$$(4)\ Y \rightarrow Z + X, \quad \text{branching}$$
$$(5)\ B + Z \rightarrow X, \quad \text{propagation}$$

In this scheme, X, Y, and Z are intermediate species; A and B are the initial species; P and T are reaction products. This mechanistic scheme involves two simple submechanisms, $N_1\langle 1,2\rangle$ and $N_2\langle 2,3,4,5\rangle$. Suppose that step 1 is catalyzed with a catalyst κ:

$$(1.1)\ A + \kappa \rightarrow A\kappa,$$
$$(1.2)\ A\kappa \rightarrow X\kappa,$$
$$(1.3)\ X\kappa \rightarrow X + \kappa.$$

Steps (1.1)–(1.3) produce the overall stoichiometry of step 1. Other pseudoelementary steps can be decomposed into simpler ones as well. As can be seen, at least two levels exist. On a macrolevel, the reaction is branched chain and contains two simple submechanisms (N_1 and N_2). On a microlevel the number of simple submechanisms is greater, although the mechanism preserves its type. In other instances, the type of the mechanism is not preserved. Thus, steps 1–3 of mechanism (3.13) can be viewed as a single pseudoelementary reaction with the overall equation

$$(1\text{–}3)\ Cl_2 + CH_4 \rightarrow {}^{\bullet}Cl + HCl + {}^{\bullet}CH_3.$$

Then, the branched chain reaction transforms into a nonbranched chain one having two simple submechanisms: $C\langle 4,5\rangle$ and $N_1\langle (1\text{–}3),6\rangle$. Several examples of catalysis of initiation, propagation, branching and termination steps can be extracted from textbooks and review papers.[18,26,27]

A similar situation can be observed in the case of catalysis when pseudoelementary steps of a catalytic process are themselves catalytic reactions.[27,28] Mechanisms of this sort can be termed "supercatalytic". The somewhat idealized reaction schemes given below serve as an illustration.

Suppose a reaction occurs via the catalytic mechanism:

$$(1)\ A + Z \rightarrow X,$$

$$(2)\ X + B \rightarrow Y,$$

$$(3)\ Y \rightarrow T + Z.$$

The overall reaction is as follows:

$$A + B \rightarrow T.$$

Species X, Y, and Z are different catalyst entities. Then, suppose that each step of the mechanism is catalyzed with "hidden" catalysts. For instance,

$$(1.1)\ A + \kappa_1 \rightarrow A\kappa_1$$
$$(1.2)\ A\kappa_1 + Z \rightarrow X + \kappa_1$$
$$\overline{}$$
$$(1)\ Z + A \rightarrow X$$

$$(2.1)\ X + \kappa_2 \rightarrow X\kappa_2$$
$$(2.2)\ X\kappa_2 + B \rightarrow Y + \kappa_2$$
$$\overline{}$$
$$(2)\ X + B \rightarrow Y$$

$$(3.1)\ Y + \kappa_3 \rightarrow Y\kappa_3$$
$$(3.2)\ Y\kappa_3 \rightarrow T + Z\kappa_3$$
$$(3.3)\ Z\kappa_3 \rightarrow Z + \kappa_3$$
$$\overline{}$$
$$(3)\ Y \rightarrow T + Z$$

There are catalytic entities of two levels.[27] On the first level, κ_1, κ_2, and κ_3 catalyze pseudoelementary steps. On the second level the catalyst Z catalyzes the overall reaction.

3.3.5. Polyfunctional Catalytic Systems

From the above discussion it follows that a catalyst may act on different levels and has particular kinetic functions; that is, it accelerates the overall process or any of its stages. The presence of several catalytic submechanisms within a single mechanism and the consideration of different levels of catalysis (as this takes place in the case of supercatalysis) allows one to regard some of the catalytic systems as *polyfunctional* ones. The polyfunctionality of a catalytic system can be due to its organization or to the structure of the overall mechanism.

Polyfunctionality of the first kind is caused by joining together different catalysts within a single active center of enzyme or a complex chemical catalyst that acts at all stages of the catalytic process. For instance, such an active center can serve both acidic and basic functions.

If several catalysts accelerate one or several steps of a multistage process, then each of these catalysts serves its own *kinetic function* (KF): it catalyzes part of the process having its overall equation. The polyfunctionality of this sort is closely related to the structure of the mechanism. Polyfunctional catalytic systems of this sort were discussed in a number of publications.[15,28-35]

Ref. 29 reports the attempts to classify polyfunctional catalytic systems using kinetic graphs. The use of bipartite graphs provides a stronger basis for the classification of these systems because nonlinearity of a mechanism is one of the key features of the polyfunctionality.

Polyfunctional catalytic systems can be characterized by (1) the number of linearly independent routes M; (2) the number of simple submechanisms N_{SS}; (3) the number of circuits (i.e., *catalytic cycles*) in the *SI* of the overall mechanism and by the manner in which they are conjugated; (4) by the number of linearly independent *kinetic functions*.

Of special interest for this characterization are kinetic functions. Each circuit in the *SI* has its overall equation, also called a kinetic function. Some of this circuits are balanced in the sense that appropriate stoichiometric numbers can be found for its *u*-vertexes (steps) such that all intermediates (*w*-vertexes) involved in the circuit (but not necessarily all intermediates involved in these steps) vanish. Other circuits may be unbalanced. Kinetic functions of balanced circuits can be uniquely defined. In the case of unbalanced circuits, a method for the canonical choice of stoichiometric numbers (and respective overall equation of the circuit) can be found. The list of the overall equations of circuits can be transformed into the stoichiometric matrix of circuits $\mathbf{B}_C$ in which rows stand for circuits and columns stand for species involved in the overall equations. Entries of the matrix are stoichiometric coefficients of species in these equations. The rank of the $\mathbf{B}_C$ matrix is equal to the number of linearly independent kinetic functions.

Analysis of the circuit matrix associated with the *SI* as a graph is also of interest.

An example of the polyfunctional catalytic system is the mechanism of acetylene hydration discussed in Chapter 2 (mechanism (2.4), Figure 2.6). The mechanism has one route (and, correspondingly, one simple submechanism). Three catalysts are initially inputted into the system: CuCl (denoted by M), RSH, and HCl. The number of circuits is seven:

(1) $X_8 \rightarrow$ *step 1* $\rightarrow X_1 \rightarrow$ *step 2* $\rightarrow X_2 \rightarrow$ *step 3* $\rightarrow X_8$,

(2) $X_8 \rightarrow$ *step 1* $\rightarrow X_1 \rightarrow$ *step 2* $\rightarrow X_7 \rightarrow$ *step 3* $\rightarrow X_8$,

(3) $X_3 \rightarrow$ *step 4* $\rightarrow X_4 \rightarrow$ *step 5* $\rightarrow X_6 \rightarrow$ *step 2* $\rightarrow X_2 \rightarrow$ *step 3* $\rightarrow X_3$,

(4) $X_4 \rightarrow$ *step 5* $\rightarrow X_6 \rightarrow$ *step 2* $\rightarrow X_7 \rightarrow$ *step 4* $\rightarrow X_4$,

(5) $X_3 \rightarrow$ *step 4* $\rightarrow X_4 \rightarrow$ *step 5* $\rightarrow X_5 \rightarrow$ *step 6* $\rightarrow X_7 \rightarrow$ *step 3* $\rightarrow X_3$,

(6) $X_4 \rightarrow$ *step 5* $\rightarrow X_5 \rightarrow$ *step 6* $\rightarrow X_7 \rightarrow$ *step 4* $\rightarrow X_4$,

(7) $X_3 \rightarrow$ *step 4* $\rightarrow X_4 \rightarrow$ *step 5* $\rightarrow X_6 \rightarrow$ *step 2* $\rightarrow X_7 \rightarrow$ *step 3* $\rightarrow X_3$.

However, not all of them are linearly independent. It is known in advance that overall equations of circuits 1 and 2 as well as 3 and 7 are the same. Therefore, rows for circuits 2 and 7 should not be included in the matrix $\mathbf{B}_C$. Circuits 1, 3, 4,

5, and 6 remain for the further consideration. The overall equations are obtained by adding up the elementary steps involved in each circuit having all stoichiometric numbers equal unity. These equations are as follows:

$$\text{Circuit 1: } C_2H_2 + X_6 \rightarrow X_3,$$
$$\text{Circuit 3: } H_2O + X_1 + X_7 \rightarrow X_5 + X_8,$$
$$\text{Circuit 4: } H_2O + X_1 + X_3 \rightarrow X_5 + X_2,$$
$$\text{Circuit 5: } H_2O + X_2 + X_7 \rightarrow CH_3CHO + X_8 + X_6,$$
$$\text{Circuit 6: } H_2O + X_3 \rightarrow CH_3CHO + X_6.$$

The matrix BC is a s follows:

	X_1	X_2	X_3	X_5	X_6	X_7	X_8	C_2H_2	H_2O	CH_3CHO
C 1	0	0	1	0	-1	0	0	-1	0	0
C 3	-1	0	0	1	0	-1	1	0	-1	0
C 4	-1	1	-1	1	0	0	0	0	-1	0
C 5	0	-1	0	0	1	-1	1	0	-1	1
C 6	0	0	-1	0	1	0	0	0	-1	1

The rank of this matrix is 4 and, therefore, there are four linearly independent kinetic functions.

More details about the classification of polyfunctional catalytic systems will be reported in our future publications.

3.4. Partially Balanced Mechanisms

A partially balanced mechanism is that containing at least one defect. The defect may consist of one elementary reaction or several reactions. The position of the defect determines its type. If deletion of the **D**-vertex from the submechanism graph increases the number of its components, the respective defects constitute a *bridging* structure in the mechanism. In this case, the defect links two or more submechanisms, which have no common steps or species. If deletion of the **D**-vertex does not affect the number of components of the submechanism graph, the respective defect is termed *pendant*. Partially balanced mechanisms in many respects resemble classes of balanced mechanisms and can be classified in a similar manner. Thus, if all simple submechanisms are of **C**-type (even if defects are present), the mechanism should be categorized as catalytic. If the submechanisms contains only N_1-submechanisms and there are at least two submechanisms that have common species or steps, the mechanism describes conjugated noncatalytic reaction. Generally speaking, if a pair of submechanisms have no common species but they are linked through a bridging defect, the overall reactions of these submechanisms are not conjugated. In all cases, the submechanism graph determines the structure of the mechanism better than commonly accepted terms in mechanistic chemistry.

Consider an example of the partially balanced mechanism with a bridging step that resembles a chain process. The mechanism of ethylene dimerization in solution of Ni complexes (X is an atom of halogen; QH is the donor of hydride) is as follows:[17]

$$(1)\ NiX_2 + QH \rightarrow HNiX + QX,$$
$$(2)\ HNiX + C_2H_4 \rightarrow C_2H_5NiX,$$
$$(3)\ C_2H_5NiX + C_2H_4 \rightarrow C_4H_9NiX, \qquad\qquad (3.15)$$
$$(4)\ C_4H_9NiX + C_2H_4 \rightarrow C_2H_5NiX + C_4H_8,$$
$$(5)\ HNiX \rightarrow HX + Ni(0).$$

Steps 1 and 5 constitute a simple N_1-submechanism having the overall equation

$$(1,5)\ NiX_2 + QH \rightarrow QX + HX + Ni(0). \qquad\qquad (3.16)$$

Steps 3 and 4 constitute a catalytic simple submechanism having the overall equation

$$(3,4)\ 2C_2H_4 \rightarrow C_4H_8. \qquad\qquad (3.17)$$

Step 2 is the defect of the mechanism. Thus, the *SISS*s of N_1- and C-submechanisms do not have a common subgraph; that is, reactions (3.16) and (3.17) are not conjugated. However, the N_1-submechanism supplies an intermediate to the C-submechanism through step 2, which is bridging. The *SI* of the overall mechanism and the submechanism graph are shown in Figure 3.15.

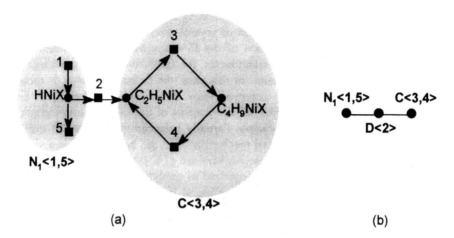

(a)   (b)

Figure 3.15. (a) The subnetwork of intermediates and (b) the submechanism graph for mechanism (3.15).

An example of a catalytic mechanism with bridging steps (defects) was discussed in Ref. 5. The catalytic acetylene hydrochlorination in solutions of Cu(I) complexes

$$C_2H_2 + HCl \rightarrow CH_2=CHCl \qquad (3.18)$$

occurs via several submechanisms, which have identical structures:

$$M + C_2H_2 \rightarrow (\eta^2\text{-}C_2H_2)M \qquad (\pi\text{-complex}),$$

$$(\eta^2\text{-}C_2H_2)M + Cl^- \rightarrow ClCH=CHM^- \qquad (\sigma\text{-complex}), \qquad (3.19)$$

$$ClCH=CHM^- + H^+ \rightarrow ClCH=CH_2 + M.$$

M represents different catalytic centers, such as $CuCl_2^-$, $CuCl_3^{2-}$, $Cu_2Cl_3^-$, etc. Different catalytic cycles (simple C-submechanisms) occur on these centers, which are linked by reversible steps:

$$CuCl_2^- + Cl^- \rightleftharpoons CuCl_3^{2-},$$

$$CuCl_3^{2-} + Cu^+ \rightleftharpoons Cu_2Cl_3^-, \qquad (3.20)$$

$$2CuCl_2^- \rightleftharpoons Cu_2Cl_3^- + Cl^-.$$

These steps are bridging and induce a defect of the mechanism. Thus, the submechanism graph of reaction (3.18) on three centers, each being a carrier of a catalytic cycle (3.19), is a star graph with three C-vertexes and a central D-vertex (Figure 3.16).

Both bridging and pendant defects may consist of several steps. To illustrate, consider the following set of reactions (CuCl, X_1 = PhC≡CCu, X_2 = PhC≡CCu[CuCl₂], and X_3 = PhC≡CPdCl are intermediates):[36]

(1) PhC≡CH + CuCl → X_1 + HCl,

(2) X_1 + CuCl₂ → X_2,

(3) X_2 + CuCl₂ → PhC≡CCl + 3CuCl,   (3.21)

(4) X_1 + PdCl₂ → X_3 + CuCl,

(5) X_3 + CO + MeOH → PhC≡CCOOMe + Pd(0) + HCl.

This mechanism contains a simple N_2-submechanism ($N_2\langle 1,4,5 \rangle$) and a defect ($D\langle 2,3 \rangle$). They are joined by two common species (CuCl and X_2). The overall reaction of N_2-submechanism is as follows:

$$PhC≡CH + CO + MeOH + PdCl_2 \rightarrow PhC≡CCOOMe + Pd(0) + 2HCl.$$

The SI of N_2-submechanism (Figure 3.17) contains a circuit corresponding to the catalytic species CuCl: $w_{CuCl} \longrightarrow u_1 \longrightarrow w_{X_1} \longrightarrow u_4 \longrightarrow w_{CuCl}$, while palladium is a stoichiometric rather than catalytic species. If the step of palladium(0) oxidation is added to the set of reactions (3.21):

$$(6)\ Pd(0) + 2CuCl_2 \rightarrow PdCl_2 + 2CuCl,$$

the N_2-submechanism transforms into a C-submechanism and becomes "closed" with respect to palladium as well. The defect of the mechanism is part of the auto-catalytic set of reactions having the overall equation that involves intermediate CuCl:

$$PhC\equiv CH + 2CuCl_2 \rightarrow PhC\equiv CCl + 2CuCl + HCl.$$

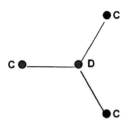

Figure 3.16. The submechanism graph for the acetylene hydrochlorination reaction having three catalytic submechanisms and a bridging defect between them (see reactions (3.19) and (3.20)).

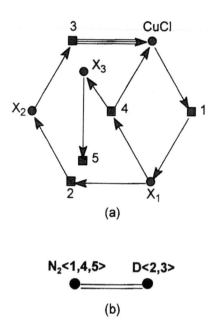

Figure 3.17. (a) The subnetwork of intermediates and (b) the submechanism graph for mechanism (3.20).

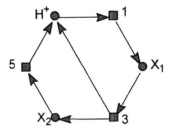

Figure 3.18. The simplified subnetwork of intermediates for mechanism (3.23)

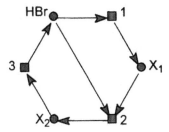

Figure 3.19. The subnetwork of intermediates for the simplified mechanism (3.24).

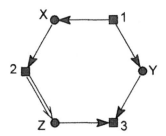

Figure 3.20. The subnetwork of intermediates for the hypothetical mechanism (3.25).

3.5. Unbalanced Mechanisms

If a mechanism does not contain simple submechanisms, it is termed *unbalanced*. The choice of the overall equation for the unbalanced mechanism is always arbitrary, because different sets of stoichiometric numbers produce different overall equations. Autocatalysis is a typical example of an unbalanced mechanism. The autocatalytic reaction produces the species that play the role of a catalyst for the very steps which produce this species. This is reflected in the structure of a mechanism. For instance, the acetone iodination reaction was found to occur via several steps of the acid-catalyzed enolization of acetone:[37]

$$CH_3C(O)CH_3 + H^+ \rightleftharpoons [CH_3C^+(OH)CH_3],$$

$$[CH_3C^+(OH)CH_3] \rightleftharpoons CH_2=C(OH)CH_3 + H^+,$$

$$CH_2=C(OH)CH_3 + I_2 \longrightarrow [CH_2IC^+(OH)CH_3] + I^-, \qquad (3.22)$$

$$[CH_2IC^+(OH)CH_3] \longrightarrow CH_2IC(O)CH_3 + H^+.$$

The following species are intermediates in this set of reactions: H^+, $X_1 = [CH_3C^+(OH)CH_3]$, $X_2 = CH_2=C(OH)CH_3$, and $X_3 = [CH_2IC^+(OH)CH_3]$. The other species are terminal. To simplify the reaction scheme, we can rewrite this mechanism using the above designations and adding the fourth equation to the third (this will not change the structure of the mechanism). Reversible first and second steps are written as four irreversible elementary reactions:

$$(1)\ CH_3C(O)CH_3 + H^+ \rightarrow X_1,$$

$$(2)\ X_1 \rightarrow CH_3C(O)CH_3 + H^+,$$

$$(3)\ X_1 \rightarrow X_2 + H^+, \qquad (3.23)$$

$$(4)\ X_2 + H^+ \rightarrow X_1,$$

$$(5)\ X_2 + I_2 \longrightarrow CH_2IC(O)CH_3 + H^+ + I^-.$$

The submechanism graph for the set of reactions (3.23) cannot be constructed because no simple submechanism can be found in this system. Figure 3.18 presents the subnetwork of intermediates for this reaction. To show the typical structure of the autocatalytic reaction, we simplified this graph by the deletion of vertexes corresponding to reactions (2) and (4). As can be seen, the branching step 3 is not compensated by steps that reduce the number of intermediates.

A similar situation is observed in the set of reactions discussed as variants 1 and 2 in Chapter 2 (Section 2.3.5). The subnetwork in Figure 2.48 is similar to that shown in Figure 3.18 except for the type of the branching step.

Another type of unbalanced mechanism is associated with the effect of *auto-inhibition*, which is structurally opposite to autocatalysis.[5] It can be illustrated by the example of the alkene hydrobromination in liquid nonpolar solvents occurring via multicentered transition states. The mechanism is somewhat simplified:

$$(1)\ HBr + RCH=CH_2 \rightarrow X_1,$$

$$(2)\ X_1 + HBr \rightarrow X_2, \qquad (3.24)$$

$$(3)\ X_2 \rightarrow RCH(Br)CH_3 + HBr.$$

In this mechanism, HBr, X_1, and X_2 are intermediates. The reaction network is shown in Figure 3.19. By comparing the subnetworks shown in Figures 3.18 and

3.19, one can see that they are completely identical except for vertex labels and direction of the diagonal arc. In the subnetwork of this autoinhibition reaction, step 2 consumes more intermediates than can be produced by the system.

Subnetworks of intermediates of autocatalytic and autoinhibition reactions always contain circuits. Due to this fact, they resemble N_2-submechanisms and branched chain reactions. However, the presence of circuits is not the necessary condition for the mechanism to be unbalanced. A third type of subnetwork of an unbalanced mechanism can be circuit-free and resemble N_1-submechanisms. An example was discussed in Chapter 2:

$$(1)\ A + B \rightarrow X + Y,$$
$$(2)\ B + X \rightarrow 2Z, \qquad\qquad (3.25)$$
$$(3)\ Y + Z \rightarrow T.$$

The subnetwork of intermediates for mechanism (3.25) is shown in Figure 3.20.

An intriguing example of a certain type of autocatalytic system was presented in Eigen and Schuster's book *The Hypercycle*.[38] The subnetwork of intermediates is highly symmetric and contains four circuits:

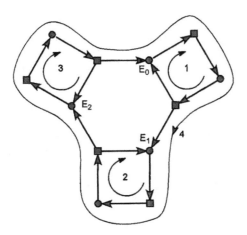

Three of them represent autocatalytic (self-instructive) cycles of enzymes E_0, E_1, and E_2. They catalyze enzymatic reactions and reproduce themselves. The fourth circuit joins them together to produce a hypercycle. More generally, a catalytic hypercycle connects several autocatalytic self-replicative and self-instructive units via linkage of circuits. This is an idealized pattern of a hypercycle. Examples of particular natural processes, which, according to the authors, match the idea of a hypercycle are somewhat contradictory to this idealized structure of *SI*. Therefore, we do not present them here.

References and Notes

1. Temkin, O. N.; Zeigarnik, A. V.; Bonchev, D. G. Application of Graph Theory to Chemical Kinetics. Part 2. Topological Specificity of Single-Route Reaction Mechanisms. *J. Chem. Inf. Comput. Sci.* **1995**, *35*, 729–737.
2. Temkin, M. I. Kinetics of Complex Reactions. In *Proceedings of USSR Conference on Chemical Reactors*; Nauka: Novosibirsk, 1966, Vol. 4, pp 628–646.
3. Christiansen, J. A. The Elucidation of Reaction Mechanisms by the Method of Intermediates in Quasi-Stationary Concentrations. In *Advances in Catalysis*; Frankenburg, W. G.; Komarewsky, V. I.; Rideal, E. K., Eds.1953, Academic: New York, Vol. 5, pp 311–353.
4. Kharash, M. S.; Fields, E. K. Factors Determining the Course and Mechanisms of Grignard Reactions. IV. The Effect of Metallic Halides on the Reaction of Aryl Grignard Reagents and Organic Halides. *J. Am. Chem. Soc.* **1941**, *63*, 2316–2320.
5. Zeigarnik, A. V.; Temkin, O. N.; Bonchev, D. G. Application of Graph Theory to Chemical Kinetics. Part 3. Topological Specificity of Multiroute Reaction Mechanisms. *J. Chem. Inf. Comput. Sci.* (in press)
6. Digraph is a directed graph. Circuit of a digraph is "regular" cycle. Digraph is termed a circuit-free digraph if all cycles are not regular.
7. Harary, F., *Graph Theory*; Addison-Wesley: Reading, MA, **1969**
8. Temkin, M. I. Kinetics of Complex Steady-Steady State Reactions. In *Mechanism and Kinetics of Complex Catalytic Reactions*; Isagulyants, G. V.; Tret'yakov, I. I., Ed.; Nauka: Moscow, 1970, p. 57–71 (in Russian).
9. Cadogan, C. C. The Möbius Function and Connected Graphs. *J. Combinatorial Theory, Sec. B* **1971**, *11*, 193–200.
10. Harary, F.; Palmer, E. M. *Graphical Enumeration*; Academic: New York, 1973.
11. See, for example: (a) Kvasnička, V.; Pospíchal, J. Canonical Indexing and Constructive Enumeration of Molecular Graphs. *J. Chem. Inf. Comput. Sci.* **1990**, *30*, 99–105; (b) Hendrickson, J. B.; Parks, C. A. Generation and Enumeration of Carbon Skeletons. *J. Chem. Inf. Comput. Sci.* **1991**, *31*, 101–107;
12. Yablonskii, G. S.; Bykov, V. I.; Elokhin, V. I. *Kinetics of Model Reactions of Heterogeneous Catalysis*; Nauka: Novosibirsk, 1984 (in Russian).
13. Reutov, O. A.; Beletskaya, I. P.; Artamkina, G. A.; Kashin, A. N. *Reactions of Organometallic Compounds as Redox Processes*; Nauka: Moscow, 1981 (in Russian).
14. Krylov, O. V. Some Analogies between Catalysis and Chain Reactions. *Kinet. Katal.* **1985**, *26*, 263–274; *Kinet. Catal. Engl. Transl.* **1985**, *26*, 223–233.
15. Moiseev, I. I. *π-Complexes in the Liquid-Phase Alkene Oxidation*; Nauka: Moscow, 1970 (in Russian).
16. Emanuel', N. M.; Knorre, D. G. *Chemical Kinetics (Homogeneous Reactions)*; Vyshaya Shkola: Moscow, 1984, 4th ed. (in Russian).
17. Temkin, O. N.; Brailovskii, S. M.; Bruk, L. G. Chain Mechanisms in Catalysis with Metal Complexes. In *Proceedings of USSR Conference on Mechanisms of Catalytic Reactions*; Nauka: Moscow, 1978, pp 74–81 (in Russian).
18. Shul'pin, G. B. *Organic Reactions Catalyzed by Metal Complexes*; Nauka: Moscow, 1988 (in Russian).
19. Denisov, E. T. *Kinetics of Homogeneous Chemical Reactions*; Vyshaya Shkola: Moscow, 1978 (in Russian).
20. Shilov, A. E. The Role of Chemical Activation in Branched Chain Reactions. In *Chemical Kinetics and Chain Reactions*; Nauka: Moscow, 1966 (in Russian).
21. Semenov, N. N. *Chain Reactions*; Nauka: Moscow, 1986, 2nd ed. (in Russian). First edition was published in 1934.

22. Semenov, N. N. *Some Problems of Chemical Kinetics and Reactivity*; USSR Academy of Sciences: Moscow, 1954 (in Russian).

23. Aglulin, A. G.; Bakshi, Yu. M.; Gel'bshtein, A. I.; Ivanova, R. A.; Snagovskii, Yu. S. Kinetics and Mechanism of Methane Chlorination in the Presence of Salt Catalysts. *Kinet. Katal.* 1985, *26*, 94–100; *Kinet. Catal. Engl. Transl.* 1985, *26*, 80–86.

24. Azatyan, V. V.; Gagarin, S. G.; Zahar'in, V. I.; Kalkanov, V. A.; Kolbanovskii, Yu. A. A Branched Chain Reaction between Chlorine and Phosphine. *Kinet. Katal.* 1985, *26*, 222–226.

25. Manzurov, V. D. *Liquid-Phase Catalytic Oxidation of Alkylnaphthalenes*; NIITEKhim: Moscow, 1993 (in Russian).

26. Kovtun, G. A.; Moiseev, I. I. Transition Metal Complexes in Catalysis of Chain Termination in Oxidation Reactions. *Koord. Khim.* 1983, *9*, 1155–1181.

27. Imyanitov, N. S.; Bogoradskii, N. M.; Semenova, T. A. An Accelerating Action of Pyridine on the Reaction between Acetyltetracarbonylcobalt and Methanol: Catalysis of the "Second Level". *Kinet. Katal.* 1978, *19*, 573–578 and references therein; Imyanitov, N. S. *Design of Homogeneous Catalytic Systems. Part 2. The Joint Action of Catalysts*. The manuscript is deposited in and available from NIITEKhim, Moscow, 1992.

28. Temkin, O. N.; Kaliya, O. L.; Shestakov, G. K.; Flid, R. M. The Mechanism for the Action of Multicomponent Metal Complex Catalysts in Solutions. *Dokl. Akad. Nauk SSSR* 1970, *190*, 398–401.

29. Temkin, O. N.; Brailovsky, S. M. The Mechanism of Catalysis in Homogeneous Polyfunctional Catalytic Systems. In *Fundamental Research in Homogeneous Catalysis*; Gordon & Breach: London, 1968, Vol. 2, pp 621–633;. The Mechanism for the Action of Polyfunctional Homogeneous Catalysts. In *Proc. Fourth Int. Symp. on Homogeneous Catalysis*; Nauka: Leningrad, 1984, Vol. 1, pp 30–32.

30. Temkin, O. N.; Bruk, L. G.; Bonchev, D. Topological Structure of Complex Reaction Mechanisms. *Teor. Eksp. Khim.* 1988, *24*, 282–291.

31. Likholobov, V. A.; Ermakov, Yu. I. Some Aspects of the Choice of Catalytic Systems Based on Metal Complexes. *Kinet. Katal.* 1980, *21*, 904–914

32. See a series of papers of Imyanitov, published by and available from NIITEKhim (Part 2 is cited in Ref. 27): Imyanitov, N. S. *Design of Homogeneous Catalytic Systems*; NIITEKhim: Moscow, 1992 (in Russian). See also Russian journal of abstracts: *Ref. Zh. Khim.* 1983, abstracts nos. 7B1167, 7B1169, 7B1178, 7B1179.

33. Golodov, V. A. The Character of Catalytic Action in Homogeneous Redox Systems: A Classification of Levels of Catalysis. In *Fundamental Research in Homogeneous Catalysis*; Gordon & Breach: London, 1968, Vol. 3, pp 1131–1139.

34. Sokol'skii, D. V.; Dorfman, Ya. A. *Catalysis by Ligands in Aqueous Solutions*; Nauka: Alma-Ata, 1975 (in Russian).

35. Heck, R. F. *Organotransition Metal Chemistry: A Mechanistic Approach*; Academic: New York, 1974.

36. Zung, T. T.; Bruk, L. G.; Temkin, O. N. A New Catalytic Reaction: Oxidative Carbonylation of Alkynes to Alkynylcarboxylic Acid Esters. *Mendeleev Commun.* 1994, 2–3.

37. Benson, S. W. *The Fundamentals of Chemical Kinetics*; McGraw Hill: New York, 1960.

38. Eigen, M.; Schuster, P. *The Hypercycle*; Springer: Berlin, 1979.

Chapter 4

Complexity
of Reaction Mechanisms

4.1. Introduction

If the mechanistic study begins with the computer-assisted generation of mechanistic hypotheses, which are further verified by experiments, a need arises in complexity measure for reaction mechanisms. First, the number of hypotheses may be huge, if at all finite. Therefore, it is reasonable to adopt some upper limit of complexity within which hypotheses will be considered. Second, if the preliminary data about the reaction is small, the combinatorial search space is rather large. Then, a complexity criterion can be a factor that restricts the hypothesis generation. Third, having a set of hypotheses, it is reasonable to start their testing from those which are simpler because more complex hypotheses more readily account for the experimental observations.

Computer assistance in handling the kinetic data revealed the problem, which remained more or less masked in the manual handling of kinetic models. Rather frequently, several kinetic models based on different mechanisms describe equally well the totality of experimental observations and agree with the background knowledge about reaction mechanisms. However, for practical purposes one needs to select one of these mechanisms as a working hypothesis; generally, this is *the simplest* mechanism. Inevitably, questions arise of how to make this choice and what to use as a criterion of simplicity (or complexity). These questions are not always easy to answer. This points to the necessity of introducing a quantitative measure of mechanistic complexity.

Another possible application of a complexity index might be in developing computer algorithms for the constructive enumeration of chemical graphs and, more specifically, graphs representing chemical reaction networks. Complexity indices and other topological invariants allow one to reduce the generation of isomorphic graphs in the cases when there is no other convenient method to do this.

Perhaps, indices like those discussed in this chapter could also be used in the computer planning of organic syntheses as a characteristic of the synthesis complexity or as a measure of structural similarity (the larger the structural similarity of the precursor and target molecules, the shorter the set of intermediate reaction steps transforming the first molecule into the second one). Several such indices have been proposed, the best known one being the *chemical distance*.[1-16] However, it is important to realize that simplicity measures used in synthesis planning cannot be applied in assessments of mechanistic complexity. This situation stems from the fact that quantitative measures used in synthesis planning serve to find the best route (or several equally acceptable routes) from the starting materials to the target ones. Conversely, the measure of mechanistic complexity should characterize the totality of all routes in a mechanism.

Background. Since the two types of chemical reaction networks, the linear and nonlinear ones, can be represented by two types of graphs (termed in the foregoing *kinetic graphs* and *bipartite graphs*), we examine first whether the known complexity measures of graphs could provide a relevant mechanism hierarchy (see Section 4.2). Complexity of graphs,[17,18] and particularly that of acyclic graphs known as *molecular branching*,[19-26] is well studied. Various graph invariants,[27-34] called *topological indices*, as well as their information-theoretic analogues,[35-40] have been used in this area.

Combined complexity measures based on several graph features have also been developed for a more complete description of graph topology.[41,42] The theory of *molecular complexity* was also developed[43-46] in a close connection with complexity of molecular graphs. The development of molecular complexity indices was greatly facilitated (as opposed to reaction mechanisms) by the available experimental evidence for various physical and chemical properties to which these indices could be compared and correlated. Not surprisingly, molecular complexity indices, as well as other topological invariants of molecules, find a broad application.[47,48]

Since the beginning of the 1980s, Bonchev and Temkin introduced a complexity index K for linear reaction mechanisms and studied its properties.[49-51] This is a kinetic index, which characterizes the complexity of linear networks from a kinetic viewpoint. This index is based on the count of step rate constants in the steady-state rate law of a route expressed in terms of fractional-rational functions. Section 4.3 presents this approach in detail.

The index K proved to match well the idea of complexity of kinetic graphs; however, this approach cannot be extended to nonlinear networks. A more general approach could be based on the stoichiometric matrix of the reaction mechanism. As mentioned earlier,[52,53] this matrix contains the entire information needed to characterize any possible mechanism. This information could be extracted by making use of the information-theoretic formalism. Such a very recent approach is discussed in Section 4.4.

4.2. Complexity of Chemical Graphs

4.2.1. Shannon's Entropy Measures of Graph Complexity

Information Theory Formalism. One of the basic formulas in Shannon's information theory[35] determines the mean entropy $H(P)$ of the probability distribution $(p_1, p_2, ..., p_k)$ of all possible outcomes (k in number) in a given situation:

$$H(P) = -\sum_{i=1}^{k} p_i \text{lb}(p_i), \tag{4.1}$$

where "lb" stands for the logarithm to base two, used for measuring entropy in bits (binary digits).

Mowshowitz[17] pointed out that, when applied to a structure, equation (4.1) measures the information content in relation to the system of transformations that leaves the structure invariant. Mowshowitz also presented a finite probability scheme that can be applied to any structure of N elements partitioned into classes of N_k equivalent elements according to a specified equivalence relation:

Equivalence classes	$1, 2, ..., k$
Element partition	$N_1, N_2, ..., N_k$
Probability E-distribution	$p_1, p_2, ..., p_k$

$$\tag{4.2}$$

In this scheme, $p_i = N_i/N$ is the probability for a randomly chosen element to belong to the class i having N_i elements, and $\sum_i N_i = N$. The probability distribution is denoted as E-distribution to mark the equivalence criterion used to partition the system elements.

According to this interpretation, equation (4.1) specifies the mean quantity of information $\bar{I}$ contained in each element of the system:

$$\bar{I} = -\sum_{i=1}^{k} p_i \text{lb}(p_i) = \sum_{i=1}^{k} \left[\left(\frac{N_i}{N}\right) \text{lb}\left(\frac{N_i}{N}\right) \right], \text{ bits/element.} \tag{4.3}$$

Bonchev and Trinajstić[21] generalized the Mowshowitz finite probabilistic scheme by ascribing "weights" or "magnitudes" $w_1, w_2, ..., w_k$ to the elements of each equivalence class:

Equivalence classes	$1, 2, ..., k$
Element partition	$N_1, N_2, ..., N_k$
Probability E-distribution	$p_1, p_2, ..., p_k$
Element weights	$w_1, w_2, ..., w_k$
Probability M-distribution	$p_1^M, p_2^M, ..., p_k^M$

$$\tag{4.4}$$

When each weight is simply the number of class elements N_i, (4.4) reduces to (4.2); i.e., the magnitude (M-)distribution reduces to the equivalence

(E-)distribution of the structure elements. The M-distribution describes the partitioning of some total magnitude $M = \sum_i N_i w_i$, characterizing the overall structure, into w_i weights. Then, $p_i^M = w_i/M$, and $\sum_i N_i p_i^M = 1$ hold for magnitude probabilities.

In this way, each chemical or mathematical structure can be characterized quantitatively by two types of information indices, I^E and I^M, calculated by equation (4.1) for the two probability distributions. Whether certain information-theoretic indices could be convenient measures of structural complexity depends on the selected equivalence relation.[39,40]

Information Indexes for Graph Orbits. Each graph is associated with a certain *automorphism group*. Graph vertexes that interchange upon the symmetry operations belonging to the automorphism group, preserving graph adjacency, are termed *topologically equivalent*; they form equivalence classes called *graph orbits*. Proceeding from these equivalence relations, Rashevsky[36] and Trucco[37] made use of equation (4.1) to introduce what they called *topological information content*. Trucco[37] also mentioned the possibility to define a similar information index proceeding from the automorphism group of graph edges. To avoid possible ambiguity, caused by the topological nature of all information-theoretic indices that can be defined for graphs, later[39] these indices were renamed *vertex orbit information index* and *edge orbit information index*, respectively. An example illustrating these first information indices defined for graphs is given below.

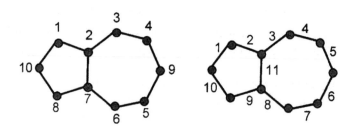

Vertex orbits: {1, 8}, {2, 7}, {3, 6}, {4, 5}, {9}, {10}
Vertex orbit partitioning: $^v P_{orb} = 10(1, 1, 2, 2, 2, 2)$
$^v \bar{I}_{orb}(1) = -4[(^2/_{10})lb(^2/_{10})] - 2[(^1/_{10})lb(^1/_{10})] = 2.522$
Edge orbits: {11}, {1, 10}, {2, 9}, {3, 8}, {4, 7}, {5, 6}
Edge orbit partition: $11(1, 2, 2, 2, 2, 2)$
$^e \bar{I}_{orb}(1) = -[(^1/_{11})lb(^1/_{11})] - 5[(^2/_{11})lb(^2/_{11})] = 2.550$

Mowshowitz[17] applied the vertex orbit information index as a measure of graph complexity and proved a number of theorems concerning patterns of increasing complexity and patterns of constant complexity. He also introduced another information index for graphs, related to the chromatic decomposition of graphs, called *chromatic information content*.[38]

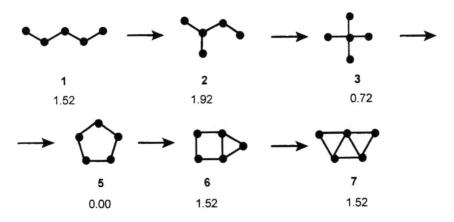

Figure 4.1. The failure of the vertex orbit information index to match the trend of increasing complexity of graphs 2–7.

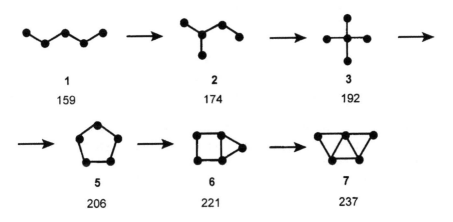

Figure 4.2. The trend of increasing complexity of graphs 2–7, illustrated by the reciprocal values of the graph distance information index (equation 4.1.6), $(1/W \bar{I}_d^M) \times 10^4$.

Unfortunately, graph complexity specified by the vertex orbit information index (as well as by the chromatic information index) contradicts common intuition, as first mentioned by Bonchev and Polansky.[44] Logically, one may expect at constant number of vertexes a trend of increasing graph complexity when more relations (edges) appear between graph vertexes, i.e., when going from non-branched linear graphs (path graphs) to branched acyclic graphs to monocyclic, bicyclic, etc. graphs. As shown in Figure 4.1 for graphs with five vertexes, $^v\bar{I}_{orb}$ classifies graph 2 (acyclic) and the much more complex graphs 6 (bicyclic) and 7 (tricyclic) as equally complex. Even more absurdly, the complexity index of 2, 6 and **7** is intermediate in value between the minimum one (the monocyclic graph **5**) and the maximum one (the branched acyclic graph **3**):

$$^{v}\bar{I}_{orb}(5) < {}^{v}\bar{I}_{orb}(4) < {}^{v}\bar{I}_{orb}(2) = {}^{v}\bar{I}_{orb}(6) = {}^{v}\bar{I}_{orb}(7) < {}^{v}\bar{I}_{orb}(3).$$

Evidently, the topological equivalence of graph vertexes and, in general, the equivalence of any other graph elements (edges, two-edge subgraphs, etc.) rather poorly matches the graph complexity. The probability M-distribution is much more discriminating than the E-distribution and provides a better basis for complexity assessments.

Graph Distance Information Index. The *distance* $d(ij)$ between two vertexes i and j in a simple connected graph is defined as the number of edges along the shortest path between these two vertexes. Thus, graph theoretical distances are integers, the adjacent vertexes being at distance one. The distance matrix $\mathbf{D}(G)$ of a graph G is a quadratic matrix symmetric about its main diagonal. The diagonal entries are zeros, $d(ij) = 0$, whereas the off-diagonal ones are the distances $d(ij)$. The sum of all entries of this matrix is a graph invariant termed *graph distance* (it does not depend on how the graph is depicted or how its vertexes are labeled):

$$D(G) = \sum_{ij} d(ij). \tag{4.5}$$

As shown by Hosoya[54] the graph distance is twice the Wiener number $W(G)$, which was introduced earlier empirically by Wiener[22] as the sum of C–C bonds in alkanes.

Bonchev and Trinajstić introduced the *graph distance information index* $\bar{I}^{M}_{dist}(G)$ by constructing the finite probability scheme for graph distances.[21] The probability for a randomly chosen distance to be equal to d was thus defined as $p_d = d/W$ and the Shannon equation (4.1) was transformed to

$$\bar{I}^{M}_{dist}(G) = -\sum_{d=1}^{d_{max}} (d/W) \mathrm{lb}(d/W) . \tag{4.6}$$

As seen in Figure 4.2, the reciprocal total distance information index I mirrors well the pattern of increasing complexity of graphs **2–7**:

$$I' = 1/I^{M}_{dist}(G) = 1/(W \bar{I}^{M}_{dist}(G)). \tag{4.7}$$

4.2.2. Topological Indexes as Potential Complexity Measures

The graph distance information index may be regarded as an information-theoretic analogue of the topological index of Wiener.[22] The Wiener index itself can match the trends of increasing complexity almost the same way as I^{M}_{dist}. Thus, the question arises of to what extent topological indices could identify and manifest complexity features.

Topological indices are numbers (integers or real numbers) that can be uniquely derived from the graph.[27-34] They are largely used in quantitative structure–property and structure–activity relationships (QSPR and QSAR).[47,48] The term *topological index* was first introduced in 1971 by Hosoya[23] who used it for his *graph nonadjacency index* $Z(G)$. The latter is defined as the sum of the *nonadjacency numbers* $p(G, k)$ which count the number of ways in which k edges could be selected from the graph G, so that no two of them are adjacent:

$$Z(G) = \sum_{k=0}^{[N/2]} p(G,k). \qquad (4.8)$$

In this equation, $N/2$ in Gauss' square brackets stands for the nearest integer not exceeding the real number in them. By definition, $P(G, 0) = 1$ and $P(G, 1) = E$ (the number of graph edges). For example, in graph 3 (see Figures 4.1 and 4.2), $p(G, 2) = 2$ (the pairs of edges 1,3 and 3,4), and $Z(3) = 1 + 4 + 2 = 7$

Another very popular topological index, called *molecular connectivity*, $\chi(G)$ was devised by Randić in 1975.[24] It is based on the vertex degrees a_i and a_j of the neighboring graph vertexes i and j (the vertex degree a_i is equal to the number of vertexes adjacent to vertex i):

$$\chi(G) = \sum_{edges} (a_i a_j)^{-1/2} = \sum \chi_{ij}. \qquad (4.9)$$

For example, the Randić index for the star-graph 3 is calculated as follows:

$$\chi(3) = 2\chi_{13} + \chi_{23} + \chi_{12} = 2\times0.577 + 0.408 + 0.707 = 2.270$$

The Randić molecular connectivity has been generalized by Kier *et al.*[25] from edges (paths of length one) to paths of length $h = 2, 3$, etc.

$$^h\chi(G) = \sum_{paths} (a_i a_j ... a_{h+1}). \qquad (4.10)$$

In (4.10), $a_i, a_j, ..., a_{h+1}$ are the degrees of all vertexes included in the path of length h, and the summation is taken over all paths.

Another generalization was proposed by Balaban[26] whose index J includes distance vertex degrees d_i (the sum of the distances from vertex i to all other graph vertexes) instead of the vertex degrees a_i:

$$J(G) = [E/(C + 1)] \sum_{edges} (d_i d_j)^{-1/2}, \qquad (4.11)$$

where E and C in the normalizing factor denote the total number of graph edges and cycles, respectively.

Many other topological indices have been devised to quantify the graph structure. However, they all share two major pitfalls when applied as complexity measures: (i) Being based on a single graph-invariant, they cannot adequately characterize the overall graph structure; (ii) They are (sometimes highly) degenerate, i.e., two or more nonisomorphic graphs can be associated with the same value of the topological index. The remedy was searched in devising either more discriminating indices (the index J was the first example of this sort) or combined topological or information-theoretic indices.

4.2.3. Combined Graph Complexity Measures

Gordon and Kennedy[19] examined the complexity of acyclic graphs by combining two criteria: (i) size factor, which was taken to be the *graph diameter* (the largest distance in the graph) and, at the same graph size, (ii) the complexity factor, which counts the subgraphs of all sizes. Though highly discriminating and matching the pattern of increasing complexity in acyclic graphs, this measure soon faces combinatorial explosion when moving to larger graphs and, particularly, in case of polycyclic graphs.

The *combinatorial complexity function CCF(G)* of Minoli[20] combines three graph invariants, the vertex, edge, and path counts, N, E, and P, respectively:

$$CCF(G) = \frac{N \cdot E \cdot P}{N + E}. \tag{4.12}$$

All paths between all pairs of vertexes are counted in P. (A path is a sequence of adjacent edges which never traverses any vertex twice.) As an illustration, the calculation of Minoli's index is shown for graph 6. This graph has $N = 5$, $E = 6$, and $P = 34$, including here six paths of length one, nine paths of length two, twelve paths of length three, and seven paths of length four. Hence, one obtains $CCF(G) = 92.7$.

Albeit based on three topological features, the Minoli index is very degenerate, particularly for acyclic graphs. Its discrimination capacity was increased considerably by Bonchev[45] who made use of the total length of all paths in the graph L_p, rather than of their count only:

$$CCF'(G) = \frac{N \cdot E \cdot L_p}{N + E}. \tag{4.13}$$

For the above example, $L_p = 6.1 + 9.2 + 12.3 + 7.4 = 88$, and $CCF'(G) = 240$.

In 1984, Randić[33] advanced the idea of *molecular identification number ID(G)* as a generalization of molecular connectivity indices. He uses weighted path count where the weight of each path is a product of the connectivities of the constituent edges χ_{ij}, as defined in equation (4.9):

$$ID(G) = N + \sum_l \sum_k \prod_{\{ij\} \in \text{path} l,k} \chi_{ij}. \tag{4.14}$$

In this equation, the summation is over all paths k with all feasible lengths l. The size factor (the total number of vertexes N) is also included as a summand in the formula.

The *ID* index was made even more discriminating by Carter, Trinajstić, and Nikolić,[34] who replaced the partial molecular connectivities χ_{ij} with distance connectivities $\chi_{d(ij)} = (d_i d_j)^{-1/2}$ and used different normalization factors. The formula for the *weighted identification number WID(G)* is as follows:

$$WID(G) = N - (1/N) + (1/N)^2 \cdot ID^*(G). \tag{4.15}$$

Combined information-theoretic indices have also been used. In his index of molecular complexity, Bertz[43] made use of the information index on the graph connections (two-edge subgraphs), I_{conn}, to which he added a graph size term, and a third term for the elemental composition I_{size} (which in case of labeled graphs corresponds to an information index on the vertex labels):

$$I_{Bertz} = I_{AC} + I_{size} + I_{conn}. \tag{4.16}$$

A more complete combined topological index, called *topological information superindex*, $SI(G)$, was devised by Bonchev, Mekenyan, and Trinajstić.[42] While remaining an open system, the superindex in its first version incorporated six terms which represent the graph vertex orbits,[36] vertex chromaticity,[38] vertex centric ordering,[55] graph edges,[42] distances,[21] and nonadjacent numbers:[21]

$$SI(G) = {}^v I^E_{orb} + {}^v I^E_{chr} + {}^v I^E_{centr} + I^E_{edge} + {}^v I^M_{dist} + I_z + \dots . \tag{4.17}$$

The summation of the individual indices is certainly not the best way to produce a combined complexity index because some of the information the individual indices contain is lost upon the summation. For that reason, the topological information superindex has been redefined[42] as a vector, composed of the individual indices:

$$SI'(G) = ({}^v I^E_{orb}, {}^v I^E_{chr}, {}^v I^E_{centr}, I^E_{edge}, {}^v I^M_{dist}, I_z, \dots). \tag{4.18}$$

Although very discriminating, the topological information superindex is a rather heterogeneous mixture of terms reflecting different graph complexity features. What is still lacking in this model of graph complexity is a certain hierarchy of topological factors. Such a hierarchical concept of topological complexity was presented in 1987 by Bonchev and Polansky.[44] The mathematical form of this concept is a hierarchically constructed multicomponent vector C_{TOP} with as many components as the number of hierarchical levels:

$$C_{TOP} = (C_{CONN}, C_{ADG}, C_{FRAG}, C_{SYM}, C_{DIST}, C_{PATHS}, \dots). \tag{4.19}$$

As seen in Figure 4.3, this is a very general scheme which starts on the first level with the *graph connectedness*, thus enabling the discrimination of disconnected, connected, and nonplanar graphs.

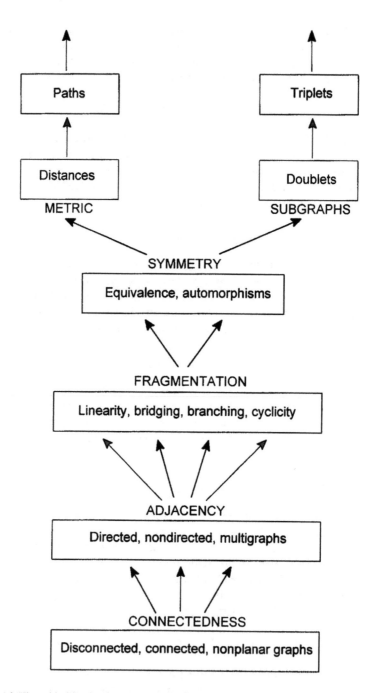

Figure 4.3. Hierarchical levels of topological complexity.

Adjacency, the second hierarchical level, distinguishes directed graphs, nondirected graphs, and multigraphs. Four basic types of molecular fragments specify on level 3 such complexity features as linearity (the least complex), bridging, branching, and cyclicity (the most complex feature). Graph symmetry, as handled by the automorphism group of the graph, comes next on level 4. Higher complexity levels (if needed) may include graph metric (distances, paths, walks, etc.).

4.3. Kinetic Complexity Index for Linear Reaction Networks

4.3.1. Complexity of Kinetic Models

Albeit well studied, complexity of graphs is only the first level of complexity of linear mechanisms as presented by kinetic graphs (KGs). Mechanistic complexity manifests itself best in kinetics. Therefore, a second level of mechanistic complexity must be considered, that of the kinetic model.

In developing the ideas of mechanistic complexity, in 1979–81 we faced a dilemma. The attempts to directly apply some of the well-known measures of graph complexity (such as the first-generation topological indices of Wiener[22] or Randić[24] or the complexity measure of Minoli[18]) to kinetic graphs failed, due to the high degeneracy of these indices for kinetic graphs. The molecular complexity of Bertz[43] and the topological superindex[42] were at the development stage and there was little hope that they would be in a state to reflect the second level of mechanistic complexity, that of the kinetic model. (All other highly sophisticated measures of graph complexity[33,34,44] appeared considerably later.)

An alternative approach was to construct a new complexity measure to mirror both levels of mechanistic complexity. In proceeding along that avenue, we developed a complexity measure based on the complexity of the steady-state kinetic model,[49–51] which can readily be obtained for linear mechanisms by making use of graph theory. We introduced a complexity index K, based on the fractional-rational form of the rate laws for reaction routes within the framework of the Vol'kenshtein-Gol'dshtein algorithm.[56,57] More specifically, K was defined as the total number of weights (rate constants) of the elementary steps (KG edges) included in the kinetic laws for all M routes of a multiroute reaction.

Our choice was based on the simple and intuitively relevant assumption that the reaction mechanism (as well as its kinetic model) will be more complex when it includes more reaction routes, with more elementary steps and reaction intermediates in each route. This trend runs also parallel with the increase in complexity of the respective kinetic graphs with the increasing number of their cycles, edges, and vertexes. Moreover, as will be shown in this section, the index proposed satisfactorily matches the enhanced mechanistic complexity of KGs with stronger interdependence between reaction routes (KG cycles). Thus, for example, complexity increases in the following order:

routes having no common steps or intermediates →
routes having a common intermediate →
routes having one common step →
routes having more than one common step.

Finer aspects of the mechanistic topology are also mirrored by our complexity index.

Although matching both the kinetic and topological features of reaction mechanisms, the complexity index K is, indeed, not in a state to account for all complexity features of chemical reactions. Thus, complexity of the reacting species (the nature of the KG vertexes), as well as the types of elementary reaction steps, such as coupling, substitution, etc. (the nature of the KG edges), are not taken into consideration. However, being related to the kinetic graph, these features could later be incorporated as components of a conveniently generalized complexity index K.

According to the Vol'kenstein-Gol'dstein algorithm,[56,57] the rate of a route p in a reaction with M routes is

$$r_p = [X_i] \sum_{k=0}^{k \le M-1} (C_{pk}^{+} - C_{pk}^{-})D_{pk}/D_i, \tag{4.20}$$

where D_i is the vertex determinant; C_{pk} is the length of the cycle corresponding to route p ($k=0$) or the length of the cycles encompassing the cycle with $k = 0$ (the cases with $k > 0$). The cycle length is the product of the elementary step weights constituting the kth cycle; D_{pk} is the base determinant of the subgraph formed after contracting the pk cycle to a vertex (D_{pk} is also called the algebraic complement of cycle pk). [X_i] is the reagent (catalyst) concentration in vertex i. If the substance is a 0-species, then [X_0] = 1.

Equation (4.20) can be rewritten in a form which is very convenient for catalytic reactions, because it accounts for the material balance with respect to the catalyst:

$$r_p = \frac{[X]_\Sigma}{\sum_i D_i} \sum_{k=0}^{k \le M-1} (C_{pk}^{+} - C_{pk}^{-}). \tag{4.21}$$

In equation (4.21), [X]$_\Sigma$ is the total concentration of all catalyst species. In the case of surface reactions, [X]$_\Sigma$ =1.

The complexity index K can now be defined for mechanisms containing only reversible steps by the equation:

$$K = MN(N-1)T_i + 2N \sum_{p=1}^{M} \sum_{k=0}^{k_{max}} D_{pk}, \tag{4.22}$$

where T_i is the number of spanning trees having a root at vertex i (this number is the same for every vertex in the KG; a *spanning tree* is any tree containing all vertexes of a cyclic graph); the double sum counts the number of spanning trees of the KG subgraphs obtained after subsequently contracting each of the graph cycles p, and after contracting each of the encompassing cycles pk, to a vertex. As seen from equation (4.22), the calculation of complexity index K is reduced to the calculation of the spanning trees of the kinetic graph, as well as of those of some specific subgraphs. Evidently, though directly related to the kinetic model, our complexity index reflects the complexity of kinetic graphs as well, and may be of use for complexity analyses of any cyclic graphs.

4.3.2. Formulas for the Spanning Trees and Algebraic Complements in Kinetic Graphs

The Spanning Tree Count. General methods are known in graph theory for generating and enumerating spanning trees.[57] Alternatively, we derived[49-51] explicit formulas for the number of KG spanning trees, which enable the detailed analysis of mechanistic complexity of linear networks with two, three, and four reaction routes:

$$T_2 = N_1 N_2 - E_{12}^2, \qquad (4.23)$$

$$T_3 = N_1 N_2 N_3 - E_{12}^2 N_3 - E_{13}^2 N_2 - E_{23}^2 N_1 - 2 E_{12} E_{13} E_{23}, \qquad (4.24)$$

$$
\begin{aligned}
T_4 = {} & N_1 N_2 N_3 N_4 - (E_{12}^2 N_3 N_4 + E_{13}^2 N_2 N_4 + E_{14}^2 N_2 N_3 + \\
& E_{23}^2 N_1 N_4 + E_{24}^2 N_1 N_3 + E_{34}^2 N_1 N_2) - (2 E_{12} E_{13} E_{23} N_4 + \\
& 2 E_{12} E_{14} E_{24} N_3 + 2 E_{13} E_{14} E_{34} N_2 + 2 E_{23} E_{24} E_{34} N_1) - \\
& (2 E_{13} E_{14} E_{23} E_{34} + 2 E_{12} E_{13} E_{24} E_{34} + 2 E_{12} E_{14} E_{23} E_{34}) + \\
& (E_{12}^2 E_{34}^2 + E_{13}^2 E_{24}^2 + E_{14}^2 E_{23}^2).
\end{aligned}
\qquad (4.25)
$$

Equations (4.23)–(4.25) are derived with the assumption that no edge is common for more than two cycles. The equations provide fast calculation of the spanning trees count and the complexity index directly from the mechanism linear code (described in Section 2.2.2), where one can find both the cycle size N_p and the number of edges two cycles have in common E_{ij}. The latter is obviously zero for classes **A** and **B**, while for class **C** it is equal to the subclass subscript (1 for $\mathbf{C} = \mathbf{C_1}$, 2 for $\mathbf{C_2}$, etc.). Thus, for tricyclic graphs, equation (4.24) in its complete form is valid for the $\mathbf{4\text{-}C^3}$ class only. The other classes are treated by the following simplified equations:

$$T_3 = N_1 N_2 N_3 - E_{12}^2 N_3 - E_{13}^2 N_2 \qquad (4.24')$$

for classes $\mathbf{4\text{-}C^2}$, $\mathbf{2\text{-}BC^2}$, $\mathbf{3\text{-}BC^2}$;

$$T_3 = N_1 N_2 N_3 - E_{12}^2 N_3 \qquad (4.24'')$$

for classes **5-AC, 4-BC, 4-A^{2}C, 3-B^{2}C, 5-B^{2}C**; and

$$T_3 = N_1 N_2 N_3 \tag{4.24'''}$$

for classes **3-A^2, 4-A^2, 4-B^2, 5-AB, 7-A^3, 4-B^3**, and **5-A^{2}B**.

It will be shown below that the double sum in the basic equation (4.22) can be expressed as the number of spanning trees in some specific subgraphs. We will consider in detail bicyclic and tricyclic KGs; however, the result can be readily generalized for larger cyclic graphs.

Algebraic Complements in Bicyclic Kinetic Graphs. Expand the double sum in equation (4.22) as follows:

$$\sum_{p=1}^{M=2} \sum_{k=0}^{1} D_{pk} = D_p + D_{p \cup k} + D_k + D_{k \cup p}. \tag{4.26}$$

For bicyclic graphs, however,

$$D_p = N_k - E_{pk},$$
$$D_k = N_p - E_{pk}, \tag{4.27}$$

and

$$D_{p \cup k} = D_{k \cup p} = E_{pk}.$$

Hence, one obtains:

$$\sum_{p=1}^{M=2} \sum_{k=0}^{1} D_{pk} = N_p + N_k. \tag{4.28}$$

Algebraic Complements in Tricyclic Kinetic Graphs. Similarly, the double sum in (4.22) is expanded as follows:

$$\sum_{p=1}^{M=2} \sum_{k=0}^{3} D_{pk} = D_p + D_k + D_l + D_{p \cup k} +$$

$$D_{p \cup l} + D_{k \cup l} + 3D_{p \cup k \cup l} = \tag{4.29}$$

$$(N_p N_k - E_{pk}^2) + (N_p N_l - E_{pl}^2) + (N_k N_l - E_{kl}^2).$$

By comparing equations (4.29) and (4.23), one can conclude that in tricyclic graphs the algebraic complement of an arbitrary cycle l and its encompassing cycles equals the number of spanning trees in the bicyclic subgraph containing the other two cycles p and k:

$$\sum_{k=0}^{3} D_{pk} = T_{p \cup k}. \qquad (4.30)$$

For bicyclic graphs, equation (4.28) can be interpreted analogously:

$$\sum_{k=0}^{1} D_{pk} = T_{p}. \qquad (4.31)$$

These results can easily be generalized for multiroute reaction mechanisms.

It should be noted that the linear code of kinetic graphs contains all the information needed for the calculation of complexity index according to equations (4.22)–(4.31). Included here are the cycle sizes N_i, the type of linking of two cycles (**A**, **B** or **C**), the number of cycles M, and vertexes N. Thus, the KG code described in Section 2.2.2 is not only a convenient hierarchical description of these graphs, but is associated directly with their complexity, which is essential in the computer handling of linear reaction networks.

4.3.3. Standard Tables with the Complexities of All Topologically Distinct Four-Route Mechanisms Having Two to Six Intermediates

Before proceeding with a complexity analysis, we present here in Figure 4.4 and Table 4.1 all 390 four-route linear networks having two to six intermediates and their complexity indices. The K values of the five single-route networks, 24 two-route networks, and 104 three-route networks were given in Chapter 2 (Table 2.6), to illustrate the codes of these linear mechanisms.

The KGs shown in Figure 4.4 were obtained[58] by the program KING (KINetic Graphs), which generates exhaustively all nonredundant KGs for a given number of cycles and vertexes. The KING program is written in C language, under MS-DOS. It runs on an IBM PC or compatible machine and is very inexpensive since it requires relatively little RAM and hard drive space. The combinatorial algorithm used for KG enumeration is similar to that used in the GENESIS program,[58] and it employs an approach to graph enumeration developed by Faradzhev et al.[59]

The linear networks shown in Figure 4.4 contain reversible elementary steps only. However, each of them can be used to generate the possible networks with irreversible steps (directed KGs), as well as an additional number of networks incorporating intermediates that are involved only in an equilibrium elementary step (KGs with pendant vertexes).

4.3.4. Trends Increasing Mechanistic Complexity

The complexity analysis we performed for one-route, two-route, three-route and four-route mechanisms[52,69] confirmed our intuitive choice of the hierarchy of classificational factors, reflected in the linear network code:

(i) The strongest complexity factor is the number of reaction routes M. For a constant number of intermediates N, the complexity index K increases rapidly with M. Thus, for $N = 5$, K is 110 for $M = 1$, it is 200–560 for $M = 2$, 470–1740 for $M = 3$, and 1240–4440 for $M = 4$.

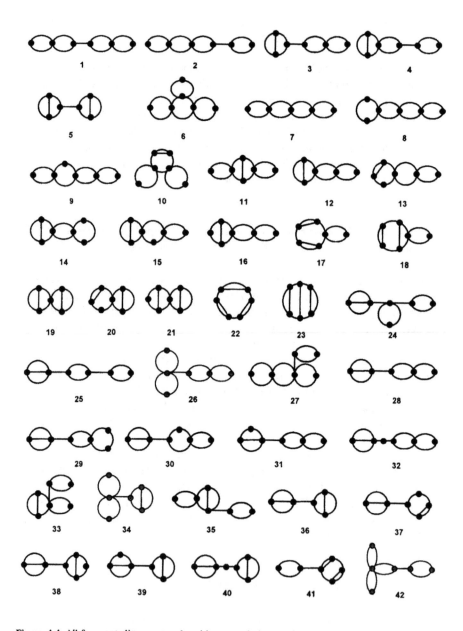

Figure 4.4. All four-route linear networks with two to six intermediates.

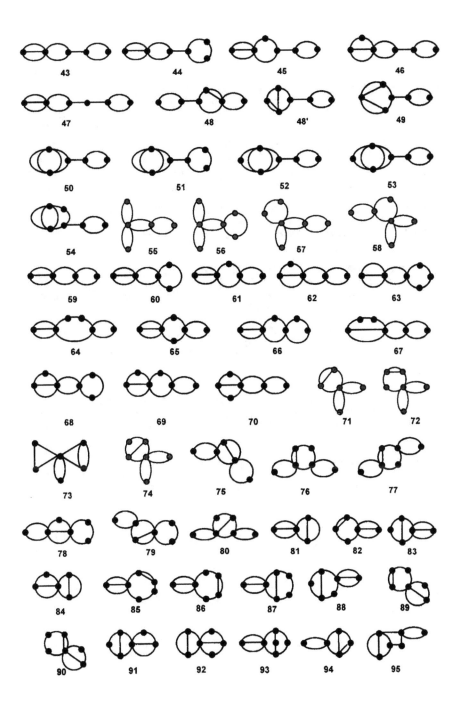

Figure 4.4. (*Continued*).

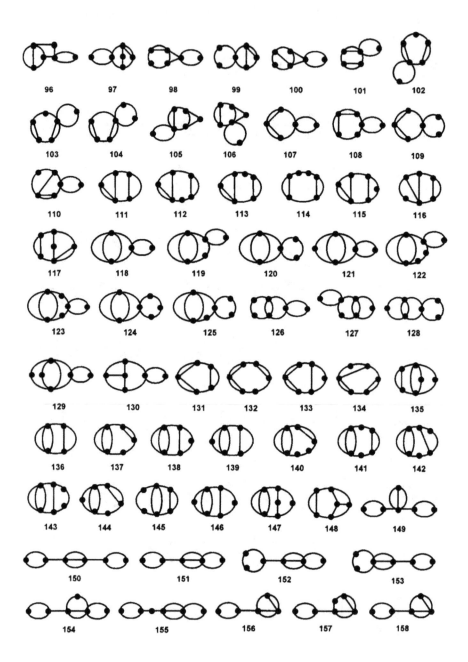

Figure 4.4. (*Continued*).

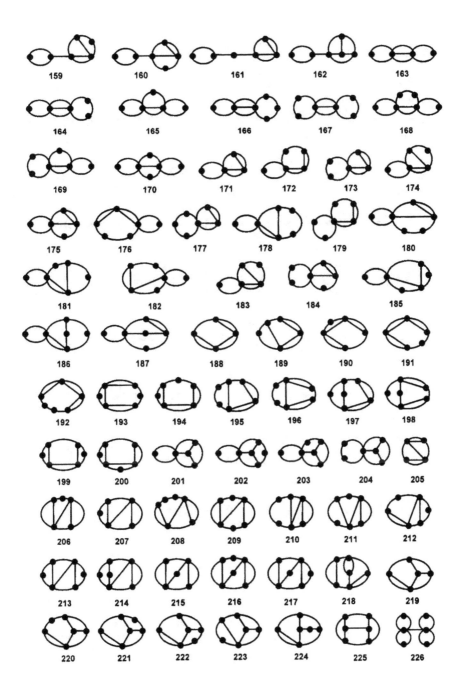

Figure 4.4. (*Continued*).

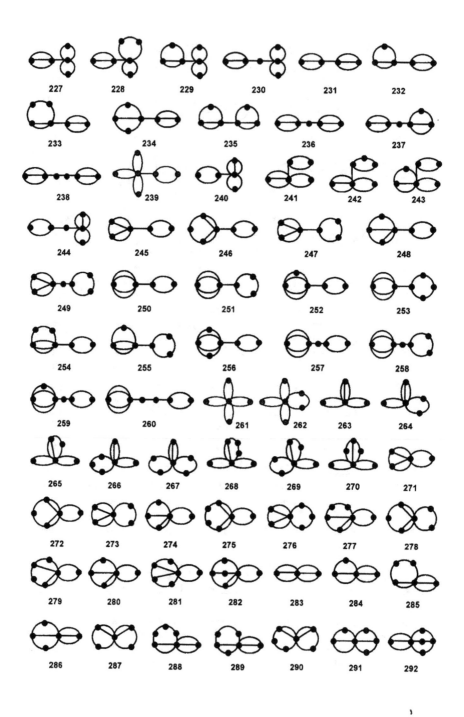

Figure 4.4. (*Continued*).

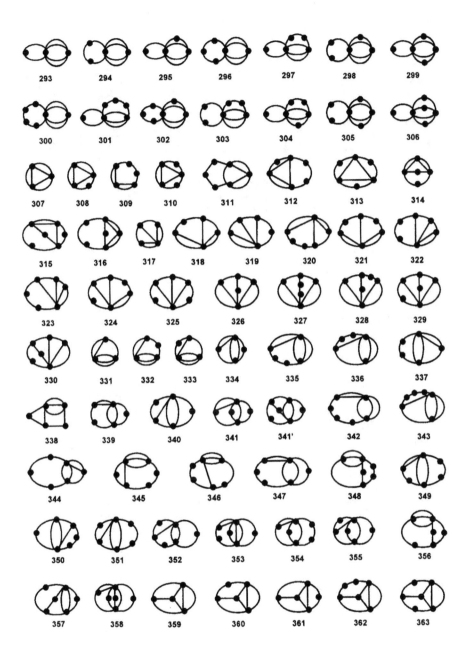

Figure 4.4. (*Continued*).

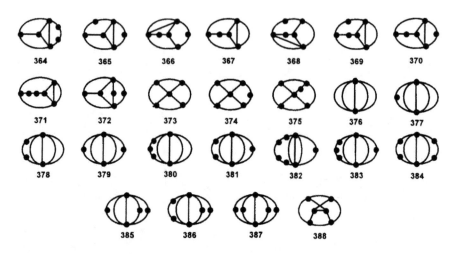

Figure 4.4. (*Continued*).

Table 4.1. Classification, codes, and complexity idexes of four-route mechanisms with two to six intermediates (Reprinted with permission from *J. Chem. Inf. Comput. Sci.* **1994**, *34*, 436–445 © 1994 American Chemical Society)

No.	Code	Index	No.	Code	Index
	$L = 3$			*Class 4-BCZ^{2}B*	
			11	4-6-4-BCZ^{2}Z$_2$B-3,2,3,2	4512
	Generalized Class AB^2			*Class 4-B^{2}Z^{2}CZ*	
	Class 4-ABZ^{2}BZ		12	4-5-4- B^{2}Z^{2}CZ$_2$-2,3,2,2	2000
1	4-6-4-ABZ^{2}BZ$_2$-2,2,2,2	2304	13	4-6-4- B^{2}Z^{2}CZ$_2$-2,4,2,2	3984
	Class 4-ABZ^{2}B		14	4-6-4- B^{2}Z^{2}CZ$_2$-2,3,3,2	4260
2	4-6-4-ABZ^{2}Z$_2$B-2,2,2,2	2304	15	4-6-4- B^{2}Z^{2}CZ$_2$-3,3,2,2	4260
			16	4-6-4- B^{2}Z^{2}CZ$_2$-2,3,2,3	4512
	Generalized Class ABC				
	Class 5-ABZ^{2}CZ			**Generalized Class BC^2**	
3	4-6-5-ABZ^{2}CZ$_2$-2,3,2,2	2880		*Class 5- BC^{2}Z^3*	
	Class 5-ABZ^{3}C		17	4-6-5-BC^{2}Z^3-5,2,2,2	4560
4	4-6-5-ABZ^{2}Z$_2$C-2,2,3,2	2880		*Class 5-BCZ^{3}C*	
			18	4-6-5-BCZ^{2}Z$_2$C-3,2,4,2	5364
	Generalized Class AC^2			*Class 5-BCZ^{2}CZ*	
	Class 6-ACZ^{2}CZ		19	4-5-5-BCZ^{2}CZ$_2$-3,3,2,2	2500
5	4-6-6-ACZ^{2}CZ$_2$-3,3,2,2	3600	20	4-6-5-BCZ^{2}CZ$_2$-3,4,2,2	4980
			21	4-6-5-BCZ^{2}CZ$_2$-3,3,2,3	5640
	Generalized Class B^3				
	Class 3-B^{3}Z^3			**Generalized Class C^3**	
6	4-6-3-B^{3}Z^3-3,2,2,2	3408		*Class 6-C^{3}Z^3*	
	Class 3-B^{2}Z^{2}BZ		22	4-6-6-C^{3}Z^3-6,2,2,2	5136
7	4-5-3-B^{2}Z^{2}BZ$_2$-2,2,2,2	1600		*Class 6-C^{2}Z^{2}CZ*	
8	4-6 3-B^{2}Z^{2}BZ$_2$-2,2,2,3	3405	23	4-6-6-C^{2}Z^{2}CZ$_2$-4,4,2,2	6360
9	4-6-3-B^{2}Z^{2}BZ$_2$-2,3,2,2	3408			
				$L=4$	
	Generalized Class B^{2}C		24	4-6-4-A^{3}CZ$_0$^{2}-2,2,2,2	1776
	Class 4-B^{2}CZ^3			*Class 5-A^{3}CZ^2*	
10	4-6-4-B^{2}CZ^3-4,2,2,2	3984	25	4-6-5-A^{3}CZ^2-2,2,2,2	1776

Table 4.1. (*Continued*)

No.	Code	Index
	Generalized Class A^2B^2	
	Class 3-$A^2B^2Z^2$	
26	4-6-3-$A^2B^2Z^2$-2,2,2,2	2304
	Class 3-A^2ZB^2Z	
27	4-6-3-A^2ZB^2Z-2,2,2,2	2304
	Generalized Class A^2BC	
	Class 4-A^2BCZ^2	
28	4-5-4-A^2BCZ^2-2,2,2,2	1240
29	4-6-4-A^2BCZ^2-2,2,2,3	2628
30	4-6-4-A^2BCZ^2-3,2,2,2	2628
31	4-6-4-A^2BCZ^2-2,2,3,2	2880
32	4-6-4-A^2BCZ^2-2,2,2,2	1776
	Class 4-A^2ZBCZ	
33	4-6-4-A^2ZBCZ-2,3,2,2	2880
	Class 4-A^2CBZ^2	
34	4-6-4-A^2CBZ^2-3,2,2,2	2880
	Class 4-A^2ZCBZ	
35	4-6-4-A^2ZCBZ-2,3,2,2	2880
	Generalized Class A^2C^2	
	Class 5-$A^2C^2Z^2$-	
36	4-5-5-$A^2C^2Z^2$-3,2,2,2	1550
37	4-6-5-$A^2C^2Z^2$-4,2,2,2	3072
38	4-6-5-$A^2C^2Z^2$-3,2,2,3	3480
39	4-6-5-$A^2C^2Z^2$-3,2,3,2	3600
40	4-6-5-$A_2C^2Z^2$-3,2,2,2	2220
	Class 5-A^2ZC^2Z	
41	4-6-5-A^2ZC^2Z-2,4,2,2	3456
	Generalized Class AB^3	
	Class 3-AB^2Z^2B	
42	4-6-3-AB^2Z^2B-2,2,2,2	2304
	Generalized Class AB^2C	
	Class 4-AB^2Z^2C	
43	4-5-4-AB^2Z^2C-2,2,2,2	1240
44	4-6-4-AB^2Z^2C-2,3,2,2	2628
45	4-6-4-AB^2Z^2C-3,2,2,2	2628
46	4-6-4-AB^2Z^2C-2,2,2,3	2880
47	4-6-4-$A_2B^2Z^2C$-2,2,2,2	1776
	Class 4-$ABCZ^2B$	
48	4-6-4-$ABCZ^2B$-3,2,2,2	2880
	Generalized Class ABC^2	
	Class 5-$ABCZ^2C$	
48'	4-6-5-$ABCZ^2$-3,2,2,3	3732
	Class 5-AC^2Z^2B	
49	4-6-5-AC^2Z^2B-4,2,2,2	3456
	Generalized Class AC^3	
	Class 4-AC^2Z^2C	
50	4-5-4-AC^2Z^2C-3,2,2,2	1470
51	4-6-4-AC^2Z^2C-3,3,2,2	3108
52	4-6-4-AC^2Z^2C-3,2,2,3	3480
53	4-6-4-$A_2C^2Z^2C$-3,2,2,2	2100
54	4-6-4-AC^2Z^2C-4,2,2,2	2952
	Generalized Class B^4	
	Class 2-B^4Z^2	
55	4-5-2-B^4Z^2-2,2,2,2	1600
56	4-6-2-B^4Z^2-2,2,2,3	3408
57	4-6-2-B^4Z^2-2,2,3,2	3408
58	4-6-2-B^4Z^2-3,2,2,2	3408
	Generalized Class B^3C	
	Class 3-B^3CZ^2	
59	4-4-3-B^3CZ^2-2,2,2,2	800
60	4-5-3-B^3CZ^2-2,2,2,3	1830
61	4-5-3-B^3CZ^2-3,2,2,2	1830
62	4-5-3-B^3CZ^2-2,2,3,2	2000
63	4-6-3-B^3CZ^2-2,2,2,4	3480
64	4-6-3-B^3CZ^2-4,2,2,2	3480
65	4-6-3-B^3CZ^2-4,2,2,2	3480
66	4-6-3-B^3CZ^2-3,2,2,3	3888
67	4-6-3-B^3CZ^2-2,2,4,2	3984
68	4-6-3-B^3CZ^2-2,2,3,3	4260
69	4-6-3-B^3CZ^2-3,2,3,2	4260
70	4-6-3-B^3CZ^2-2,3,3,2	4512
	Class 3-B^2CBZ^2	
71	4-5-3-B^2CBZ^2-3,2,2,2	2000
72	4-6-3-B^2CBZ^2-4,2,2,2	3984
73	4-6-3-B^2CBZ^2-3,2,3,2	4260
74	4-6-3-B^2CBZ^2-3,2,2,3	4512
	Class 3-B^2CZBZ	
75	4-5-3-B^2CZBZ-3,2,2,2	2000
76	4-6-3-B^2CZBZ-4,2,2,2	3984
77	4-6-3-B^2CZ_2BZ-4,2,2,2	3984
78	4-6-3-B^2CZBZ-3,2,2,2	4260
79	4-6-3-B^2CZBZ-3,3,2,2	4260
80	4-6-3-B^2CZBZ-3,2,2,3	4512
	Generalized Class B^2C^2	
	Class 4-$B^2C^2Z^2$	
81	4-4-4-$B^2C^2Z^2$-3,2,2,2	1000
82	4-5-4-$B^2C^2Z^2$-4,2,2,2	2140
83	4-5-4-$B^2C^2Z^2$-3,2,2,3	2420
84	4-5-4-$B^2C^2Z^2$-3,2,3,2	2500
85	4-6-4-$B^2C^2Z^2$-5,2,2,2	3924
86	4-6-4-$B^2C^2Z^2$-5,2,2,2	3924
87	4-6-4-$B^2C^2Z^2$-3,2,2,4	4740
88	4-6-4-$B^2C^2Z^2$-4,2,2,3	4740
89	4-6-4-$B^2C^2Z^2$-4,2,3,2	4980
90	4-6-4-$B^2C^2Z^2$-3,2,4,2	4980
91	4-6-4-$B^2C^2Z^2$-3,2,3,3	5640
92	4-6-4-$B^2C^2Z^2$-3,3,3,2	5640
93	4-6-4-$B^2C_2CZ^2$-4,2,2,4	5184
	Class 4-B^2CZCZ	
94	4-5-4-B^2CZCZ-3,2,23	2590
95	4-6-4-B^2CZCZ-4,2,2,3	5112
96	4-6-4-B^2C_2CZ-4,2,2,3	5112
97	4-6-4-B^2C_2ZCZ-4,2,2,4	5592

Table 4.1. (*Continued*)

No.	Code	Index
98	4-6-4-B^2CZCZ-3,2,2,4	5364
99	4-6-4-B^2CZCZ-3,2,3,3	5520
100	4-6-4-B^2CZCZ-3,3,2,3	5892
	Class 4-B^2ZC^2Z	
101	4-5-4-B^2ZC^2Z-2,4,2,2	2400
102	4-6-4-B^2ZC^2Z-2,5,2,2	4560
103	4-6-4-B^2ZC^2Z-2,5,2,2	4560
104	4-6-4-B^2ZC^2Z-3,4,2,2	5112
105	4-6-4-B^2ZC^2Z-2,4,2,3	5364
106	4-6-4-B^2ZC^2Z-2,4,3,2	5364
	Class 4-BC^2Z^2B	
107	4-5-4-BC^2Z^2B-4,3,2,2	2400
108	4-6-4-BC^2ZZ_2B-5,2,2,2	4560
109	4-6-4-BC^2Z^2B-4,3,2,2	5112
110	4-6-4-BC^2Z^2B-4,2,2,3	5364
	Generalized Class BC^3	
	Class 5-BC^3Z^2	
111	4-5-5-BC^3Z^2-4,2,3,2	3090
112	4-6-5-BC^3Z_2Z-5,2,3,2	5832
113	4-6-5-BC^3Z^2-5,2,3,2	5832
114	4-6-5-BC^3Z^2-4,2,4,2	6360
115	4-6-5-BC^3Z^2-4,2,3,3	6900
116	4-6-5-BC^3Z^2-4,3,3,2	7020
117	4-6-5-$BC_2C^2Z^2$-5,2,4,2	6684
	Class 3-BC^2Z^2C	
118	4-4-3-BC^2Z^2C-3,2,2,2	952
119	4-5-3-BC^2Z^2C-4,2,2,2	2060
120	4-5-3-BC^2Z^2C-3,3,2,2	2170
121	4-5-3-BC^2Z^2C-3,2,2,3	2420
122	4-6-3-BC^2Z^2C-5,2,2,2	3504
123	4-6-3-$BC^2Z_2^2C$-5,2,2,2	3804
124	4-6-3-BC^2Z^2C-3,4,2,2	4116
125	4-6-3-BC^2Z^2C-4,3,2,2	4368
126	4-6-3-BC^2Z^2C-3,2,2,4	4860
127	4-6-3-BC^2Z^2C-4,2,2,3	4860
128	4-6-3-BC^2Z^2C-3,3,2,3	5148
129	4-6-3-BC^2Z^2C-3,2,3,3	5712
	Class 5-BC^2Z^2C	
130	4-6-5-BC^2Z^2C-4,3,3,3	6768
	Class 5-BCZCZC	
131	4-5-5-BCZCZC-2,2,5,2	2800
132	4-6-5-BCZCZC_2-2,2,6,2	5136
133	4-6-5-BCZCZC-2,2,5,3	6216
134	4-6-5-BCZCZC-2,3,5,2	6216
	Generalized Class C^4	
	Class 4-C^4Z^2	
135	4-6-4-C^4Z^2-4,3,3,2	6852
136	4-4-4-C^4Z^2-4,2,2,2	1152
137	4-5-4-C^4Z^2-5,2,2,2	2370
138	4-5-4-C^4Z^2-4,2,2,3	2760
139	4-5-4-C^4Z^2-4,2,3,2	2920
140	4-6-4-C^4Z^2-6,2,2,2	4248
141	4-6-4-$C^4Z_2^2$-6,2,2,2	4248
142	4-6-4-C^4Z^2-5,2,2,3	5220

No.	Code	Index
143	4-6-4-C^4Z^2-4,2,2,4	5376
144	4-6-4-C^4Z^2-5,2,3,2	5580
145	4-6-4-C^4Z^2-4,2,4,2	5856
146	4-6-4-C^4Z^2-4,2,3,3	6528
147	4-6-4-$C^2C_2CZ^2$-5,2,2,4	5820
	Class 6-C^4Z	
148	4-6-6-C^4Z^2-5,3,3,2	7896
	L = 5	
	Generalized Class A^4C	
	Class 4-A^2CZA^2	
149	4-6-4-$A^2CZ_0A^2$-2,2,2,2	1776
150	4-6-4-A^2CZA^2-2,2,2,2	1776
	Generalized Class A^2B^2C	
	Class 3-A^2CZB^2	
151	4-5-3-A^2CZB^2-2,2,2,2	1240
152	4-6-3-A^2CZB^2-3,2,2,2	2628
153	4-6-3-A^2CZB^2-2,2,2,3	2680
154	4-6-3-A^2CZB^2-2,2,3,2	2880
155	4-6-3-$A_2^2CZB^2$-2,2,2,2	1776
	Generalized Class A^2BC^2	
	Class 4-A^2ZCBC	
156	4-5-4-A^2ZCBC-2,2,3,2	1640
157	4-6-4-A^2ZCBC-2,2,4,2	3456
158	4-6-4-A^2ZCBC-3,2,3,2	3480
159	4-6-4-A^2ZCBC-2,2,3,3	3732
160	4-6-4-A^2ZCBC-2,3,3,2	3732
161	4-6-4-A_2^2ZCBC-2,2,3,2	2352
	Generalized Class A^2C^3	
	Class 5-A^2ZC^3	
162	4-6-5-A^2ZC^3-2,3,3,3	4608
	Generalized Class B^4C	
	Class 2-B^2CZB^2	
163	4-4-2-B^2CZB^2B-2,2,2,2	800
164	4-5-2-B^2CZB^2B-2,2,2,2	1830
165	4-5-2-B^2CZB^2B-2,2,2,3	2000
166	4-6-2-B^2CZB^2B-2,2,4,2	3480
167	4-6-2-B^2CZB^2B-2,3,3,2	3888
168	4-6-2-B^2CZB^2B-2,2,2,4	3984
169	4-6-2-B^2CZB^2B-2,2,3,3	4260
170	4-6-2-B^2CZB^2B-3,2,2,3	4512
	Generalized Class B^3C^2	
	Class 3-B^2CZBC	
171	4-4-3-B^2CZBC-2,2,2,3	1056
172	4-5-3-B^2CZBC-2,2,2,4	2400
173	4-5-3-B^2CZBC-2,3,2,3	2420
174	4-5-3-B^2CZBC-2,2,3,3	2590
175	4-5-3-B^2CZBC-3,2,2,3	2590
176	4-6-3-B^2CZBC-2,2,2,5	4560
177	4-6-3-B^2CZBC-2,4,2,3	4608

Table 4.1. (*Continued*)

No.	Code	Index
178	4-6-3-B^{2}CZBC-2,2,4,3	5112
179	4-6-3-B^{2}CZBC-2,3,2,4	5112
180	4-6-3-B^{2}CZBC-4,2,2,3	5112
181	4-6-3-B^{2}CZBC-2,2,3,4	5364
182	4-6-3-B^{2}CZBC-3,2,2,4	5364
183	4-6-3-B^{2}CZBC-2,3,3,3	5520
184	4-6-3-B^{2}CZBC-3,3,2,3	5892
185	4-6-3-B^{2}CZBC-3,2,3,3	5892
186	4-6-3-B^{2}CZBC$_2$-2,2,4,4	5592
187	4-6-3-B^{2}C$_2$ZBC-4,2,2,4	5592
	Ceneralized Class B^{2}C^3	
	Class 4-B^{2}CZC^2	
188	4-4-4-B^{2}CZC^2-2,2,2,4	1312
189	4-5-4-B^{2}CZC^2-2,2,3,4	3180
190	4-5-4-B^{2}CZC^2-3,2,2,4	3180
191	4-5-4-B^{2}CZC^2-2,2,2,5	2800
192	4-6-4-B^{2}CZC^2-2,2,2,6	5136
193	4-6-4-B^{2}CZC^2-3,2,2,5	6216
194	4-6-4-B^{2}CZC^2-2,2,3,5	6216
195	4-6-4-B^{2}CZC^2-2,2,4,4	6240
196	4-6-4-B^{2}CZC^2-4,2,2,4	6240
197	4-6-4-B^{2}CZCC$_2$-2,2,4,5	6816
198	4-6-4-B^{2}C$_2$ZC^2-4,2,2,5	6816
199	4-6-4-B^{2}CZC^2-2,3,3,4	7152
200	4-6-4-B^{2}CZC^2-3,2,3,4	7152
	Class 4-B^{2}ZC^3	
201	4-5-4-B^{2}ZC^3-2,3,3,3	3200
202	4-6-4-B^{2}ZC^3-2,3,3,4	6768
203	4-6-4-B^{2}ZC^3-2,3,4,3	6768
204	4-6-4-B^{2}ZC^3-3,3,3,3	6816
	Class 4-BC^{3}ZB	
205	4-4-4-BC^{3}ZB-3,2,3,2	1376
206	4-5-4-BC^{3}ZB-3,2,4,2	3090
207	4-5-4-BC^{3}ZB-3,2,3,3	3370
208	4-6-4-BC^{3}ZB-3,2,5,2	5832
209	4-6-4-BC^{3}ZB-4,2,4,2	6360
210	4-6-4-BC^{3}ZB-3,2,3,4	6648
211	4-6-4-BC^{3}ZB-3,3,4,2	6900
212	4-6-4-BC^{3}ZB-3,2,4,3	7020
213	4-6-4-BC^{3}ZB-3,3,3,3	7680
214	4-6-4-BC^{2}C$_2$ZB-3,4,4,2	7368
215	4-6-4-BC$_2$C^{2}ZB-4,2,4,2	3320
216	4-6-4-BC$_2$C^{2}ZB-4,2,5,2	6684
217	4-6-4-BC$_2$C^{2}ZB-4,2,4,3	7300
218	4-6-4-BC$_2$C^{2}ZB-5,2,5,2	6480
	Generalized Class BC^4	
	Class 5-BC^{3}ZC	
219	4-5-5-BC^{3}ZC-3,2,4,3	3980
220	4-6-5-BC^{3}ZC-3,2,5,3	7896
221	4-6-5-BC^{3}ZC-3,2,4,4	8304
222	4-6-5-BC^{3}ZC-4,2,4,3	8304
223	4-6-5-BC^{3}ZC-3,3,4,3	8976
224	4-6-5-BCC$_2$CZC-4,2,4,4	9024

No.	Code	Index
	Generalized Class C^5	
	Class 6-C^{5}Z	
225	4-6-6-C^{5}Z-4,4,3,3	10560
	L = 6	
	Generalized Class A^{4}B^2	
	Class 2-A^{2}B^{2}A^2	
226	4-6-2-A^{2}B^{2}A^2-2,2,2,2	2304
	Generalized Class A^{4}BC	
	Class 3-A^{2}BCA^2	
227	4-5-3-A^{2}BCA^2-2,2,2,2	1240
228	4-6-3-A^{2}BCA^2-2,2,2,3	2628
229	4-6-3-A^{2}BCA^2-2,2,3,2	2880
230	4-6-3-A$_2$^{2}BCA$_2$-2,2,2,2	1776
	Generalized Class A^{4}C^2	
	Class 4-A^{2}C^{2}A^2	
231	4-4-4-A^{2}C^{2}A^2-2,2,2,2	624
232	4-5-4-A^{2}C^{2}A^2-2,2,2,3	1550
233	4-6-4-A^{2}C^{2}A^2-2,2,2,4	3072
234	4-6-4-A^{2}C^{2}A^2-2,3,3,2	3480
235	4-6-4-A^{2}C^{2}A^2-2,2,3,3	3600
236	4-6-4-A$_2$^{2}C^{2}A$_2$^{2}-2,2,2,2	960
237	4-6-4-A$_2$^{2}C^{2}A$_2$^{2}-2,2,2,3	2220
238	4-6-4-A$_2$^{2}C^{2}A$_2$^{2}-2,2,2,2	1368
	Generalized Class A^{3}C^3	
	Class 2-A^{3}B^3	
239	4-6-2-A3B3-2,2,2,2	2304
	Generalized Class A^{3}B^{2}C	
	Class 3-A^{3}B^{2}C	
240	4-5-3-A^{3}B^{2}C-2,2,2,2	1240
241	4-6-3-A^{3}B^{2}C-2,3,2,2	2628
242	4-6-3-A^{3}B^{2}C-3,2,2,2	2628
243	4-6-3-A^{3}B^{2}C-2,2,2,3	2880
244	4-6-3-A$_2$^{3}B^{2}C-2,2,2,2	1776
	Generalized Class A^{3}BC^2	
	Class 4-A^{3}BC^2	
245	4-5-4-A^{3}BC^2-2,2,2,3	1640
246	4-6-4-A^{3}BC^2-2,2,2,4	3456
247	4-6-4-A^{3}BC^2-3,2,2,3	3480
248	4-6-4-A^{3}BC^2-2,2,3,3	3732
249	4-6-4-A$_2$^{3}BC^2-2,2,2,3	2325
	Generalized Class A^{3}C^3	
	Class 3-A^{3}C^3	
250	4-4-3-A^{3}C^3-2,2,2,2	576
251	4-5-3-A^{3}C^3-3,2,2,2	1300
252	4-5-3-A^{3}C^3-2,2,2,3	1470
253	4-6-3-A^{3}C^3-4,2,2,2	2448
254	4-6-3-A^{3}C^3-2,2,2,4	2952

Table 4.1. (*Continued*)

No.	Code	Index		No.	Code	Index
255	4-6-3-A^{3}C^3-3,2,2,3	3108		297	4-5-2-B^{3}C^3-2,2,2,4	2060
256	4-6-3-A^{3}C^3-2,2,3,3	3480		298	4-5-2-B^{3}C^3-3,2,2,3	2170
257	4-6-3-A$_2$^{3}C^3-2,2,2,2	880		299	4-5-2-B^{3}C^3-2,2,3,3	2420
258	4-6-3-A$_2$^{3}C^3-3,2,2,2	1848		300	4-6-2-B^{3}C^3-5,2,2,2	3048
259	4-6-3-A$_2$^{3}C^3-2,2,2,3	2100		301	4-6-2-B^{3}C^3-2,2,2,5	3804
260	4-6-3-A$_3$^{3}C^3-2,2,2,2	1248		302	4-6-2-B^{3}C^3-4,2,2,3	4116
				303	4-6-2-B^{3}C^3-3,2,2,4	4368
	Generalized Class B^6			304	4-6-2-B^{3}C^3-2,2,3,4	4860
	Class 1-B^6			305	4-6-2-B^{3}C^3-3,2,3,3	5148
261	4-5-1-B^6-2,2,2,2	1600		306	4-6-2-B^{3}C^3-2,3,3,3	5712
262	4-6-1-B^6-2,2,2,3	3408			*Class 3-B^{2}CBC^2*	
				307	4-3-3-B^{2}CBC^2-2,2,2,3	480
	Generalized Class B^{6}C			308	4-4-3-B^{2}CBC^2-2,2,3,3	1328
	Class 2-B^{5}C			309	4-5-3-B^{2}CBC^2-2,2,4,3	2840
263	4-4-2-B^{5}C-2,2,2,2	800		310	4-5-3-B^{2}CBC^2-2,3,3,3	3290
264	4-5-2-B^{5}C-2,3,2,2	1830		311	4-6-3-B^{2}CBC^2-2,2,5,3	5208
265	4-5-2-B^{5}C-2,2,2,3	2000		312	4-6-3-B^{2}CBC^2-2,3,4,3	6528
266	4-6-2-B^{5}C-2,4,2,2	3480		313	4-6-3-B^{2}CBC^2-3,3,3,3	7560
267	4-6-2-B^{5}C-3,3,2,2	3888		314	4-5-3-B^{2}CBCC$_2$-2,2,4,4	3240
268	4-6-2-B^{5}C-2,2,2,4	3984		315	4-6-3-B^{2}CBCC$_2$-2,3,4,4	7380
269	4-6-2-B^{5}C-2,3,2,3	4260		316	4-6-3-B^{2}CBCC$_2$-2,2,5,4	6312
270	4-6-2-B^{5}C-2,2,3,3	4512			*Class 4-B^{2}C^{2}BC*	
				317	4-4-4-B^{2}C^{2}BC-2,2,3,3	1376
	Generalized Class B^{4}C^2			318	4-5-4-B^{2}C^{2}BC-2,2,3,4	3090
	Class 3-B^{4}C^2			319	4-5-4-B^{2}C^{2}BC-2,3,3,3	3370
271	4-4-3-B^{4}C^2-2,2,2,3	1056		320	4-6-4-B^{2}C^{2}BC-2,2,3,5	5832
272	4-5-3-B^{4}C^2-2,2,2,4	2400		321	4-6-4-B^{2}C^{2}BC-2,2,4,4	6360
273	4-5-3-B^{4}C^2-3,2,2,3	2420		322	4-6-4-B^{2}C^{2}BC-2,4,3,3	6648
274	4-5-3-B^{4}C^2-2,2,3,3	2590		323	4-6-4-B^{2}C^{2}BC-2,3,4,3	6900
275	4-6-3-B^{4}C^2-2,2,2,5	4560		324	4-6-4-B^{2}C^{2}BC-2,3,3,4	7020
276	4-6-3-B^{4}C^2-4,2,2,3	4608		325	4-6-4-B^{2}C^{2}BC-3,3,3,3	7680
277	4-6-3-B^{4}C^2-2,2,4,3	5112		326	4-6-4-B^{2}C^{2}BC$_2$-2,2,4,4	3320
278	4-6-3-B^{4}C^2-3,2,2,4	5112		327	4-6-4-B^{2}C^{2}BC$_3$-2,2,5,5	6480
279	4-6-3-B^{4}C^2-2,2,3,4	5364		328	4-6-4-B^{2}C^{2}BC$_2$-2,2,4,5	6684
280	4-6-3-B^{4}C^2-3,2,3,3	5520		329	4-6-4-B^{2}C^{2}BC$_2$-2,3,4,4	7500
281	4-6-3-B^{4}C^2-2,3,3,3	5892		330	4-6-4-B^{2}CC$_2$BC-2,4,4,3	7368
282	4-6-3-B^{4}CC$_2$-2,2,4,4	5592				
	Class 3-B^{2}C^{2}B^2				**Generalized Class B^{2}C^4**	
283	4-3-3-B^{2}C^{2}B^2-2,2,2,2	360			*Class 3-B^{2}C^4*	
284	4-4-3-B^{2}C^{2}B^2-2,2,2,3	1000		331	4-3-3-B^{2}C^4-2,2,2,3	450
285	4-5-3-B^{2}C^{2}B^2-2,2,2,4	2140		332	4-4-3-B^{2}C^4-2,2,2,4	1152
286	4-5-3-B^{2}C^{2}B^2-2,3,3,2	2420		333	4-4-3-B^{2}C^4-3,2,2,3	1224
287	4-5-3-B^{2}C^{2}B^2-2,2,3,3	2500		334	4-4-3-B^{2}C^4-2,2,3,3	1272
288	4-6-3-B^{2}C^{2}B^2-2,2,2,5	3924		335	4-5-3-B^{2}C^4-2,2,2,5	2370
289	4-6-3-B^{2}C^{2}B^2-2,3,4,2	4740		336	4-5-3-B^{2}C^4-4,2,2,3	2590
290	4-6-3-B^{2}C^{2}B^2-2,2,3,4	4980		337	4-5-3-B^{2}C^4-2,2,4,3	2750
291	4-6-3-B^{2}C^{2}B^2-2,3,3,3	5640		338	4-5-3-B^{2}C^4-3,2,2,4	2760
292	4-6-3-B^{2}C^{2}B^2-2,4,4,2	5184		339	4-5-3-B^{2}C^4-2,2,3,4	2920
				340	4-5-3-B^{2}C^4-3,2,3,3	3120
	Generalized Class B^{3}C^3			341	4-5-3-B^{2}C^4-2,3,3,3	3250
	Class 2-B^{3}C^3			341'	4-5-3-B^{2}C$_2$C^3-4,2,2,4	2900
293	4-3-2-B^{3}C^3-2,2,2,2	336		342	4-6-3-B^{2}C^4-2,2,2,6	4248
294	4-4-2-B^{3}C^3-3,2,2,2	848		343	4-6-3-B^{2}C^4-5,2,2,3	4716
295	4-4-2-B^{3}C^3-2,2,2,3	952		344	4-6-3-B^{2}C^4-2,2,5,3	5076
296	4-5-2-B^{3}C^3-4,2,2,2	1720		345	4-6-3-B^{2}C^4-3,2,2,5	5220

Table 4.1. (*Continued*)

No.	Code	Index	No.	Code	Index
346	$4\text{-}6\text{-}3\text{-}B^2C^4\text{-}4,2,2,4$	5376	369	$4\text{-}6\text{-}4\text{-}B^2CC_2C^2\text{-}2,4,5,3$	7776
347	$4\text{-}6\text{-}3\text{-}B^2C^4\text{-}2,2,3,5$	5580	370	$4\text{-}6\text{-}4\text{-}B^2CC_2C^2\text{-}3,4,4,3$	8208
348	$4\text{-}6\text{-}3\text{-}B^2C_2C^3\text{-}5,2,2,4$	5568	371	$4\text{-}6\text{-}4\text{-}B^2CC_2C^2\text{-}2,5,5,3$	8220
349	$4\text{-}6\text{-}3\text{-}B^2C^4\text{-}2,2,4,4$	5856	372	$4\text{-}6\text{-}4\text{-}B^2C_2C^3\text{-}4,3,3,4$	9120
350	$4\text{-}6\text{-}3\text{-}B^2C^4\text{-}4,2,3,3$	6156		*Class 5-BC⁴B*	
351	$4\text{-}6\text{-}3\text{-}B^2C^4\text{-}3,2,4,3$	6276	373	$4\text{-}5\text{-}5\text{-}BC^4B\text{-}3,3,3,3$	4440
352	$4\text{-}6\text{-}3\text{-}B^2C^4\text{-}3,2,3,4$	6528	374	$4\text{-}6\text{-}5\text{-}BC^4B\text{-}3,3,3,4$	9228
353	$4\text{-}6\text{-}3\text{-}B^2C^4\text{-}2,3,3,4$	6852	375	$4\text{-}6\text{-}5\text{-}BC^3C_2B\text{-}3,4,3,4$	9744
354	$4\text{-}6\text{-}3\text{-}B^2C^4\text{-}2,3,4,3$	6840		*Class 2-C⁶*	
355	$4\text{-}6\text{-}3\text{-}B^2C^4\text{-}3,3,3,3$	7404	376	$4\text{-}2\text{-}2\text{-}C^6\text{-}2,2,2,2$	120
356	$4\text{-}6\text{-}3\text{-}B^2C_2C^3\text{-}4,2,2,5$	5820	377	$4\text{-}3\text{-}2\text{-}C^6\text{-}2,2,2,3$	396
357	$4\text{-}6\text{-}3\text{-}B^2C_2C^3\text{-}4,2,3,4$	6876	378	$4\text{-}4\text{-}2\text{-}C^6\text{-}2,2,2,4$	944
358	$4\text{-}6\text{-}3\text{-}B^2CC_2^3\text{-}2,4,4,4$	7632	379	$4\text{-}4\text{-}2\text{-}C^6\text{-}2,2,3,3$	1120
	Class 4-B²C⁴		380	$4\text{-}5\text{-}2\text{-}C^6\text{-}2,2,2,5$	1860
359	$4\text{-}4\text{-}4\text{-}B^2C^4\text{-}2,3,3,3$	1616	381	$4\text{-}5\text{-}2\text{-}C^6\text{-}2,2,3,4$	2420
360	$4\text{-}5\text{-}4\text{-}B^2C^4\text{-}2,3,4,3$	3730	382	$4\text{-}6\text{-}2\text{-}C^6\text{-}2,2,2,6$	3240
361	$4\text{-}5\text{-}4\text{-}B^2C^4\text{-}3,3,3,3$	4020	383	$4\text{-}6\text{-}2\text{-}C^6\text{-}2,2,3,5$	4464
362	$4\text{-}6\text{-}4\text{-}B^2C^4\text{-}2,3,5,3$	7152	384	$4\text{-}6\text{-}2\text{-}C^6\text{-}2,2,4,4$	4872
363	$4\text{-}6\text{-}4\text{-}B^2C^4\text{-}2,4,4,3$	7812	385	$4\text{-}6\text{-}2\text{-}C^6\text{-}2,3,3,3$	2960
364	$4\text{-}6\text{-}4\text{-}B^2C^4\text{-}4,3,3,3$	7992	386	$4\text{-}6\text{-}2\text{-}C^6\text{-}2,3,3,4$	5928
365	$4\text{-}6\text{-}4\text{-}B^2C^4\text{-}3,3,4,3$	8604	387	$4\text{-}6\text{-}2\text{-}C^6\text{-}3,3,3,3$	7056
366	$4\text{-}6\text{-}4\text{-}B^2C^3C_2\text{-}2,4,4,4$	8700			
367	$4\text{-}6\text{-}4\text{-}B^2CC_2C^2\text{-}2,4,4,3$	3700		*Class 6-C⁶*	
368	$4\text{-}6\text{-}4\text{-}B^2C^3C_2\text{-}2,3,5,4$	8028	388	$4\text{-}6\text{-}6\text{-}CC_2^4C\text{-}4,4,4,4$	11616

(ii) For a constant number of routes M, the number of intermediates N is a stronger complexity factor than route connectedness and intermediate connectedness for networks with 2 to 4 intermediates for which no complicated topology could arise. However, for $N = 5$, the two factors are competing, whereas for $N = 6$ and $N > 6$, the enhanced intermediate connectedness produces generally a higher complexity than that of a network with one more intermediate, but with a weaker KG connectedness. As an illustration, the ranges of K values for three-route and four-route networks are presented in Table 4.2. Indeed, for a constant number of routes M, and for a constant network type and class, K increases in parallel with the number of intermediates.

Table 4.2. Ranges of the complexity index K for three-route and four-route linear networks at different number of intermediates N

N	Complexity index range	N	Complexity index range
	M = 3		*M = 4*
2	64	2	120
3	174–228	3	336–480
4	304–768	4	576–1616
5	600–1740	5	880–4440
6	864–3744	6	1248–11616

Table 4.3. Complexity index K for different types and classes of linear networks

No.	Code* M–L/Class**	Complexity index K		
		$N = 4$	$N = 5$	$N = 6$
	2–1			
1	**A**	128	290	552–612
2	**B**	184	380–420	684–804
3	**C**	216–240	430–510	756–996
	3–2			
4	$\mathbf{A^2}$	–	–	864
5	**AB**	–	600	1272
6	**AC**	–	750	1488–1680
7	$\mathbf{B^2}$	384	880	1680–1872
8	**BC**	480	1030–1160	1896–2472
9	$\mathbf{C^2}$	576	1180–1360	2112–3072
	3–3			
10	$\mathbf{A^2B}$	–	600	1272
11	$\mathbf{A^2C}$	304	690–750	1308–1680
12	$\mathbf{B^3}$	384	880	1680–1872
13	$\mathbf{B^2C}$	444–480	910–1160	1626–2280
14	$\mathbf{2\text{-}BC^2}$	504–584	1000–1480	1752–2520
15	$\mathbf{3\text{-}BC^2}$	576–620	1180–1510	2112–3072
16	$\mathbf{C^3}$	768	1740	3312–3594
	4–3			
17	$\mathbf{4\text{-}AB^2}$	–	–	2304
18	$\mathbf{5\text{-}AB^2}$	–	–	2880
19	**ABC**	–	–	2880
20	$\mathbf{AC^2}$	–	–	3600
21	$\mathbf{B^3}$	–	1600	3408
22	$\mathbf{B^2C}$	–	2000	3984–4512
23	$\mathbf{BC^2}$	–	2500	4980–5640
24	$\mathbf{C^3}$	–	–	5136–6360
	4–4			
25	$\mathbf{A^3C}$	–	–	1776
26	$\mathbf{A^2B^2}$	–	–	2304
27	$\mathbf{AB^3}$	–	1240	2628–2880
28	$\mathbf{A^2BC}$	–	1240	2628–2880
29	$\mathbf{A^2C^2}$	–	1550	3072–3600
30	$\mathbf{ABC^2}$	–	–	3456–3732
31	$\mathbf{AC^3}$	–	1470	2952–3480
32	$\mathbf{B^4}$	–	1600	3408
33	$\mathbf{B^3C}$	800	1830–2000	3408–4512
34	$\mathbf{B^2C^2}$	1000	2140–2590	3924–5892
35	$\mathbf{3\text{-}BC^3}$	952	2060–2420	3804–5712
36	$\mathbf{5\text{-}BC^3}$	–	2800–3090	5136–7020
37	$\mathbf{C^4}$	1152	2370–2920	4248–7896

Table 4.3. (*Continued*)

No.	Code* M–L/Class**	Complexity index K		
		$N = 4$	$N = 5$	$N = 6$
	4–5			
38	A^4C	–	–	1776
39	A^2B^2C	–	1240	1776–2880
40	A^2BC^2	–	1640	3456–3732
41	A^2C^3	–	–	4608
42	B^4C	800	1830–2000	3480–4512
43	B^3C^2	1056	2400–2590	4560–5892
44	B^2C^3	1312–1376	3180–3370	5136–7680
45	BC^4	–	3980	7896–8976
46	C^5	–	–	10560
	4–6			
47	A^4B^2	–	–	2304
48	A^4BC	–	1240	2628–2880
49	A^3B^2C	–	1240	2628–2880
50	A^3B^3	–	–	2304
51	B^6	–	1600	3408
52	B^5C	800	1830–2000	3480–4512
53	A^4C^2	624	1550	3072–3600
54	A^3BC^2	–	1640	3456–3732
55	A^3C^3	576	1300–1470	2448–3480
56	B^4C^2	1000–1056	2140–2590	3924–5892
57	2-B^3C^3	848–952	1720–2420	3048–5712
58	3-B^3C^3	1328	2480–3290	5208–7360
59	4-B^3C^3	1376	3090–3370	5832–7680
60	3-B^2C^4	1152–1272	2370–3250	4248–7404
61	4-B^2C^4	1616	3730–4020	7152–8604
62	5-B^2C^4	–	4440	9228
63	2-C^6	944–1120	1860–2960	3240–7056
64	6-C^6	–	–	11616

* The class prefix-number is given only when there are classes that differ by this number only.
** The class ranges do not include the values for subclasses A_2, A_3, C_2, and C_3, to enable the
   evaluation of the **class** topology impact on the complexity index K.

(iii) For a constant number of routes and intermediates, changing the network
type by increasing the route interconnectedness (the number L showing the number
of connected pairs of routes) generally increases network complexity when the
additional pair of routes is connected by a common step (class **C**), and keeps constant the network complexity when the supplementary connection is either by an
intermediate (class **B**) or by an equilibrium step (class **A**). These trends can be
traced in Table 4.3.

Some specific examples of increased complexity with the higher ranking of the network type are given below: For $M = 3$ ($L = 2 \rightarrow L = 3$): $\mathbf{A}^2 \rightarrow \mathbf{A}^2\mathbf{C}$; $\mathbf{B}^2 \rightarrow \mathbf{B}^2\mathbf{C}$; $\mathbf{BC} \rightarrow \mathbf{BC}^2$, $\mathbf{C}^2 \rightarrow \mathbf{C}^3$, etc. For $M = 4$ ($L = 3 \rightarrow L = 4 \rightarrow L = 5 \rightarrow L = 6$): $\mathbf{C}^3 \rightarrow \mathbf{C}^4 \rightarrow \mathbf{C}^5 \rightarrow \mathbf{C}^6$; $\mathbf{BC}^2 \rightarrow \mathbf{BC}^3 \rightarrow \mathbf{BC}^4 \rightarrow \mathbf{BC}^5$; $\mathbf{B}^2\mathbf{C} \rightarrow \mathbf{B}^2\mathbf{C}^2 \rightarrow \mathbf{B}^2\mathbf{C}^3 \rightarrow \mathbf{B}^2\mathbf{C}^4$, etc. Examples of constant complexity (isocomplexity) are given in Section 4.3.5.

(iv) For a constant number of routes and intermediates, and a constant number of route interconnections (network type L), complexity index K increases with the stronger type of route interconnectedness, i.e., in the sequence of classes

$$\mathbf{A} < \mathbf{B} < \mathbf{C}. \qquad (4.32)$$

This general trend is illustrated by the data in Table 4.3 and by Figures 4.5 and 4.6. It is seen that the two alterations in the KG topology that enhance the network complexity are the contraction of a bridge between two cycles into a common vertex ($\mathbf{A} \rightarrow \mathbf{B}$) and the change from a common vertex to a common edge ($\mathbf{B} \rightarrow \mathbf{C}$).

Another trend of increasing mechanistic complexity is related to the increase in the class prefix n in the linear network code. Since n, by definition, is equal to the number of vertexes in the smallest homomorphic image of the KGs from a certain class of mechanisms, then the larger this number, the more complex the mechanism. This trend can be illustrated by comparing in Table 4.1 the three specific classes belonging to the same generalized class $\mathbf{B}^3\mathbf{C}^3$: $2\text{-}\mathbf{B}^3\mathbf{C}^3 \rightarrow 3\text{-}\mathbf{B}^2\mathbf{CBC}^2 \rightarrow 4\text{-}\mathbf{B}^2\mathbf{C}^2\mathbf{BC}$ (KGs **293** to **306**, **307** to **316**, and **317** to **330**, respectively).

(v) For a constant number of routes and intermediates, as well as for a constant type and class of the network, the complexity index increases when the number of equilibrium steps connecting two routes decreases,

$$... < \mathbf{A}_3 < \mathbf{A}_2 < \mathbf{A}, \qquad (4.33)$$

as well as when the number of steps two routes have in common increases

$$\mathbf{C} < \mathbf{C}_2 < \mathbf{C}_3 < .... \qquad (4.34)$$

Some examples illustrating these patterns are shown in Figure 4.7.

(vi) For a constant number of reaction routes and intermediates, and for a constant type, class and subclass, K increases with more uniform distribution of intermediates over the reaction routes, thus manifesting an entropy-like behavior. As an illustration, compare KGs **123**, **126**, and **129** from Figure 4.4. They have six intermediates and belong to the same class $3\text{-}\mathbf{BC}^2\mathbf{Z}^2\mathbf{C}$ but differ in cycle sizes, which are, respectively, 5,2,2,2; 3,2,2,4; and 3,2,3,3. The cycle-size equalizing results in an increase in the complexity index from 3804 to 4860 to 5712, respectively.

4.3.5. Isocomplexity

Complexity Index Degeneracy. Albeit closely related to the unique linear code, the complexity index of the linear mechanisms is not entirely discriminating. The number of distinct KGs with the same value of the K index increases rapidly with the increase in the number of reaction routes. This is illustrated in Table 4.4, where the degree of degeneracy of the complexity index is calculated as the ratio of the total number of networks to that of the networks with different K values.

The high degeneracy found for four-route networks reflects the higher degree of similarity of the graphs having four cycles. The difficulties involved in discriminating the highly connected KGs parallel those involved in discriminating the kinetic hypotheses for four-route reaction mechanisms. Thus, the complexity index K helps explain why it is so difficult to discriminate some mechanisms, the reason being the high similarity in the topological structure of mechanisms.

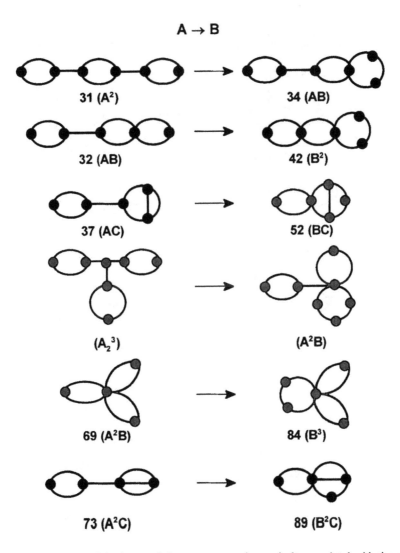

Figure 4.5. Illustration of the increased three-route network complexity associated with the class changes $A \rightarrow B$. The network numbering is that from Table 2.6.

Isocomplexity Levels. The phenomenon of isocomplexity encompasses not only mechanisms differing in minor structural details, but also covers all classification levels of mechanisms: types, generalized classes, specific classes, subclasses, and different distributions of cycle sizes.

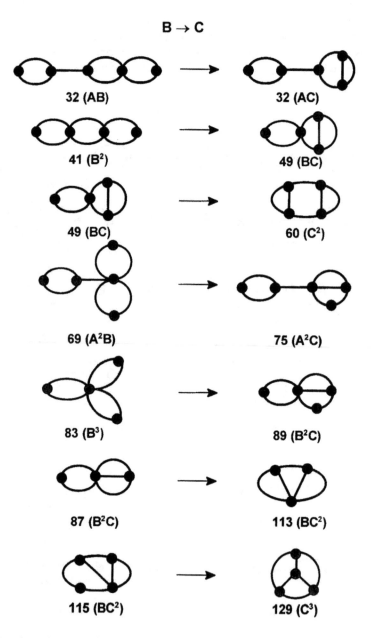

Figure 4.6. Illustration of the increased three-route network complexity associated with the class changes **B** → **C**. The network numbering is that from Table 2.6.

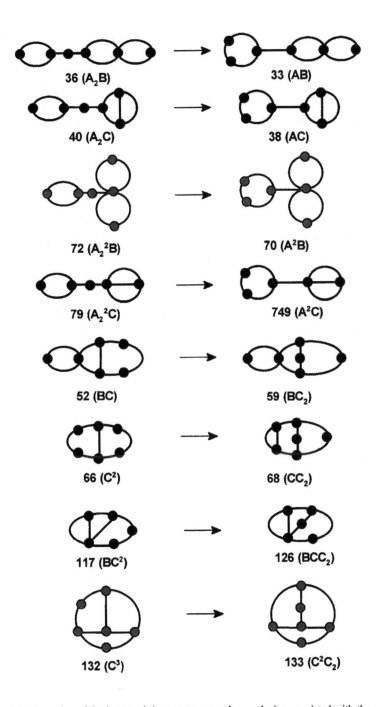

Figure 4.7. Illustration of the increased three-route network complexity associated with the subclass changes in class A (shortening of the bridge between two cycles) and class C (increasing the number of edges two cycles have in common).

An illustration is presented in Figure 4.8, where 11 KGs belonging to 11 specific classes, 6 generalized classes, and four types of linear mechanisms have the complexity index $K = 2880$ showing the highest degeneracy. For example, KGs **31** and **33-35** belong to the specific classes **4-A^{2}BCZ^2**, **4-A^{2}ZBCZ**, **4-A^{2}CBZ^2**, and **4-A^{2}ZCBZ**, respectively, all of which are included in the generalized class **A^{2}BC** and type $L = 4$ (four cycle interconnections). Another generalized class, **AB^{2}C**, of the same type $L = 4$, is also represented by KGs **46** and **48** (specific classes **4-AB^{2}Z^{2}C** and **4-ABCZ^{2}B**, respectively). KG **154** is of generalized class **A^{2}B^{2}C** and type $L = 5$, and KGs **229** and **243** are of type $L = 6$ and generalized classes **A^{4}BC** and **A^{3}B^{2}C**, respectively.

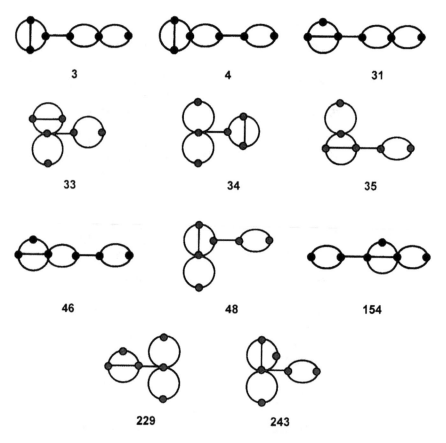

Figure 4.8. Eleven isocomplex linear networks with complexity index $K = 2880$. The network numbering corresponds to those in Figure 4.4 and Table 4.1 (Reprinted with permission from *J. Chem. Inf. Comput. Sci.* **1994**, *34*, 436–445 © 1994 American Chemical Society).

In addition to the intrinsic network isocomplexity described above, it should be mentioned that 15 cases of accidental degeneracies have been found. These are cases in which the same K index value results by chance from different summands reflecting different mechanistic topologies; no systematic graph transformations connect these KGs. An example is presented below, in which four linear mechanisms have the same complexity index ($K = 2420$):

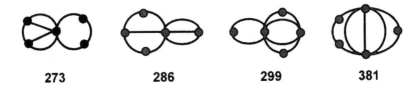

| 273 | 286 | 299 | 381 |

Graph Transformations Preserving Complexity. The analysis of equations (4.23)–(4.25) indicates that graph transformations preserving complexity are all transformations that do not change cycle sizes N_i and the number E_{ij} of the edges common for cycles i and j. Otherwise, these are different cases of "positional isomerism" that deal mainly with **A** and **B** classes (weak intercycle linkage). Upon such a graph transformation, a cycle linked by a bridge or by a common vertex is displaced so as to be connected to other cycles by any one of these weak linkages.

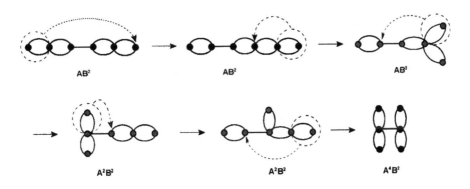

In general, the same type of transformation can be performed for subgraphs containing two or more weakly connected cycles:

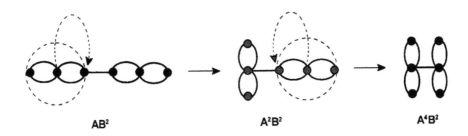

Table 4.4. Degeneracy of the complexity index k of linear reaction networks having 1 to 4 reaction routes

Number of routes	Total number of mechanisms	Total number of the different index values	Degree of degeneracy
1	5	5	1.00
2	24	23	1.04
3	104	65	1.60
4	390	171	2.28

Some transformations of strongly connected cycles (class **C**) also produce isocomplexity. These are displacements of an outer cycle sharing a common edge with a large cycle whose sites are nonequivalent:

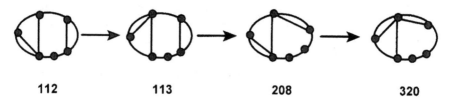

| 112 | 113 | 208 | 320 |

Complexity Flowcharts. All isocomplexity relationships found for the classes of three-route and four-route networks can be presented in flowcharts (Figures 4.9 and 4.10, respectively). The classes with the same complexity are connected there by vertical lines. The flowcharts also show the relationships of increasing complexity; these are shown by horizontal or diagonal lines for all generalized classes of the two types ($L = 1$ and $L = 2$) of three-route networks, and for all four types ($L = 3$ to $L = 6$) of the four-route networks.

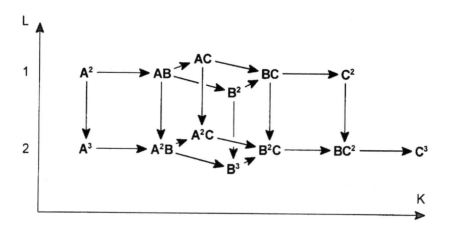

Figure 4.9. Complexity flowchart for types (L) and classes of three-route linear networks. (Reprinted with permission from *J. Chem. Inf. Comput. Sci.* **1994,** *34,* 436–445 © 1994 American Chemical Society.)

From the two flowcharts one can see that the KG transformations that increase complexity include all **B → C** transitions, as well as some of the **A → B** ones. The first trend deals with replacing the common vertex between two KG cycles with a common edge (or, otherwise, with replacing a common intermediate with a common elementary step). The second trend, the replacement of a bridge between two KG cycles with a common vertex, is weaker because both are a "weak" type of cycle linkage. The increase in complexity in such cases comes (see equations (4.24) and (4.25)) from the increase by 1 in the size of one of the KG cycles in order to preserve a constant total number of KG vertexes. However, in those cases in which the **A → B** transformation can be performed by cycle displacements only (i.e., without any alteration of the cycle sizes), the complexity index remains unchanged.

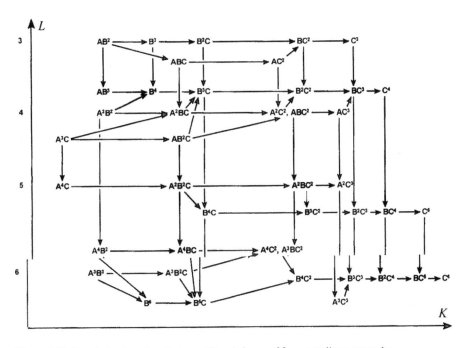

Figure 4.10. Complexity flowchart for types (L) and classes of four-route linear networks.

4.3.6. Complexity of Networks with Pendant Vertexes

The complexity analysis of networks with pendant vertexes (describing equilibrium steps) requires some modification of complexity index K. In fact, it follows from equation 4.22 that network complexity depends on the basic graph topology, i.e., it depends on the number of cycles and their interconnectedness, on the number of vertexes included in cycles, etc. All these factors remain unchanged upon addition of pendant vertexes, therefore, the network complexity does not change. However, for stationary or quasistationary reactions with linear mechanism, pendant vertexes are of importance for the kinetic model and, in particular, for the mass balance of catalyst in catalytic reactions and for that of the reagents for noncatalytic reactions.

This role of pendant vertexes may be reflected in a conveniently modified complexity index

$$K' = K + \Delta K = K + M(N-1)\sum_{i=1}^{N} T_i N_p(i), \tag{4.35}$$

where K is the complexity index as defined by equation (4.22), N is the number of vertexes in the cyclic part of the network, $N_p(i)$ is the number of pendant vertexes attached to vertex i, and T_i is the number of spanning trees for vertex i neglecting the pendant vertexes.

Example. $M = 1$, $N = 3$, $N_p = 4$, $N_p(1) = 1$, $N_p(2) = 1$, $N_p(3) = 2$. $\Delta K = 1.2 \times (3.1 + 3.1 + 3.2) = 24$.

In deriving the expression for ΔK, one proceeds from the steady-state rate law for a route accounting for the pendant vertexes. This equation differs from equation (4.21) only in its denominator, which is modified:

$$\sum_{i=1}^{N+N_p} D_i = \sum_{i=1}^{N} D_i F_i. \tag{4.36}$$

Here, D_i is the determinant of vertex i neglecting the pendant vertexes, and F_i is the function defined by equation (2.61) [see Section 2.2.4].

4.4. Stoichiometric Complexity Index for Reaction Networks

4.4.1. Background

In Section 4.3, we introduced the complexity index K specifically constructed to match the complexity features of linear reaction networks. Based on the fractional-rational rate laws, this index mirrored quite well the complexity of kinetic graphs used for depicting linear reaction mechanisms. This approach, however, could not directly be extended to nonlinear networks because these are described by a different type of graphs, the bipartite graphs. For that reason, Zeigarnik and Temkin[60] developed a more general approach, based on the stoichiometric matrix of reaction network. In this section, we present the results obtained for linear networks within the framework of this new approach. The results for the nonlinear networks will be published in the near future after the completion of the generation and enumeration of these networks.[61]

The new approach to the complexity of reaction networks is essentially based on the idea of simple submechanisms (hereinafter, submechanisms),[62,63] which was analyzed in detail in Chapters 2 and 3. In constructing the new measure of reaction networks complexity we proceeded from the intuitively clear ideas of mechanistic complexity whose increase parallels that of the number of steps, intermediates, and submechanisms, as well as that of the increase in the sizes of the simple sub-mechanisms. It was also logical to expect network complexity to increase with the increase in the number and size of the common subgraphs of the submechanism graphs or, otherwise, the stronger the interrelation between the reaction routes, the more complex the reaction mechanism.

4.4.2. Mathematical Formalism

In Section 4.2, we introduced the formula (4.3) for the mean information content per element of the structure under investigation. For complexity evaluations one may also use the derivative equation[64] for the overall information content of the structure:

$$I = \bar{I}N = N\mathrm{lb}N - \sum_{i=1}^{k} N_i \mathrm{lb} N_i, \qquad (4.37)$$

where k is the number of equivalence classes.

When the system Φ can be decomposed into r subsystems Φ_i ($\Phi = \Phi_1 \cup \Phi_2 \cup \ldots \cup \Phi_r$), and the cardinality of the jth subsystem $|\Phi_j| = N^{(j)}$, then the total information content can be expressed as follows:

$$I_{\mathrm{total}} = \left[\sum_{j=1}^{r} N^{(j)} \right] \mathrm{lb} \left[\sum_{j=1}^{r} N^{(j)} \right] - \sum_{j}\sum_{i} N_i^{(j)} \mathrm{lb} N_i^{(j)}. \qquad (4.38)$$

The summation in equation (4.38) is taken over all equivalence classes i and over all subsystems j. This equation accounts for the nonequivalence of elements included in different subsets.

Recall that different graph elements have been used to apply equations (4.1), (4.3), and (4.37) for complexity assessments of chemical systems (molecules, reaction mechanisms). Use was made of equivalence classes based on distributions of vertexes,[36] edges,[37] two-edge subgraphs,[41] distances,[21] etc., over the respective orbits of the automorphism group of the graph. All these elements can be taken with different weights[21,40] which makes the respective information index much more discriminating. The individual complexity measures can also be used to construct combined complexity indices.[42]

Consider now a reaction mechanism which includes r simple submechanisms. The information content of the ith simple submechanism will be presented by three terms:

$$I(\mathbf{B}_i) = I_{\mathrm{int}}(\mathbf{B}_i) + I_{\mathrm{st}}(\mathbf{B}_i) + I_{\mathrm{sub}}(\mathbf{B}_i), \qquad (4.39)$$

where $\mathbf{B}_i$ is the matrix of stoichiometric coefficients of the intermediates included in the simple submechanism (this matrix is homomorphic to the bipartite graph of the submechanism); $I_{int}(\mathbf{B}_i)$ is the information content of the subset of intermediates (the columns of the $\mathbf{B}_i$ matrix); $I_{st}(\mathbf{B}_i)$ is the information content of the subset of steps (the rows of the $\mathbf{B}_i$ matrix); $I_{sub}(\mathbf{B}_i)$ is the information content of the sub-mechanism as a whole (the set of all $\mathbf{B}_i$ matrix entries). The incorporation of the third term in (4.39) makes $I(\mathbf{B}_i)$ less degenerate. The first two terms in this equation are calculated by equation (4.38), whereas the third term is found by equation (4.37).

The elements of the simple submechanism whose equivalence is counted in applying equations (4.37) and (4.38) are the stoichiometric coefficients of intermediates in the different steps. The equivalence relations are as follows: (i) In calculating $I_{int}(\mathbf{B}_i)$, the stoichiometric coefficients are nonequivalent when they belong to different intermediates or have different numerical values; (ii) In calculating $I_{st}(\mathbf{B}_i)$, the coefficients are regarded nonequivalent when they belong to intermediates from different steps or when they have different numerical values; (iii) In calculating $I_{sub}(\mathbf{B}_i)$, the different coefficient values are the only criterion of nonequivalence.

Consider now the information content of the reaction mechanism as a whole, I_{mech}. It might be calculated by equation (4.39), where $\mathbf{B}_i$ is the full stoichiometric matrix of intermediates; however, this would produce a strongly degenerate complexity index $I(\mathbf{B})$. The degeneracy can be diminished by adding a second term, $\sum_i I(\mathbf{B}_i)$, which adds up the information contents of all simple submechanisms:

$$I_{mech} = I(\mathbf{B}) + \sum_{i=1}^{r} I(\mathbf{B}_i), \qquad (4.40)$$

where $\mathbf{B}$ is the matrix of the stoichiometric coefficients of intermediates in the overall mechanism.

Example. Consider the reaction mechanism of ammonia synthesis, proposed by M. I. Temkin.[65] This mechanism includes one simple submechanism only. For simplicity, all steps will be regarded irreversible. If all steps were taken reversible, then two submechanisms would result, those for the forward and reverse overall reaction.

$$(1)\ Z + N_2 \rightarrow ZN_2, \qquad (4.41)$$

$$(2)\ ZN_2 + H_2 \rightarrow ZN_2H_2, \qquad (4.42)$$

$$(3)\ ZN_2H_2 + Z \rightarrow 2ZNH, \qquad (4.43)$$

$$(4)\ ZNH + H_2 \rightarrow Z + NH_3. \qquad (4.44)$$

The resulting overall reaction is

$$N_2 + 3H_2 \rightarrow 2NH_3. \qquad (4.45)$$

The matrix of the stoichiometric coefficients of the intermediates is as follows:

$$
\mathbf{B} = \begin{array}{c} \\ (1) \\ (2) \\ (3) \\ (4) \end{array}
\begin{array}{cccc}
Z & ZN_2 & ZN_2H_2 & ZNH \\
-1 & 1 & 0 & 0 \\
0 & -1 & 1 & 0 \\
-1 & 0 & -1 & 2 \\
1 & 0 & 0 & -1
\end{array}
$$

Calculate first $I_{sub}(\mathbf{B})$. The $\mathbf{B}$ matrix has $4 \times 4 = 16$ entries: seven zeros, three "1"s, five "−1"s, and one "2". Hence, the distribution of the matrix elements into the four equivalence classes is $16\{6, 3, 5, 1\}$, and

$$I_{sub}(\mathbf{B}) = 16\,lb\,16 - 7\,lb\,7 - 3\,lb\,3 - 5\,lb\,5 - 1\,lb\,1 = 27.984.$$

In calculating $I_{st}(\mathbf{B})$, one counts as equivalent two zeros in the first row, two zeros in the second row, two zeros in the fourth row, and two "−1" entries in the third row. Hence,

$$I_{st}(\mathbf{B}) = 16\,lb\,16 - 4 \times 2\,lb\,2 = 56.$$

In calculating $I_{int}(\mathbf{B})$, equivalent are two zeros in each of the second, third, and fourth columns, and two "−1"s in the first column. Therefore, $I_{int}(\mathbf{B}) = I_{st}(\mathbf{B}) = 56$, and

$$I(\mathbf{B}) = I_{int}(\mathbf{B}) + I_{st}(\mathbf{B}) + I_{sub}(\mathbf{B}) = 56 + 56 + 27.984 = 139.98.$$

Since the mechanism of ammonia synthesis includes one simple submechanism only, I_{mech} will be simply twice as much (the basic term and the supplementary term in equation (4.40) are equal).

4.4.3. Calculation Methodology for Linear Networks

General Remarks. Calculations were performed for all kinetic graphs (KGs) with 2–6 vertexes and 1–3 independent cycles, given in Section 2.2; i.e., for all linear networks with 1–3 independent routes and 2–6 intermediates. Complexity analysis of these networks makes also use of the classification we introduced in Chapter 2, according to the topologies of the homomorphically irreducible graphs.[66] We recall here that the procedure of homomorphic reduction of a graph includes deleting all vertexes of degree two and their incident edges, and adding an edge between any pair of vertexes, which have been incident to deleted edges.

It is convenient to perform the calculations proceeding from the stoichiometric matrix of intermediates $\mathbf{B}$. The $\mathbf{B}$ matrix is readily obtained as the incidence matrix of the directed KG. In the general case, KG is a mixed graph; some of its edges are directed, whereas others are not. In constructing **B**, one has to replace each nondi-rected edge by two edges (arcs) with opposite directions. The KGs, shown in Table 2.6, have only nondirected edges; therefore, each of them must be substituted by two arcs.

In calculating the complexity index we regard any two submechanisms containing the same set of steps, but having opposite direction, as formally distinct. All simple submechanisms were included in the calculations, not only the linearly independent ones, because the choice of linearly independent routes is arbitrary.

How To Deal with Complexity of Nonlinear Networks? In the case of nonlinear networks, the **B** matrix is obtained from the respective bipartite graphs. If the adjacency matrix is constructed for the subset of intermediates and if the vertexes are ordered according to their type (w-vertexes for intermediates and u-vertexes for steps), then the block matrix **A** can be written in the following form:

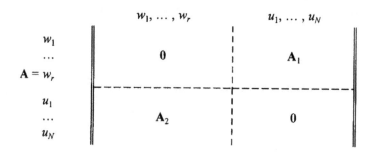

Then,

$$\mathbf{B} = \mathbf{A}_1 - \mathbf{A}_2^{\mathsf{T}}. \tag{4.46}$$

Computer Assistance. The enumeration of all simple submechanisms of the networks under examination was done by the specifically developed program **GERM** (Generic Ensembles of Reaction Mechanisms) whose algorithm was reported recently[62] (See also Section 2.3) The program was written in **C**, and was tested on a considerable number of reaction mechanisms. GERM handles the overall matrix of stoichiometric coefficients **B** (a manual input) and generate its submatrix of intermediates, **B**(*SI*). The program also calculates the complexity index, according to equations (4.1) and (4.2).

4.4.4. Influence of Principal Complexity Factors and Properties of the Index

Degeneracy. The I_{mech} index is considerably less degenerate than the K index. As seen in Table 4.5, for the 136 linear networks with 2–6 intermediates and 1–3 independent routes, we found only 13 cases of degeneracy (less than 0.15%) versus 67 cases for K. In most cases, the isocomplex networks have similar structures. Two pairs of networks (**30, 36**) and (**55, 78**) show accidental degeneracy. With one exception (**55, 78**), the degeneracy of I_{mech} is associated with that of K.

The Number of Simple Submechanisms as the Strongest Complexity Factor.
Graphs **83** and **129** (numbering is as in Figure 2.33) illustrate well the point. Each
of these graphs contains four vertexes and six edges; however, the first graph has
only three cycles, whereas the second one has seven.

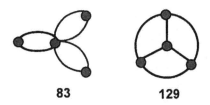

83   **129**

Hence, at rather close values of the first terms of equation (4.40) ($I(\mathbf{B})$ = 469.14
and 484.08, respectively), the overall I_{mech} values differ drastically (589.14 and
1870.7, respectively).

In order to examine this trend in detail, we singled out sets of KGs with cardi-
nalities larger than one, such that each set $A_{p/q}$ incorporated graphs with the same
number of vertexes p and edges q. Examples of such sets are $A_{4/6}$, $A_{5/6}$, $A_{6/7}$, and
$A_{6/8}$. (We recall here that p and q are related with the *cyclomatic number* ϕ of a
graph; $q = p - 1 + \phi$.) Each set of KGs was partitioned into subsets, according to
the total number of cycles (the number of simple submechanisms).

For example, the set $A_{4/6}$ = {**41, 49, 60, 73, 83, 88, 89, 103, 104, 114, 115, 129**}
was partitioned into four subsets (the superscript in the subset notation stands for
the number of cycles (simple submechanisms)):

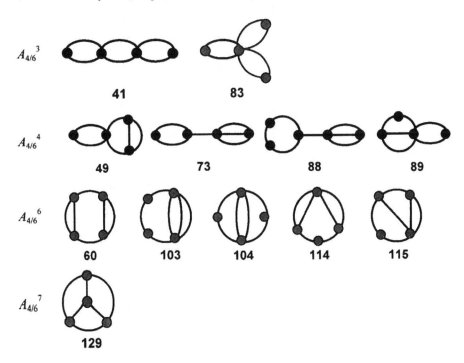

Table 4.5. The I_{mech} complexity index of linear reaction networks and related network characteristics (kinetic graphs are numbered as in Figure 2.33)

KG	N_{int}	s	r	Number of cycles of length n					K	$I(\mathbf{B})$	I_{mech}
				2	3	4	5	6			
Monocyclic kinetic graphs											
1	2	2	2	1	0	0	0	0	8	48.00	88.00
2	3	3	2	0	1	0	0	0	24	160.65	303.29
3	4	4	2	0	0	1	0	0	56	304.00	576.00
4	5	5	2	0	0	0	1	0	110	487.84	925.67
5	6	6	2	0	0	0	0	1	192	714.59	1357.2
Class A											
6	4	5	4	2	0	0	0	0	128	391.72	471.72
7	5	6	4	1	1	0	0	0	290	599.99	782.63
8	6	7	4	1	0	1	0	0	552	851.15	1163.1
9	6	7	4	0	2	0	0	0	612	851.15	1136.4
10	5	6	4	2	0	0	0	0	200	599.99	679.99
11	6	7	4	1	1	0	0	0	420	851.15	1033.8
12	6	7	4	2	0	0	0	0	288	851.15	931.15
Class B											
13	3	4	4	2	0	0	0	0	72	218.12	298.12
14	4	5	4	1	1	0	0	0	184	389.22	571.87
15	5	6	4	1	0	1	0	0	380	598.03	910.03
16	5	6	4	0	2	0	0	0	420	598.03	883.32
17	6	7	4	1	0	0	1	0	684	849.44	1327.3
18	6	7	4	0	1	1	0	0	804	849.44	1264.1
Class C											
19	2	3	6	3	0	0	0	0	28	79.02	199.02
20	3	4	6	1	2	0	0	0	90	223.10	548.39
21	4	5	6	1	0	2	0	0	216	391.72	975.72
22	4	5	6	0	2	1	0	0	240	391.72	949.01
23	5	6	6	1	0	0	2	0	430	599.99	1515.7
24	5	6	6	0	1	1	1	0	510	599.99	1452.5
25	6	7	6	1	0	0	0	2	756	851.15	2176.3
26	6	7	6	0	1	0	1	1	936	851.15	2074.2
27	6	7	6	0	0	2	0	1	996	851.15	2037.7
28	5	6	6	0	0	3	0	0	560	599.99	1416.0
29	6	7	6	0	0	1	2	0	1068	851.15	1998.8
Class 3-A^2											
30	6	8	6	3	0	0	0	0	864	989.95	1110.0

Table 4.5. (*Continued*)

KG	N_{int}	s	r	\multicolumn{5}{c}{Number of cycles of length n}	K	$I(\mathbf{B})$	I_{mech}
				2	3	4	5	6			
\multicolumn{12}{c}{Class 4-A^2}
31	6	8	6	3	0	0	0	0	864	991.51	1111.5
\multicolumn{12}{c}{Class AB}
32	5	7	6	3	0	0	0	0	600	714.53	834.53
33	6	8	6	2	1	0	0	0	1272	989.95	1212.6
34	6	8	6	2	1	0	0	0	1272	989.95	1212.6
35	6	8	6	2	1	0	0	0	1272	989.95	1212.6
36	6	8	6	3	0	0	0	0	864	989.95	1110.0
\multicolumn{12}{c}{Class AC}
37	5	7	8	2	2	0	0	0	750	716.24	1081.5
38	6	8	8	1	3	0	0	0	1590	991.51	1459.5
39	6	8	8	1	2	1	0	0	1680	991.51	1588.8
40	6	8	8	2	2	0	0	0	1080	991.51	1356.8
136	6	8	8	2	0	2	0	0	1488	989.95	1614.0
\multicolumn{12}{c}{Class B^2}
41	4	6	6	3	0	0	0	0	384	480.16	600.16
42	5	7	6	2	1	0	0	0	880	712.82	935.47
43	5	7	6	2	1	0	0	0	880	712.82	935.47
44	6	8	6	2	0	1	0	0	1680	988.39	1340.4
45	6	8	6	2	0	1	0	0	1680	988.39	1340.4
46	6	8	6	2	1	0	0	0	1680	988.39	1340.4
47	6	8	6	1	2	0	0	0	1872	988.39	1313.7
48	6	8	6	1	2	0	0	0	1872	988.39	1313.7
\multicolumn{12}{c}{Class BC}
49	4	6	8	2	2	0	0	0	480	482.12	847.41
50	5	7	8	2	0	2	0	0	1030	714.53	1338.5
51	5	7	8	1	3	0	0	0	1100	714.53	1182.5
52	5	7	8	1	2	1	0	0	1160	714.53	1311.8
53	6	8	8	2	0	0	2	0	1896	989.95	1945.6
54	6	8	8	2	0	0	2	0	1896	989.95	1587.2
\multicolumn{12}{c}{Class BC}
55	6	8	8	1	2	1	0	0	2100	989.95	1587.2
56	6	8	8	1	1	2	0	0	2190	989.95	1716.6
134	6	8	8	1	1	1	1	0	2220	989.95	1882.4
57	6	8	8	1	1	1	1	0	2280	989.95	1882.4
58	6	8	8	0	3	1	0	0	2472	989.95	1689.9
59	6	8	8	1	0	3	0	0	2496	989.95	1846.0

Table 4.5. (*Continued*)

KG	N_{int}	s	r	Number of cycles of length n					K	$I(\mathbf{B})$	I_{mech}
				2	3	4	5	6			
Class C^2											
60	4	6	12	2	0	4	0	0	576	484.08	1652.1
61	5	7	12	2	0	0	4	0	1180	716.24	2547.6
62	5	7	12	1	1	2	2	0	1380	716.24	2318.6
63	6	8	12	2	0	0	0	4	2112	991.51	3601.9
64	6	8	12	2	0	0	0	4	2112	991.51	3641.9
65	6	8	12	1	1	0	2	2	2598	991.51	3335.0
66	6	8	12	1	0	3	0	2	2700	991.51	3132.7
67	6	8	12	0	2	1	2	1	3072	991.51	3067.1
68	6	8	12	1	0	1	4	0	2916	991.51	3054.9
Class A^2B											
69	5	7	6	3	0	0	0	0	600	711.11	831.11
70	6	8	6	2	1	0	0	0	1272	986.93	1209.6
71	6	8	6	2	1	0	0	0	1272	986.93	1209.6
72	6	8	6	3	0	0	0	0	864	986.93	1106.9
Class A^2C											
73	4	6	8	4	0	0	0	0	304	482.12	642.12
74	5	7	8	3	1	0	0	0	690	714.53	977.18
75	5	7	8	2	2	0	0	0	750	714.53	1079.8
76	6	8	8	3	0	1	0	0	1308	989.95	1382.0
77	6	8	8	1	3	0	0	0	1590	989.95	1457.9
78	6	8	8	1	2	1	0	0	1680	989.95	1587.2
79	5	7	8	4	0	0	0	0	470	714.53	874.53
80	6	8	8	3	1	0	0	0	990	989.95	1252.6
81	6	8	8	2	2	0	0	0	1080	989.95	1355.2
82	6	8	8	4	0	0	0	0	672	989.95	1150.0
135	6	8	8	2	0	4	0	0	1488	991.51	1615.5
Class B^3											
83	4	6	6	3	0	0	0	0	384	469.14	589.14
84	5	7	6	2	1	0	0	0	880	705.60	928.25
85	6	8	6	2	0	1	0	0	1680	982.35	1334.4
86	6	8	6	1	2	0	0	0	1872	982.35	1307.6
Class B^2C											
87	3	5	8	4	0	0	0	0	174	281.72	461.72
88	4	6	8	3	1	0	0	0	444	477.92	740.56
89	4	6	8	2	2	0	0	0	480	477.92	843.21
90	5	7	8	3	0	1	0	0	910	711.11	1103.1

Table 4.5. (*Continued*)

KG	N_{int}	s	r	Number of cycles of length n					K	I(**B**)	I_{mech}
				2	3	4	5	6			
				Class B^2C							
91	5	7	8	2	0	2	0	0	1030	711.11	1335.1
92	5	7	8	1	3	0	0	0	1100	711.11	1179.1
93	5	7	8	1	2	1	0	0	1160	711.11	1308.4
94	6	8	8	3	0	0	1	0	1626	986.93	1544.8
95	6	8	8	2	0	0	2	0	1896	986.93	1942.6
96	6	8	8	1	2	1	0	0	2100	986.93	1584.2
97	6	8	8	1	1	2	0	0	2190	986.93	1713.6
98	6	8	8	1	1	1	1	0	2280	986.93	1879.4
99	6	8	8	0	3	1	0	0	2472	986.93	1686.9
100	6	8	8	1	0	3	0	0	2496	986.93	1842.9
				Class 2-BC^2							
101	2	4	12	6	0	0	0	0	64	112.00	352.00
102	3	5	12	3	3	0	0	0	210	286.45	834.39
103	4	6	12	3	0	3	0	0	504	480.16	1416.2
104	4	6	12	1	4	1	0	0	584	480.16	1362.7
105	5	7	12	3	0	0	3	0	1260	712.82	2146.3
106	5	7	12	1	2	2	1	0	1260	712.82	2019.9
107	5	7	12	0	3	3	0	0	1480	712.82	1956.8
108	6	8	12	3	0	0	0	3	1752	988.39	3036.2
109	6	8	12	1	2	0	2	1	2328	988.39	2831.9
110	6	8	12	1	0	4	0	1	2520	988.39	2759.0
				Class 2-BC^2							
111	6	8	12	0	2	2	2	0	2490	988.39	2693.4
112	6	8	12	0	0	6	0	0	3360	988.39	2620.4
				Class 3-BC^2							
113	3	5	12	2	4	0	0	0	228	288.94	939.53
114	4	6	12	2	0	4	0	0	576	482.12	1650.1
115	4	6	12	1	3	2	0	0	620	482.12	1494.1
116	5	7	12	2	0	0	4	0	1180	714.53	2545.9
117	5	7	12	1	2	1	2	0	1320	714.53	2187.5
118	5	7	12	1	1	2	2	0	1380	714.53	2316.8
119	5	7	12	0	3	3	0	0	1510	714.53	2124.3
120	6	8	12	2	0	0	0	6	2112	989.95	3640.3
121	6	8	12	1	2	0	1	2	2418	989.95	3038.3
122	6	8	12	1	1	0	2	2	2598	989.95	3333.4
123	6	8	12	1	0	3	0	2	2700	989.95	3131.1

Table 4.5. (*Continued*)

KG	N_{int}	s	r	Number of cycles of length n					K	$I(\mathbf{B})$	I_{mech}
				2	3	4	5	6			
124	6	8	12	0	2	2	1	1	2982	989.95	2899.7
125	6	8	12	0	2	1	2	1	3072	989.95	3065.5
126	5	7	12	1	0	5	0	0	1470	714.53	2114.5
127	6	8	12	1	0	1	4	0	2916	989.95	3053.3
128	6	8	12	0	1	3	2	0	3330	989.95	2824.3
						Class $\mathbf{C}^3$					
129	4	6	14	0	4	3	0	0	768	484.08	1870.7
130	5	7	14	0	2	3	2	0	1740	716.24	2693.2
131	6	8	14	0	2	1	2	2	3312	991.51	3709.7
132	6	8	14	0	1	2	3	1	3594	991.51	3634.3
133	6	8	14	0	0	5	0	2	3744	991.51	3636.7

The calculation showed that, for any set, the ranges of I_{mech} values for different subsets do not overlap at all or overlap only marginally. Thus, the following ranges of I_{mech} resulted for the set $A_{4/6}$ and its subsets with 3, 4, 6, and 7 cycles: 589–600, 642–847, 1363–1652, and 1870 (a single value). A minor overlap was observed in the $A_{5/7}$ and $A_{6/8}$ subsets; however, it does not change the general trend toward an increase in I_{mech} in parallel with the increase in the number of simple submechanisms. Apparently, this trend is due to the second term in equation (4.40), since the contribution of the first term $I(\mathbf{B})$ is generally less than that of the minimal simple submechanism.

The Number of Intermediates and the Number of Steps as Complexity Factors. These two factors are not independent, as follows from their relation to the cyclomatic number; for any class of KGs, they vary in parallel. Indeed, for a constant number of simple submechanisms, and the same basic graph topology (same *KG* class), network complexity increases with the increase in the number of intermediates p or steps q. Thus, one can find in Table 4.5 that for the **BC** class and $\phi = 8$ (eight simple submechanisms), and for the p/q pairs 4/6, 5/7, and 6/8, the I_{mech} ranges are 847, 1183–1338, and 1587–1946, respectively. For the **2-BC²** class, and $\phi = 12$, $I_{mech} = 352$; 834; 1362–1416; 1957–2146; and 2693–3036, respectively. The number of intermediates and the number of reaction steps thus appear as the second strongest complexity factor. However, as indicated by these two examples, the basic network topology (the network class) is also of importance.

Complexity and the Basic Network Topology (the KG Class as a Complexity Factor). Basic network topology is the one determined by the homomorphically irreducible graph, *HIG*. All graphs that can be transformed to a *HIG* by a homomorphic reduction, form a mechanism *class*.[66,67]

The importance of the class for network complexity follows mainly from the fact that the class determines the number of simple submechanisms which is essential for the I_{mech} calculation. However, even at the same number of simple submechanisms, intermediates and steps, the higher degree of network connectedness, as mirrored by the class, increases the network complexity index. This can be illustrated in Table 4.5 by the comparison of classes $2\text{-}BC^2$ and $3\text{-}BC^2$, all KGs of which have 12 cycles. For constant p/q pairs the following I_{mech} values (ranges) result: for 3/5, 834 versus 940; for 4/6, 1362–1416 versus 1494–1650; for 5/7, 1957–2146 versus 2115–2346; and for 6/8, 2620–2832 versus 2824-3640, respectively.

KG Subclasses as Complexity Factors. Subtle differences in KG topology, determined by their *subclass*, are also matched by the I_{mech} index. Thus, for constant number of cycles, vertexes, and edges, and for a constant class, the lengthening of a bridge between cycles changes the subclass $A_1 = A$ to A_2 and A_3 and decreases network cyclicity. An example is given by KG **76**, **80**, and **82**, for which I_{mech} values are 1382, 1257, and 1150.

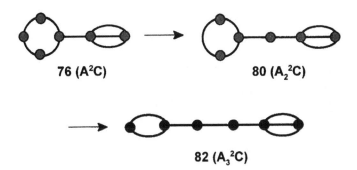

76 (A²C)   80 (A₂²C)

82 (A₃²C)

The steps, included in bridges, affect only the first term $I(B)$ in equation (4.40), because they do not belong to any simple submechanism. Due to this fact, I_{mech} increases faster in a series in which the additional edges are included in a cycle, because in these cases the more important second term of equation (4.40) increases rapidly. An extreme case of this kind is when more edges become shared by two cycles, i.e., in subclasses $C \rightarrow C_2 \rightarrow C_3$. The C_3 subclass can occur in networks larger than those included in Table 4.5. For that reason, we show below only two pairs of graphs, containing $C_1 = C$ and C_2 subclasses only. I_{mech} increases for $58 \rightarrow 59$ from 1690 to 1846, and for $99 \rightarrow 100$, from 1687 to 1843.

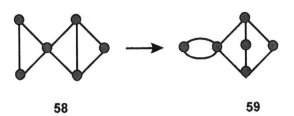

58   59

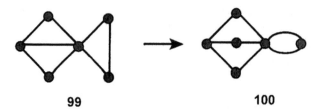

99   **100**

The subclass factor is not strong, and in some cases the ordering induced by a higher subclass can be reversed by the larger sizes of the simple subnetworks in the lower subclass, a complexity factor which is regarded next.

The Simple Submechanism Size as a Complexity Factor. As any other Shannon-type information index, I_{mech}, and particularly its second term $\sum_i I(\mathbf{B}_i)$, decreases with the more even distribution of the cycle sizes in KG. In order to single out the influence of this factor, compare KGs with the same number of cycles, vertexes, and edges, as well as with the same total number of vertexes included in graph cycles. For example, I_{mech} increases along the series **28 → 24 → 23**,

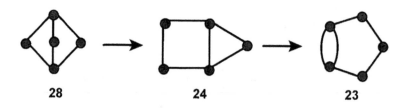

28   **24**   **23**

since the cycle size distribution is the most even one in **28** (4 + 4 + 4), and it is the least even in **23** (2 + 5 + 5). The complexity index K, which is not an information-theoretic index, displays the opposite behavior; for the same conditions, it increases with the evenness of the cycle size distribution.

4.4.5. Concluding Remarks

In Section 4.4 we presented a new complexity index I_{mech} for reaction mechanisms. In difference from the previously proposed index K (Section 4.3), the new complexity measure allows one to treat both linear and nonlinear mechanisms in a unified scale, and it is less degenerate. The index was tested on the set of linear mechanisms with 2–6 intermediates and 1–3 linearly independent routes. It was shown that the I_{mech} index matches very well the mechanistic complexity in agreement with the hierarchical classification discussed in Section 4.2. Thus, the same hierarchy of complexity factors was established as for the earlier complexity index K. More specifically, listed in a series of lesser importance are the following:

number of simple submechanisms (graph cycles) →
number of intermediates (graph vertexes) or steps (graph edges) →
degree of mechanism connectedness (the basic graph topology) →
evenness (unevenness) of the cycle size distribution.

However, the larger generality of the new approach goes beyond kinetic graphs on which the index K was defined for linear networks, and refers to bipartite graphs describing nonlinear networks as well.

In contrast to the numerous indices of molecular complexity, the I_{mech} index cannot be used for quantitative structure-property or structure-activity correlations, because reaction networks are not associated with any physical or chemical property. Nevertheless, this index may find application in comparative assessments of mechanistic complexities in modeling chemical processes, in computer-assisted generation and discrimination of mechanistic hypotheses and, generally, in creating databases in chemical kinetics. Due to its low degeneracy, it might also be of use in reducing the generation of isomorphic structures in the enumeration of reaction networks of various kinds.

References and Notes

1. Dugunji, J.; Ugi, I. An Algebraic Model of Constitutional Chemistry as a Basis for Chemical Computer Programs. *Top. Curr. Chem.* **1973**, *39*, 19–64.
2. Jochum, C.; Gasteiger, J.; Ugi, I. The Principle of Minimum Chemical Distance. *Angew. Chem. Int. Ed. Engl.* **1980**, *19*, 495–505.
3. Jochum, C.; Gasteiger, J.; Ugi, I. The Principle of the Minimum Chemical Distance and the Principle of the Minimum Structure Change. *Z. Naturforsch. B: Chem. Sci.* **1982**, *37*, 1205–1215.
4. Ugi, I.; Bauer, J.; Bley, K.; Dengler, A.; Dietz, A.; Fontain, E.; Gruber, B.; Herges, R; Bley, K.; Reitsam, K.; Stein, N. Computer-Assisted Solution of Chemical Problems – The Historical Development and the Present State of the Art of a New Discipline of Chemistry. *Angew. Chem. Int. Ed. Engl.* **1993**, *32*, 201–207.
5. Hendrickson, J. B.; Parks, C. A. Generation and Enumeration of Carbon Skeletons. *J. Chem. Inf. Comput. Sci.* **1991**, *31*, 101–107.
6. Koča, J.; Kratochvíl, M.; Kvasnička, V.; Matyska, L.; Pospíchal, J. *Synthon Model of Organic Chemistry and Synthesis Design*; Springer: Berlin, 1989 (*Lect. Notes Chem.,* Vol. 51).
7. Kvasnička, V.; Kratochvíl, M.; Koča, J. Reaction Graphs. *Collect. Czech. Chem. Commun.* **1983**, *48*, 2284–2304.
8. Koča, J. A Graph Model of the Synthon. *Collect. Czech. Chem. Commun.* **1988**, *53*, 3108–3118; The Reaction Distance. *Collect. Czech. Chem. Commun.* **1988**, *53*, 3119–3130.
9. Koča, J. A Mathematical Model of Realistic Constitutional Chemistry. A Synthon Approach. 1. An Algebraic Model of a Synthon. *J. Math. Chem.* **1989**, *3*, 73–89.
10. Kvasnička, V.; Pospíchal, J. Two Metrics for a Graph Theoretical Model of Organic Chemistry. *J. Math. Chem.* **1989**, *3*, 161–191.
11. Kvasnička, V.; Pospíchal, J. Graph Theoretical Interpretation of Ugi's Concept of the Reaction Network. *J. Math. Chem.* **1990**, *5*, 309–322.

12. Pospíchal, J.; Kvasnička, V. Graph Theory of Synthons. *Int. J. Quantum Chem.* **1990**, *38*, 253–278.

13. Bertz, S. H.; Herndon, W. C. The Similarity of Graphs and Molecules. In *Artificial Intelligence Applications in Chemistry*; Pierce, T.H., Hohne, B.A., Eds.; American Chemical Society: Washington, DC; **1986** (ACS Symp. Ser., no. 306), pp 169–175.

14. Bertz, S. H. Converegence, Molecular Complexity and Synthetic Analysis. *J. Am. Chem. Soc.* **1982**, *104*, 5801–5803.

15. For reviews on molecular similarity see: Johnson, M. A.; Maggiora, G. M., Eds.; *Concept and Applications of Molecular Similarity*; Wiley: New York, 1990.

16. Johnson, M. A. A Review and Examination of the Mathematical Spaces Underlying Molecular Similarity Analysis. *J. Math. Chem.* **1989**, *3*, 117–145.

17. Mowshowitz, A. Entropy and the Complexity of Graphs: I. An Index of the Relative Complexity of a Graph. *Bull. Math. Biophys.* **1968**, *30*, 175–204; II. The Information Content of Digraphs and Infinitr Graphs. *ibid.* **1968**, *30*, 225–240; III. Graphs with Prescribed Information Content. *ibid.* **1968**, *30*, 387–414.

18. Minoli, D. Combinatorial Graph Complexity. *Atti. Acad. Waz. Lincei Rend.* **1976**, *A59*, 651–661.

19. Gordon, M.; Kennedy, J. W. The Graph-Like State of Matter. Part 2. LCGI Schemes for Thermodynamics of Alkanes and the Theory of Inductive Inference. *J. Chem. Soc. Faraday II* **1973**, *69*, 484–504.

20. Gutman, I.; Ruscic, B.; Trinajstić, N.; Wilcox, C. W., Jr. Graph Theory and Molecular Orbitals. 12. Acyclic Polyenes. *J. Chem. Phys.* **1975**, *69*, 3399–3405.

21. Bonchev, D.; Trinajstić, N. Information Theory, Distance Matrix, and Molecular Branching, *J. Chem. Phys.* **1977**, *67*, 4517-4533.

22. Wiener, H. Structural Determination of Paraffin Boiling Points. *J. Am. Chem. Soc.* **1947**, *69*, 17–20; Correlation of the Heats of Isomerizations, and Differences in Heats of Vaporizations of Isomers, among the Paraffin Hydrocarbons, *J. Am. Chem. Soc.* **1947**, *69*, 2636–2638.

23. Hosoya, H. Topological Index. A Newly Proposed Quantity Characterizing the Topological Nature of Structural Isomers of Saturated Hydrocarbons *Bull. Chem. Soc. Jpn.* **1971**, *44*, 2332- 2339.

24. Randić, M. On Characterization of Molecular Branching. *J. Am. Chem. Soc.* **1975**, *97*, 6609–6615.

25. Kier, L. B.; Hall, L. H.; Murray, M. J.; Randić, M. Molecular Connectivity. I. Relationship to Nonspecific Local Anesthesia. *J. Pharm. Sci.* **1975**, *64*, 1971–1974.

26. Balaban, A. T. Highly Discriminating Distance-Based Topological Index. *Chem. Phys. Lett.* **1982**, *89*, 399–404.

27. Rouvray, D. H. The Modeling of Chemical Phenomena Using Topological Indexes. *J. Comput. Chem.* **1987**, *8*, 470-480; The Limit of Applicability of Topological Indexes. *J. Mol. Struct. (Theochem)* **1989**, *185*, 187–201.

28. Balaban, A. T. Topological Indexes and Their Uses: A New Approach for the Coding of Alkanes. *J. Mol. Struct. (Theochem)* **1988**, *165*, 243–253.

29. Mekenyan, O.; Bonchev, D.; Balaban, A. T. Topological Indexes for Molecular Fragments and New Graph Invariants. *J. Math. Chem.* **1988**, *2*, 347–375.

30. Stankevich, M. I.; Stankevich, I. V.; Zefirov, N. S. Topological Indexes in Organic Chemistry. *Russ. Chem. Rev.* **1988**, *57*, 337–366.

31. Randić, M. Generalized Molecular Descriptors, *J. Math. Chem.* **1991**, *7*, 155–168. Randić, M. In Search of Structural Invariants. *J. Math. Chem.* **1992**, *9*, 97–146.

32. Trinajstić, N. *Chemical Graph Theory*; 2nd revised ed.; CRC Press: Boca Raton, FL, 1992.

33. Randić, M. On Molecular Identification Numbers. *J. Chem. Inf. Comput. Sci.* **1984**, *24*, 164–175.

34. Carter, S.; Trinajstić, N.; Nikolić, S. On the Use of ID Numbers in Drug Research: A QSAR Study of Neuroleptic Pharmacophores. *Med. Sci. Res.* **1988**, *16*, 185–186.

35. Shannon, C.; Weaver, W. *Mathematical Theory of Communications*; University of Illinois: Urbana, 1949.

36. Rashevsky, N. Life, Information Theory and Topology. *Bull. Math. Biophys.* **1955**, *17*, 229–235.

37. Trucco, E. A Note on the Information Content of Graphs. *Bull. Math. Biophys.* **1956**, *18*, 129-135; On the Information Content of Graphs: Compound Symbols; Different States for Each Point. *Bull. Math. Biophys.* **1956**, *18*, 237–253.

38. Mowshowitz, A. Entropy and the Complexity of Graphs: IV. Entropy Measures and Graphical Structure. *Bull. Math. Biophys.* **1968**, *30*, 533–546.

39. Bonchev, D.; Trinajstić, N. Chemical Information Theory. Structural Aspects. *Int. J. Quantum Chem. Symp.* **1982**, *16*, 463–480.

40. Bonchev, D. *Information-Theoretic Indexes for Characterization of Chemical Structures*; Research Studies: Chichester, 1983.

41. Bertz, S. H. On the Complexity of Graphs and Molecules. *Bull. Math. Biol.* **1983**, *45*, 849-855; The Bond Graph. *J. Chem. Soc., Chem. Commun.* **1981**, 818–820.

42. Bonchev, D.; Mekenyan, O.; Trinajstić, N. Isomer Discrimination by Topological Information Approach. *J. Comput. Chem.* **1981**, *2*, 127–148.

43. Bertz, S. H. The First General Index of Molecular Complexity. *J. Am. Chem. Soc.* **1981**, *103*, 3599-3601; In *Chemical Applications of Topology and Graph Theory*; King, R. B., Ed.; Elsevier: Amsterdam, 1983, pp 206–221.

44. Bonchev, D.; Polansky, O. E. On the Topological Complexity of Chemical Systems. In *Graph Theory and Topology in Chemistry*; King, R. B.; Rouvray, D. H., Eds.; Elsevier: Amsterdam, 1987, pp 126–158.

45. Bonchev, D. The Problems of Computing Molecular Complexity, In *Computational Chemical Graph Theory*; Rouvray, D. H., Ed.; Nova Publications: New York, 1990, pp 34–67.

46. Hendrickson, J. B.; Huang, P.; Toczko, A. G. Molecular Complexity: A Simplified Formula Adapted to Individual Atoms. *J. Chem. Inf. Comput. Chem.* **1987**, *27*, 63–66.

47. Seybold, P. G., May, M.; Bagal, U. A. Molecular Structure–Property Relationship. *J. Chem. Educ.* **1987**, *64*, 575–581.

48. Kier, L. B.; Hall, L. H. *Molecular Connectivity in Structure–Activity Analysis*; Research Studies Press: Letchworth, U.K., 1986.

49. Bonchev, D.; Temkin, O. N.; Kamenski, D. On the Complexity of Linear Reaction Mechanisms. *React. Kinet. Catal. Lett.* **1980**, *15*, 119–124.

50. Bonchev, D.; Kamenski, D.; Temkin, O. N. Complexity Index for the Linear Mechanisms of Chemical Reactions. *J. Math. Chem.* **1987**, *1*, 345–388.

51. Gordeeva, E.; Bonchev, D.; Kamenski, D.; Temkin, O. N. Enumeration, Coding, and Complexity of Linear Reaction Mechanisms. *J. Chem. Inf. Comput. Sci.* **1994**, *34*, 436–445.

52. Temkin, O. N.; Bruk, L. G.; Bonchev, D. The Topological Structure of the Mechanisms of Complex Reactions. *Teor. Eksp. Khim.* **1988**, *24*, 282–291.

53. Temkin, O. H.; Bonchev, D. Classification and Coding of Chemical Reaction Mechanisms. In *Graph Theory and Its Applications to Chemistry*; Tyutyulkov, N.; Bonchev, D., Eds.; Nauka Izkustvo: Sofia, 1987 (in Bulgarian).

54. Hosoya, H.; Kawasaki, J.; Mizutani, K. Topological Index and Thermodynamic Properties. I. Empirical Rules on the Boiling Points of Saturated Hydrocarbons. *Bull. Chem. Soc. Jpn.* **1972**, *45*, 3415–3421.

55. Bonchev, D.; Balaban, A. T.; Mekenyan, O. Generalization of the Graph Center Concept, and Derived Topological Indexes. *J. Chem. Inf. Comput. Sci.* **1980**, *20*, 106–113.

56. Vol'kenshtein, M. V.; Gol'dshtein, B. N. Application of Graph Theory to Calculation of Complex Reactions. *Dokl. Akad. Nauk SSSR* **1966**, *170*, 963–965; Vol'kenshtein, M. V. *Physics of Enzymes*; Nauka: Moscow, 1967 (in Russian).

57. Christofides, N. *Graph Theory: An Algorithmic Approach*; Academic: New York, 1975 and references therein.

58. Gordeeva, E. V.; Molchanova, M.S.; Zefirov, N. S. General Methodology and Computer Program for the Exhausted Restoring of Chemical Structures by Molecular Connectivity Index. Solution of the Inverse Problem in QSAR/QSPR. *Tetrahedron Comp. Method.* **1990**, *3*, 389–415.

59. Faradzhev, I. A. Generation of Nonisomorphic Graphs with a Given Distribution of Vertex Types. In *Algorithmic Investigations in Combinatorics;* Nauka: Moscow, 1978; pp 11–19.

60. Zeigarnik, A. V.; Temkin, O. N. A Graph-Theoretical Model of Complex Reaction Mechanisms: A New Complexity Index for Reaction Mechanisms. *Kinet. Katal.* **1996**, *37*, 372–385; *Kinet. Catal. Engl. Transl.* **1996**, *37*, 347–360.

61. Zeigarnik, A. V.; Gordeeva, E. V.; Bonchev, D.; Temkin, O. N. (work in progress).

62. Zeigarnik, A. V.; Temkin, O. N. A Graph-Theoretical Model of Complex Reaction Mechanisms: Bipartite Graphs and the Stoichiometry of Complex Reactions. *Kinet. Katal.* **1994**, *35*, 702–710; *Kinet. Catal. Engl. Transl.* **1994**, *35*, 647–655.

63. Zeigarnik, A. V. A Graph-Theoretical Model of Complex Reaction Mechanisms: The Method for Classifying Complex Reaction Mechanisms. *Kinet. Katal.* **1995**, *36*, 653–657.

64. Brillouin, L. *Science and Information Theory*; Academic: New York, 1956.

65. Temkin, M. I. Kinetics of Complex Reactions. In *Proceedings of the All-Union Conference on Chemical Reactors*; Nauka: Novosibirsk, 1966, Vol. 4. pp 628–646.

66. Bonchev, D.; Temkin, O. N.; Kamenski, D. On the Classification and Coding of Linear Reaction Mechanisms. *React. Kinet. Catal. Lett.* **1980**, *15*, 113–118.

67. Bonchev, D.; Temkin, O. N.; Kamenski, D. Graph-Theoretical Classification and Coding of Chemical Reactions with a Linear Mechanism. *J. Comput. Chem.* **1982**, *3*, 95–110.

Chapter 5

Topological Structure of a Mechanism and Its Kinetic Analysis

In Chapter 3, we discussed the topological structure underlying specific features of reaction mechanisms. It is of interest to analyze how different elements of this structure can reveal themselves in kinetics and which methods can be used to recognize the key elements of the topological structure using kinetic methods. Thus, proceeding from a particular hypothetical mechanism and its graph that depicts its topological structure, the goal of a researcher is to answer the following questions: Which type of kinetic behavior should be expected? What kind of experiments can reveal the elements of the topological structure?

Several factors stipulate the kinetic behavior. Among these are the structure (topology) of a mechanism graph and the nature of elementary steps. Feasibility of the experimental procedures to provide adequate monitoring of intermediates is also of importance.

3.1. Topological Structure of a Mechanism and the Structure of Its Kinetic Model

Analysis of the steady-state kinetic models of the three classes of two-route linear mechanisms ($\mathbf{A}$, $\mathbf{B}$, and $\mathbf{C}$) revealed that there is a correspondence between the mechanism graph structures and the kinetic model.[1,2] According to the concept of kinetic complexity discussed in Chapter 4, there is a trend toward increasing complexity from classes $\mathbf{A}^i$, $\mathbf{B}^j$, and $\mathbf{A}^i\mathbf{B}^j$ to classes $\mathbf{C}^k$, $\mathbf{Z}^v$, and $\mathbf{C}^k\mathbf{Z}^v$ because the number of step weights increases.

253

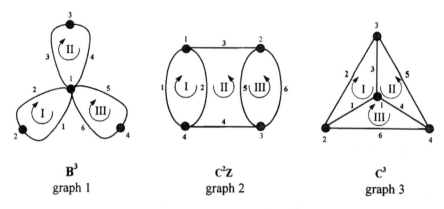

B³   **C²Z**   **C³**
graph 1   graph 2   graph 3

Figure 5.1. Kinetic graphs of linear mechanisms of classes **B³**, **C²Z** and **C³** ($M = 3$, $N = 4$). The numeration of cycles, edges, and vertexes is not canonical.

Consider as an example three mechanisms of three-route reactions each having four intermediates (Figure 5.1).[3,4] The steady-state rate law for route I of the catalytic reaction of class **B³** can be expressed by the equation (see Section 2.2.4):

$$r_I = \frac{[\mathcal{X}]_\Sigma (C_{10}^+ - C_{10}^-) D_{10}}{\sum D_i} \qquad (5.1)$$

or in more general form:[1]

$$r_I = \frac{[\mathcal{X}]_\Sigma W_I P_I}{\sum D_i}, \qquad (5.2)$$

where W_I is the cyclic characteristic of route I ($W_I = C_{10}^+ - C_{10}^-$); P_I is a *parameter of conjugation* (i.e., the root determinant of cycle I, $D_{10} = P_I$); and $[\mathcal{X}]_\Sigma$ is the total catalyst concentration;

$$W_I = \omega_{+1}\omega_{+2} - \omega_{-1}\omega_{-2},$$

$$P_I = \omega_{+4}\omega_{+6} + \omega_{+4}\omega_{+5} + \omega_{-3}\omega_{+6} + \omega_{-3}\omega_{-5} = (\omega_{+4} + \omega_{-3})(\omega_{+6} + \omega_{-5}).$$

Complexity index K (see Chapter 4) can characterize not only the entire mechanism, but any rate law that is expressed in terms of the step weights. Therefore, it is worthwhile to discuss how the topological structure (class) of the mechanism is reflected in the complexity of the rate law of a route.

The kinetic complexity index for equation (5.2) is 128 (when the index refers to the particular equation, we will denote it by K with asterisk: $K^* = 128$). If the concentration of a free catalyst [$\mathcal{X}$] (or the portion of free active sites θ_0 on the heterogeneous catalyst surface) is known, one can obtain a far simpler expression for r_1 than equation (5.2):

$$r_{\mathrm{I}} = \frac{[\mathcal{K}]W_{\mathrm{I}}}{D_1^{\mathrm{I}}} = \frac{[\mathcal{K}](\omega_{+1}\omega_{-2} - \omega_{-1}\omega_{-2})}{\omega_{+2} + \omega_{-1}}, \tag{5.3}$$

where D_1^{I} is the root determinant of vertex 1 calculated for cycle I only. In this case, the complexity index $\overline{K}^*$ for the rate law, which involves only the step weights for route I, is 6. (If index K^* refers to the rate law expressed in terms of the free catalyst concentration, we use $\overline{K}^*$ notation.)

If the network class notation contains $\mathbf{C}$, $\mathbf{C}^k$, or $\mathbf{C}^k\mathbf{Z}^\nu$, the rate law of each route will involve all steps of other routes, even if the catalyst material balance is not taken into account. This is due to the chemical conjugation of reaction routes. In the case of class $\mathbf{C}^3$ the rate law takes the following form:

$$r_1 = \frac{[\mathcal{K}](W_1 P_1 + C^*)}{D_1}, \tag{5.4}$$

where $C^* = \sum_l W_l P_l$ is the *factor of conjugation*;[1] the sum is taken over all cycles l ($l \le M - 1$) other than cycle I (route I) that include the step belonging to route I only (step 2, graph 3); W_1 is the cyclic characteristics of cycle I (route I); P_l is the algebraic complement of cycle l. In the case of equation (5.4) (class $\mathbf{C}^3$), $\overline{K}^*$ is 112 (compare to $\overline{K}^* = 6$ for equation (5.3)).

To illustrate the procedure of obtaining the index $\overline{K}^*$, consider the second route of mechanism $\mathbf{C}^2\mathbf{Z}$ (Figure 5.1). The rate law for the second route is

$$r_{\mathrm{II}} = \frac{[\mathcal{K}](W_{\mathrm{II}} P_{\mathrm{II}} + C^*)}{D_1}, \tag{5.5}$$

where

$$W_{\mathrm{II}} = \omega_{+1}\omega_{+3}\omega_{+4}\omega_{+5} - \omega_{-1}\omega_{-3}\omega_{-4}\omega_{-5},$$

$$P_{\mathrm{II}} = 1,$$

$$C^* = \omega_{+4}\omega_{+3}(\omega_{+5}\omega_{-2} + \omega_{-6}\omega_{+1} + \omega_{-6}\omega_{-2}) - \omega_{-4}\omega_{-3}(\omega_{-5}\omega_{+2} + \omega_{+6}\omega_{-1} + \omega_{+6}\omega_{+2}),$$

$$D_1 = \omega_{+2}\omega_{+3}\omega_{+5} + \omega_{+2}\omega_{+3}\omega_{-4} + \omega_{+2}\omega_{+3}\omega_{-6} + \omega_{+2}\omega_{-5}\omega_{-4} + \omega_{+2}\omega_{+6}\omega_{-4} + \omega_{+3}\omega_{+4}\omega_{-6} +$$
$$\omega_{-1}\omega_{+3}\omega_{+5} + \omega_{-1}\omega_{+3}\omega_{-4} + \omega_{-1}\omega_{+3}\omega_{-6} + \omega_{-1}\omega_{-4}\omega_{-5} + \omega_{-1}\omega_{-4}\omega_{+6} + \omega_{+3}\omega_{+4}\omega_{+5}.$$

Table 5.1 lists the values of K^* and $\overline{K}^*$ for one of the routes of mechanisms shown in Figure 5.1 and total values of K.

As can be seen from the above examples, the network class define the complexity of a rate law (index K). The complexity of particular rate equations (indices K^* and $\overline{K}^*$) depends also on which concentrations of intermediates are measured and involved in the rate law. That is, different rate laws corresponding to the mechanism of the same class may differ in their complexity.

Table 5.1. Complexity indices of kinetic models

Class	K^*	$\overline{K}^*$	K
B^3	128	6	384
C^2Z	176[†]	68	576
C^3	256	112	768

[†] Route II.

Consider one more example. Let us discuss the kinetic features of class B^j illustrated by the two-route mechanism ($j = 1$, see Figure 5.2). In the case of the catalytic network ($M = 2$) depicted by graph 4, X_i is the catalyst of both route I and route II. Suppose its steady-state concentration is known. Then, the rates r_I and r_{II} expressed in terms of $[X_i]$ will be kinetically independent (i.e., each of the rates does not contain any of the rate constants of the other):

$$r_I = \frac{[X_i]W_I}{D_{X_i}^I} \quad \text{and} \quad r_{II} = \frac{[X_i]W_{II}}{D_{X_i}^{II}}.$$

If one measures only $[X_m]$ and the reaction rate is expressed in terms of the unknown concentration $[X_i]$, then the latter can be expressed in terms of the former,

$$[X_i] = [X_m]\omega_m / \omega_0,$$

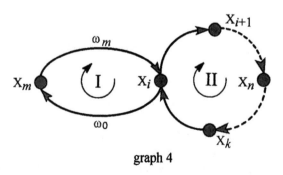

graph 4

Figure 5.2. Graph of the linear mechanism of class B.

and the rate law of the second route can be defined by the equation

$$r_{II} = \frac{[X_i]W_{II}}{D_{X_i}^{II}} = \frac{[X_m]\omega_m W_{II}}{\omega_0 D_{X_i}^{II}} = \frac{W_m}{\omega_0} \frac{W_{II}}{D_{X_i}^{II}}.$$
(5.6)

Thus, the rate of the second route depends on the rate W_m of X_i formation and on the weight of its decay ω_0; W_{II} is the cyclic characteristic of the second route. If the rate is expressed by equation (5.6), the kinetics of a catalytic reaction take on the features of a chain process kinetics, involving the initiation W_m, termination ω_0, and propagation ($W_{II} / D_{X_i}^{II}$) rates. In the case of a typical chain process, one of the routes is noncatalytic, in which the initiator is consumed.

It is important to note that in addition to the topological structure of the mechanism, the nature of elementary steps also can substantially affect the form of the rate laws and kinetic features of this structure. Let us discuss two somewhat simplified mechanisms of the branched chain reaction of hydrocarbon RH oxidation in the gas (I) and liquid (II) phases:[6]

Mechanism I: **Mechanism II:**

(1) RH $\xrightarrow{W_i}$ r$^\bullet$ RH $\xrightarrow{W_i}$ r$^\bullet$

(2) r$^\bullet$ + RH $\xrightarrow{k_2}$ P + r$^\bullet$ r$^\bullet$ + RH $\xrightarrow{k_2}$ P + r$^\bullet$

(3) P $\xrightarrow{k_3}$ r$^\bullet$ P $\xrightarrow{k_3}$ r$^\bullet$

(4) P $\xrightarrow{k_3'}$ products   P $\xrightarrow{k_3'}$ products

(5) r$^\bullet$ + $\xrightarrow{\omega_i}$ decay   2r$^\bullet$ + $\xrightarrow{k_6}$ decay

Step 2 is the generalized step of chain propagation. The subnetworks of intermediates for these mechanisms are shown in Figure 5.3.

As can be seen, the topological structures of these mechanisms are very similar. The only difference is the molecularity of step 5. However, the rate laws are quite different, although the basic features of chain mechanisms are preserved (see Table 5.2).

Although the topological structure manifests itself in the kinetics, the experimental kinetic data obtained without preliminary hypothesis can provide conclusive evidence for the topological structure of the mechanism in the very simple cases. Therefore, complexity of rate laws in the case of graphs with high degree of connectedness ($C^k Z^v$) serves as an additional argument in favor of the rational strategy of mechanistic studies[7] and the necessity of hypothesis advancing at the first stage of this study. For example, the simplest mechanism of acetylene carbonylation in alcohol solutions of Pd(I) complexes contains five routes, ten steps, and five independent routes. The products of this reaction are acrylate (A), succinate (S), maleate (M), propionate, and ethylene (E). The schematic representation of the mechanism (Figure 5.4) shows only metal-containing intermediates.

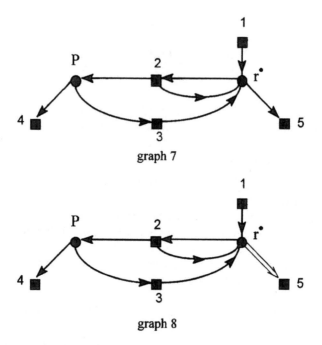

graph 7

graph 8

Figure 5.3. Subnetworks of intermediates for mechanisms I and II.

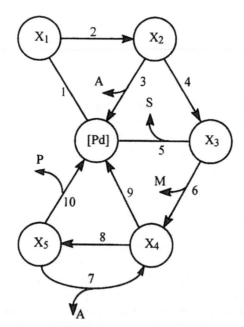

Figure 5.4. Schematic representation of the mechanism of acetylene carbonylation (X_1 = [Pd]OR, X_2 = [Pd]CH=CHCOOR, X_3 = [Pd]H(M); X_4 = [Pd]CH=CH$_2$; X_5 = [Pd]H(A); [Pd] is the complex of Pd(I)).

Five of ten step weights contain $[H^+]$: ω_{-1}, ω_{+3}, ω_{+5}, ω_{+9}, and ω_{+10}. When all concentrations except $[H^+]$ are constant, the dependence of the acrylate formation rate r_A on $[H^+]$ contains polynomials in the numerator and denominator:

$$r_A = \frac{\sum_{n=0}^{4} K_{1n}[H^+]^n}{\sum_{m=0}^{5} K_m[H^+]^m}. \tag{5.7}$$

Not having any hypothesis on the mechanism (Figure 5.4) and not knowing the form of the rate law (equation (5.7)), it is impossible to obtain this rate law from the experimental data.

Table 5.2. Formulas for rates and concentrations for mechanisms I and II

Parameter	Formula	
	Mechanism I	Mechanism II
Rate of the RH consumption at the beginning of reaction	$W_2 \sim W_i e^{\varphi t}[RH]$	$W_2 \sim \sqrt{W_i}\,[RH]$
Concentration of active centers	$[r^\cdot] = (W_i + k_3[P])/\omega_t$	$[r^\cdot] = \sqrt{(W_i + k_3[P])/k_6}$
Concentration of intermediate P	$[P] \sim W_i e^{\varphi t}[RH]/\varphi$	$[P] \sim [RH]^2 t^2$

Note: φ is the factor of acceleration; t is time; subscripts "i" and "t" denote initiation and termination, respectively.

Indeed, handling the experimental S-shaped kinetic curves on the plots of r_A versus $[H^+]$, which were obtained within the narrow limits of H^+ concentrations, resulted in the equation[8]

$$r_A = \frac{k_{10}'[H^+]^2}{1 + k_1'[H^+] + k_2[H^+]^2}. \tag{5.8}$$

This equation hardly informs a researcher about the structure of the mechanism.

A long-standing experience of mechanistic studies[9] allows us to suggest that in the case of multiroute reactions which are *stoichiometrically ambiguous* (i.e., having several different overall equations), relationships like r_i/r_j-C_k bring more information than the r_i-C_k plots.[3]

5.2. Analysis of Conjugation Nodes

One of the most common expedients in the study and identification of the topological structure of complex reaction mechanisms is the analysis of hypothetical conjugation nodes on the patways to different products. Surely, this method of studying the topological structure is meaningful only if the reaction to be studied is stoichiometrically ambiguous. The *conjugation node* is a fragment of the reaction network that involves the latest common intermediate and its transformation on the pathway to different products. If the products are denoted by $P(i)$ and $P(j)$, then the rate of transformation of the common intermediate will enter the expressions for rates $r_{P(i)}$ and $r_{P(j)}$. Conjugation nodes are branching points in the kinetic graph (Figure 5.5). When the reaction is far from equilibrium and the routes in which products are formed involve irreversible steps, ratios between the rates of product formation are much simpler functions of reactant concentrations than the rates themselves. This is also the case even when the conjugation nodes are rather complicated.

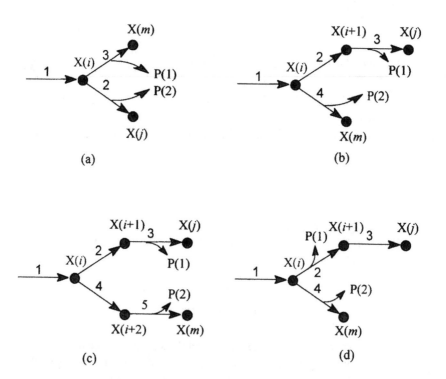

Figure 5.5. Simplest conjugation nodes of linear mechanisms.

In the case of the conjugated node shown in Figure 5.5d, the weight of step ω_{-2} involves [P(1)]. The form of $r_{P(1)}/r_{P(2)}$ expressions would not change if the first step were reversible:

$$(a) \; r_{P(1)} / r_{P(2)} = \omega_{+2} / \omega_{+3},$$

$$(b) \; r_{P(1)} / r_{P(2)} = \omega_{+2}\omega_{+3} / [\omega_{+4}(\omega_{-2} + \omega_{+3})],$$

$$(c) \; r_{P(1)} / r_{P(2)} = [\omega_{+2}\omega_{+3}(\omega_{-4} + \omega_{+5})] / [\omega_{+4}\omega_{+5}(\omega_{-2} + \omega_{+3})],$$

$$(d) \; \text{the same as case (b).}$$

(5.9)

Comparison of $r_{P(i)}/r_{P(j)}$ ratios with the results of kinetic experiments makes it possible to identify most probable conjugation nodes. Let us discuss several examples of analyzing conjugation nodes.

It is common knowledge that Cu(I) complexes catalyze reactions of acetylene in the $CuCl-CuCl_2-NH_4Cl-HCl-H_2O$ system:[9]

$$C_2H_2 + HCl \longrightarrow CH_2{=}CHCl \; (VC),$$

$$C_2H_2 + 2CuCl_2 \longrightarrow 2CuCl + ClCH{=}CHCl \; (DCE),$$

$$C_2H_2 + 2CuCl_2 \longrightarrow 2CuCl + CH_2{=}CCl_2 \; (VDC).$$

The relationships between the rates of product formation (r_{VC}, r_{DCE}, and r_{VDC}) and the concentration of free Cu^{2+} were studied at a constant concentration of copper(I) chloride complexes.[10,11] The nature of $r-[CuCl_2]$ relationships suggested the presence of at least one conjugation node, which is common for all three products.

The equation

$$r_{DCE} / r_{VC} = k_1[Cu^{2+}]^2$$

(5.10)

accounts for the simplest conjugation node, which is shown in Figure 5.6a. This structure was refined using the ratio

$$r_{DCE} / r_{VC} = k_2[Cu^{2+}]^2 / (k_3 + [Cu^{2+}]),$$

(5.11)

which supported the presence of at least one nonequilibrium step between the intermediate X_1, which is transformed to VC, and X_2, which is cleaved to produce VDC (see Figure 5.6b). Thus, the topological structure of the mechanism of the three product synthesis can be depicted by the kinetic graph (class C^3) shown in Figure 5.6c.

Having the graph of the mechanism (Figure 5.6c), one can derive the rate laws as functions of $[Cu^{2+}]$:

$$r_{VC} = (\alpha + [Cu^{2+}])/D,$$

(5.12)

$$r_{DCE} = \varphi[Cu^{2+}]^2(\alpha + [Cu^{2+}])/D,$$

(5.13)

$$r_{VDC} = \kappa[Cu^{2+}]^2/D,$$

(5.14)

where

$$D = \beta[Cu^{2+}]^3 + \gamma[Cu^{2+}]^2 + \delta[Cu^{2+}] + \varepsilon, \qquad (5.15)$$

and α, β, γ, δ, ε, φ, κ are some constants. Until the topological structure of this mechanism was revealed, one of the authors of this book attempted to derive these equations unsuccessfully applying the traditional methodology by processing the $r–[Cu^{2+}]$ data. The method based on analysis of the conjugation nodes is much simpler than analysis of expressions for selectivities as well.

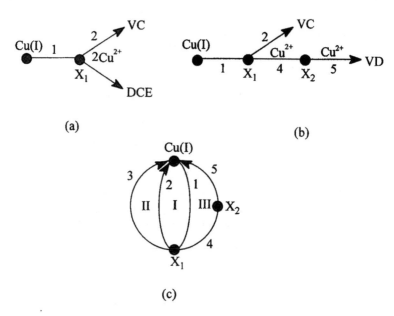

Figure 5.6. (a) and (b) Structures of conjugation nodes and (c) the kinetic graph of the mechanism of VC, DCE, and VDC formation.

Analysis of conjugation nodes was successfully applied in the discrimination of hypothetical mechanisms for the very complicated process of acetylene carbonylation, which was already mentioned here.[9] Thus, analysis of conjugation nodes of the mechanism for this reaction in the $PdBr_2$–HBr–acetone–n-butanol catalytic system made it possible to reduce all alternative variants to only two mechanisms. This system produced dibutyl maleates (M), dibutyl fumarates (F), and dibutyl succinates (S). The mechanism involved the conjugation node shown in Figure 5.7. The ratio between the rates of succinate and fumarate formation can be expressed as follows:

$$r_C / r_F = k[HBr].$$

Note that the rates of ester formation for these mechanisms take far more complicated forms (e.g., the r_A–[H$^+$] dependence, equation (5.7)).

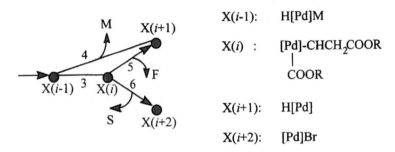

X(i-1):   H[Pd]M

X(i) :   [Pd]-CHCH$_2$COOR
           |
           COOR

X(i+1):   H[Pd]

X(i+2):   [Pd]Br

Figure 5.7. The conjugation node on the pathway to maleic (M), fumaric (F), and succinic (S) acid
esters in the PdBr$_2$–HBr–acetone–n-butanol catalytic system.

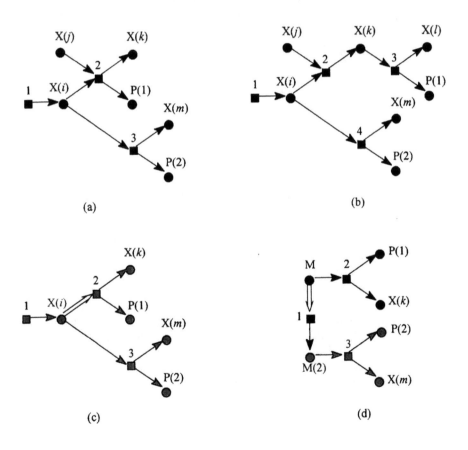

Figure 5.8. Conjugation nodes having nonlinear elementary steps (step 2 in network (b) and step 1 in
network (d) are reversible, but this is not shown in this figure).

The mechanism of formation of acrylic (A), maleic (M), succinic, and propionic acid esters in solutions of Pd(I) complexes $(Li_2Pd_2I_4)$[23] was proposed on the basis of literature data, which were available up to that moment. These data included possible pathways of formation of each product and different variants of the conjugation of oxidation and reduction reactions assuming that each product is formed by one of the possible pathways. After the preliminary discriminating experiments, 90 of 1344 mechanisms remained for the further consideration, which involved five pathways for acrylate, three pathways for maleate and succinate, and two pathways for propionate.

Analysis of conjugation nodes in 90 mechanism graphs showed that none of these 90 mechanisms is able to account for the experimental relationships between the $r_{P(i)}/r_{P(j)}$ ratios and the partial pressures of carbon monoxide and acetylene and the concentration of HCl. As a result of interfacing different pathways assuming that some products can be generated via two different mechanisms, four mechanisms were found for which conjugation nodes were not contradictory to the experimental observations. One of these mechanisms is presented in Figure 5.7.

Conjugation nodes that involve nonlinear steps are quite common in catalysis. Their analysis is a more challenging task than analysis of conjugation nodes that are parts of linear fragments of mechanisms, although helpful as well. The ratios between rates $r_{P(1)}/r_{P(2)}$ for the nodes shown in Figure 5.8 match equations:

$$\text{(a)} \quad \frac{r_{P(1)}}{r_{P(2)}} = \frac{\omega_{+2}}{\omega_{+3}}[X_j],$$

$$\text{(b)} \quad \frac{r_{P(1)}}{r_{P(2)}} = \frac{\omega_{+2}\omega_{+3}}{\omega_{+4}(\omega_{-2} + \omega_{+3})}[X_j],$$

$$\text{(c)} \quad \frac{r_{P(1)}}{r_{P(2)}} = \frac{\omega_{+2}}{\omega_{+3}}[X_i].$$

In the case of a catalytic reaction in solution of any metal complex, concentrations $[X_i]$ and $[X_j]$ will increase with increasing the total catalyst concentration $[M]_\Sigma$. Therefore, the effect of $[M]_\Sigma$ on the $r_{P(1)}/r_{P(2)}$ ratio supports the presence of nonlinear steps. In more complicated cases, it will be difficult to predict the form of the $r_{P(1)}/r_{P(2)}-[M]_\Sigma$ dependence; e.g., when both steps after the first step in the conjugation node are nonlinear. This type of a conjugation node is observed when the catalyst exists in two states: monomer and polymer (an associate of species or a polynuclear complex). Figure 5.8d shows the conjugation node with the formation of dimer. If step 1 is equilibrium, the routes of P(1) and P(2) formation are independent, but they are related to each other by the equilibrium constant K_1:

$$\text{(d)} \quad \frac{r_{P(2)}}{r_{P(1)}} = \frac{\omega_{+3}[M_2]}{\omega_{+2}[M]} = \frac{\omega_{+3}}{\omega_{+2}}K_1[M]$$

or by the material balance of the catalyst.

In the general case, step 1 can be written as

$$nM \xrightleftharpoons{K_n} M_m,$$

$$[M]_\Sigma = [M] + n[M_n].$$

If $[M] \gg [M_n]$, then $[M] \approx [M]_\Sigma$ and

$$\frac{r_{P(2)}}{r_{P(1)}} = \frac{\omega_{+3}}{\omega_{+2}}[M]_\Sigma^{n-1}.$$

If $[M] \ll [M_n]$, then $[M]_\Sigma = nK_n[M]^n$ and

$$[M] = \sqrt[n]{\frac{[M]_\Sigma}{nK_n}}.$$

Then,

$$\frac{r_{P(2)}}{r_{P(1)}} = \frac{\omega_{+3}}{\omega_{+2}}(nK_n)^{-1/n}[M]_\Sigma^{(n-1)/n}.$$

Polyfunctional catalytic systems often contain nonlinear conjugation nodes. An intriguing example of a polyfunctional catalytic system was found in the study of the mechanism of 1-chloro-1,3-butadiene (CBD) synthesis from acetylene in the $PdCl_2–HgCl_2–FeCl_3–HCl–H_2O$.[12] The reaction mechanism can be described by the following, somewhat simplified, sequence of steps (the numeration of steps is given as in the original paper):

$$(1)\ PdCl_2 + C_2H_2 \xrightleftharpoons{K_1} X_1,$$

$$(2)\ X_1 + C_2H_2 \longrightarrow X_2,$$

$$(3)\ X_2 + HgCl_2 \xrightleftharpoons{K_3} X_3 + PdCl_2,$$

$$(4)\ X_2 + C_2H_2 \xrightleftharpoons{K_4} X_4,$$

$$(6)\ X_3 + H_3O^+ \xrightarrow{k_6} P_1 + HgCl_2,$$

$$(8) \ X_4 + C_2H_2 \xrightarrow{\ k_8\ } X_5,$$

$$(9) \ X_5 + 2FeCl_3 \longrightarrow PdCl_2 + 2FeCl_2 + P_2,$$

where $X_1 = ClPdCH{=}CHCl$, $X_2 = ClPd(CH{=}CH)_2Cl$, $X_3 = ClHg(CH{=}CH)_2Cl$, $X_4 = ClPd(C_2H_2)(CH{=}CH)_2Cl$, $X_5 = ClPd(C_2H_2)(CH{=}CH)_3Cl$, $P_1 = CBD$, and $P_2 = Cl(CH{=}CH)_4Cl$.

The reaction rate increases with concentrations $[PdCl_2]_\Sigma$ and $[HgCl_2]_\Sigma$. The r_{CBD}–$[PdCl_2]_\Sigma$ dependence can be described by the equation that corresponds to the deceleration by $PdCl_2$. As can be seen from Figure 5.9, there is a conjugation node at vertex X_2 and having nonlinear reversible step 3. The $r_{P(1)}/r_{P(2)}$ ratio in the conjugation node is described by the equation

$$\frac{r_{P(1)}}{r_{P(2)}} = \frac{k_6 K_3 [H_3O^+][HgCl_2]}{k_8 K_4 [C_2H_2]^2 [PdCl_2]}.$$

Provided that $[C_2H_2]$ and $[H_3O^+]$ are constant and there exist linear proportions $[PdCl_2] = A[PdCl_2]_\Sigma$ and $[HgCl_2] = B[HgCl_2]_\Sigma$, we arrive at the following expression:

$$\frac{r_{P(1)}}{r_{P(2)}} = Const \times \frac{[HgCl_2]_\Sigma}{[PdCl_2]_\Sigma}.$$

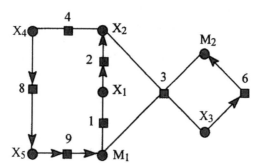

Figure 5.9. The subnetwork of intermediates for the mechanism of CBD synthesis ($M_1 = PdCl_2$ and $M_2 = HgCl_2$). For the sake of simplicity, the reversible (pseudoequilibrium) steps 1, 3, and 4 are shown without arrows. To identify their directions see the scheme of the mechanism.

If a *stoichiometrically unambiguous* reaction (i.e., having one overall equation) is studied, then it is useful to purposefully change the topological structure of the mechanism so as to create the conjugation node. This can be done by using the reactant that selectively affects some intermediates. The addition of inhibitors and activators to the enzymatic systems is successfully used in mechanistic studies.[13,14] This expedient is of particular assistance if one manages to create a new reaction route to produce a new product. In single-route reactions having a rate-limiting step or slow irreversible step, the same method serves well to receive more information about steps that occur after the irreversible one.

For instance, the well-known Nieuwland synthesis of vinylacetylene in the $CuCl-NH_4Cl-H_2O$ systems occurs according the following scheme:[9,15]

$$(1) \; C_2H_2 + Cu(I) \underset{\longleftarrow}{\overset{\longrightarrow}{}} X_1 + H_3O^+,$$

$$(2) \; X_1 + C_2H_2 \longrightarrow X_2,$$

$$(3) \; X_2 + H_3O^+ \longrightarrow C_4H_4 + Cu(I),$$

(5.16)

which corresponds to the equation

$$r_{C_4H_4} = \frac{k_{+1}k_{+2}[C_2H_2]^2[Cu(I)]}{k_{-1}[H_3O^+] + k_{+2}[C_2H_2]},$$

(5.17)

if the concentration of the active Cu(I) complexes remains constant. The addition of small amounts of $CuCl_2$ (~10^{-3} mol/l) results in the occurrence of a new route with branching at X_2 and the new overall reaction:

$$X_2 + 2CuCl_2 \longrightarrow CH_2=CCl-C\equiv CH + 2CuCl + HCl,$$

which becomes catalytic with respect to $CuCl_2$ if $CuCl$ is oxidized.[16] The following new steps emerge:

$$(4) \; X_2 + CuCl_2 \longrightarrow X_3,$$

$$(5) \; X_3 + CuCl_2 \longrightarrow CH_2=CCl-C\equiv CH + 2CuCl.$$
$$(C_4H_3Cl)$$

(5.18)

The conjugation node is shown in Figure 5.10.

The expression

$$\frac{r_{C_4H_4}}{r_{C_4H_3Cl}} = \frac{k_{+3}[H_3O^+]}{k_{+4}[CuCl_2]}$$

(5.19)

matches the experimental data. According to the reaction mechanism described by steps 1–5 (equations (5.16) and (5.18)), the sum of rates $r_{C_4H_4} + r_{C_4H_3Cl}$ will be equal to the rate of step 2, because step 2 is irreversible:

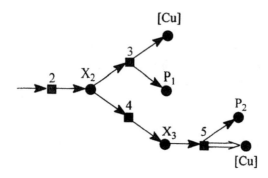

Figure 5.10. The conjugation node in the synthesis of C_4H_4 and C_4H_3Cl (X_2 = [Cu]–CH=CH–C≡CH, $P_1 = C_4H_4$, $P_2 = C_4H_3Cl$, and [Cu] and [Cu]* are active Cu(I) complexes).

$$r_{C_4H_4} + r_{C_4H_3Cl} = \frac{k_{+1}k_{+2}[C_2H_2]^2[Cu(I)]}{k_{-1}[H_3O^+] + k_{+2}[C_2H_2]}. \tag{5.20}$$

Using equations (5.19) and (5.20), we arrive at the rate law

$$r_{C_4H_4} = \frac{k_{+1}k_{+2}[C_2H_2]^2[Cu(I)]}{(k_{-1}[H_3O^+] + k_{+2}[C_2H_2])} \times \frac{k_{+3}[H_3O^+]}{(k_{+3}[H_3O^+] + k_{+4}[CuCl_2])}. \tag{5.21}$$

It can be seen from this equation that after the change in the topological structure of the mechanism, the rate law contains weights of steps that occur after the irreversible step 2, and the reaction rate dependence on $[H_3O^+]$ is the function that passes through the maximum (compare to equation (5.17)). More examples of changing the topological structure have been reported.[3,9]

5.3. Topological Structure of Mechanisms and "Dimensionless" Rate Equations

Another sort of rate equations can be used to analyze the topological structure of multiroute mechanisms, which contain conjugation nodes. These can be referred to as *dimensionless rate equations*, which describe relationships like r_i–r_Σ, r_i–r_k/k_i, and r_i/r_j–r_k/r_j, where r_i, r_j, and r_k are rates of routes or the rates of formations of different products; r_Σ is the sum of rates of all routes. These equations were termed dimensionless because each term of these equations is dimensionless; e.g., the ratio between rates. The key idea of these equations is that they describe the relationships between different rates rather than between rates and concentrations. Dimensionless rate equations of this sort can be used in the evaluation and discrimination of hypothetical mechanisms, the identification of fragments of the topological structure, and the derivation of standard rate laws (in which the rate is a function of concentrations).

Suppose a linear multiroute mechanism involves at least one route having at least one common step (class $\mathbf{C}^k$) with other routes. Then, one can choose such set of independent routes that at least one route from this set will have the same number of common steps with each of the other routes from this set. Clearly, the stoichiometric numbers of common steps v_j^M in each of M routes will be equal to unity. Let this route be route 1 having the rate $r^{(1)}$. For this route, the Temkin equation[17,18]

$$\sum_{m=1}^{M} r^{(M)} (v_{S_1}^{(M)} \overset{+}{r}_{S_2} \overset{+}{r}_{S_3} \ldots \overset{+}{r}_{S_m} + \overset{-}{r}_{S_1} v_{S_2}^{(M)} \overset{+}{r}_{S_3} + \ldots + \overset{-}{r}_{S_1} \overset{-}{r}_{S_2} \overset{-}{r}_{S_3} \ldots \overset{-}{r}_{S_{m-1}} v_{S_m}^{(M)}) =$$

$$\overset{+}{r}_{S_1} \overset{+}{r}_{S_2} \overset{+}{r}_{S_3} \ldots \overset{+}{r}_{S_m} - \overset{-}{r}_{S_1} \overset{-}{r}_{S_2} \overset{-}{r}_{S_3} \ldots \overset{-}{r}_{S_m}, \tag{5.22}$$

can be rewritten in the block form:

$$r^{(1)}(A + B) + r^{(2)}B + r^{(3)}B + \ldots r^{(M)}B = C. \tag{5.23}$$

Subscripts $S_1 \ldots S_m$ denote steps of route 1; $v_{S_m}^{M}$ is the stoichiometric number of step S_m in route M ($v_{S_m}^{M} \neq 0$ if step S_m is common for routes 1 and M). The number of terms in brackets for route 1 is equal to the number of steps m of this route. For other routes, 2 ... M, only terms with nonzero $v_{S_m}^{M}$ are put in brackets. For example, if the mechanism contains five steps and the first three steps are common for three routes and $\overset{+}{r}_{S_1} = r_{+1}$, $\overset{+}{r}_{S_2} = r_{+2}$, ..., $\overset{+}{r}_{S_5} = r_{+5}$, etc., then

$$r^{(1)}(r_{+2}r_{+3}r_{+4}r_{+5} + r_{-1}r_{+3}r_{+4}r_{+5} + r_{-1}r_{-2}r_{+4}r_{+5} + r_{-1}r_{-2}r_{-3}r_{+5} +$$

$$r_{-1}r_{-2}r_{-3}r_{-4}) + r^{(2)}(r_{+2}r_{+3}r_{+4}r_{+5} + r_{-1}r_{+3}r_{+4}r_{+5} + r_{-1}r_{-2}r_{+4}r_{+5}) + \tag{5.24}$$

$$r^{(3)}(r_{+2}r_{+3}r_{+4}r_{+5} + r_{-1}r_{+3}r_{+4}r_{+5} + r_{-1}r_{-2}r_{+4}r_{+5}) =$$

$$r_{+1}r_{+2}r_{+3}r_{+4}r_{+5} + r_{-1}r_{-2}r_{-3}r_{-4}r_{-5}.$$

Equation (5.24) divided by $r_{+1}r_{+2}r_{+3}r_{+4}r_{+5}$ gives equation

$$r^{(1)}(A + B) + r^{(2)}B + r^{(3)}B = C, \tag{5.25}$$

where

$$A = \frac{r_{-1}r_{-2}r_{-3}}{r_{+1}r_{+2}r_{+3}r_{+4}} + \frac{r_{-1}r_{-2}r_{-3}r_{-4}}{r_{+1}r_{+2}r_{+3}r_{+4}r_{+5}},$$

$$B = \frac{1}{r_{+1}} + \frac{r_{-1}}{r_{+1}r_{+2}} + \frac{r_{-1}r_{-2}}{r_{+1}r_{+2}r_{+3}},$$

$$C = 1 - \frac{r_{-1}r_{-2}r_{-3}r_{-4}r_{-5}}{r_{+1}r_{+2}r_{+3}r_{+4}r_{+5}},$$

or in more compact form:

$$r^{(1)}A + r_\Sigma B = C, \tag{5.26}$$

where $r_\Sigma = r^{(1)} + r^{(2)} + r^{(3)}$. In the general case $r_\Sigma = \sum_{i=1}^{M} r^{(i)}$.

Equation (5.26) is the simplest relationship between the rate of one of the routes and the sum of rates of all routes. This equation can be easily transformed into a dimensionless one by dividing all terms by r_Σ. Obviously, this equation is useful only when applied to stoichiometrically ambiguous reactions for which the number of linearly independent overall equations is equal to the number of linearly independent routes (i.e., empty routes are absent).

Equation (5.26) would be wrong if variations in reactant concentrations would affect the values of A, B, and C, in which case these parameters would not be constant. The graph in Figure 5.6c describes the reaction in which three routes producing VC, DCE, and VDC have the common step 1. Then, $r^{(1)} = r_{VC}$, $r^{(2)} = r_{DCE}$, $r^{(3)} = r_{VDC}$, and

$$r_{VC}\left(\frac{1}{\omega_{+1}} + \frac{\omega_{-1}}{\omega_{+1}\omega_{+2}}\right) + r_{DCE}\left(\frac{1}{\omega_{+1}}\right) + r_{VDC}\left(\frac{1}{\omega_{+1}}\right) = 1 \tag{5.27}$$

where $\omega_{+1} = r_{+1}$. Then,

$$r_{VC} = k_4 - k_5 r_\Sigma, \tag{5.28}$$

where $k_4 = \omega_{+1}\omega_{+2}/\omega_{-1}$ and $k_5 = \omega_{+2}/\omega_{-1}$; $r_\Sigma = r_{VC} + r_{DCE} + r_{VDC}$. Because the $CuCl_2$ concentration ($[Cu^{2+}]$) is not included in k_4 and k_5, and if the acetylene partial pressure $p_{C_2H_2}$, $[H_3O^+]$, and activities of Cu^+ and Cl^- ions (a_{Cu^+} and a_{Cl^-}) are constant, we arrive at the linear dependence in the form of equation (5.28).[10] This equation agrees well with the experimental data.

It can be shown that the equations (5.12)–(5.14) for rates r_{VC}, r_{DCE}, and r_{VDC} can be derived from the equations (5.10) and (5.11) for r_{DCE}/r_{VC} and r_{VDC}/r_{VC} (for conjugation nodes) and equation (5.28).

The equations for conjugation nodes (5.10) and (5.11) can be used to derive another sort of dimensionless equations, which do not involve concentration of Cu^{2+} either. These ions participate in the second and third route. It follows from equations for conjugation nodes that

$$r_{DCE} / r_{VC} = k_1[Cu^{2+}]^2,$$

$$r_{DCE} / r_{VDC} = k_3 + [Cu^{2+}].$$

Taking into account that $k_1 = k_{10} / [H_3O^+]$, we arrive at the following formula:[10]

$$r_{DCE} / r_{VDC} = k_3 + k_{11}[H_3O^+]^{\frac{1}{2}}(r_{DCE} / r_{VC})^{\frac{1}{2}}.$$

This rate equation also accounts for the totality of experimental data and allows one to derive the dependence of k_3 and k_{11} on the catalyst (CuCl) concentration, which is involved in r_{+1}.[11]

The value of some route r_n can be expressed in terms of the sum of ratios r_i / r_n ($i \neq n$) if, in the conjugation node, several products have a common intermediate that is produced by the reversible but not equilibrium step (Figure 5.11):

$$r_n = r_X^+ \left[\frac{\omega_X^-}{\omega_n^+} + \left(1 + \sum_i \frac{r_i}{r_n} \right) \right], \tag{5.29}$$

where W_X is the rate of transformation of intermediate X; ω_X^- and ω_n^+ are weights of transformations of X into its precursor and into the product P_n, respectively.

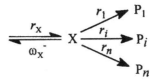

Figure 5.11. The conjugation node with the last common intermediate X.

If the form of dependencies of W_X, ω_X^-, and ω_n^+ on reactant concentrations is known from the preliminary experiments or deduced from the hypothetical mechanism, all experimental data will be described by equation (5.29). One can easily derive equation (5.26). If an intermediate is common for two products, as it was found in oxidative chlorination of chloroacetylene (CA) in CuCl–CuCl$_2$–HCl–NH$_4$Cl solutions (products are *cis*-1,2-dichloroethylene [DCE] and trichloroethylene [TCE]), then equation (5.29) can be written as follows:[19]

$$r_{DCE} = \frac{r_X^+}{\dfrac{\omega_X^-}{\omega_{DCE}^+} + 1 + \dfrac{r_{TCE}}{r_{DCE}}}. \tag{5.30}$$

The r_{TCE} / r_{DCE} ratio in the conjugation node is described by the following equation:

$$\frac{r_{TCE}}{r_{DCE}} = \frac{k_1 [CuCl_2]_\Sigma}{[H_3O^+]};$$ (5.31)

W_X is proportional to $p_{CA}[CuCl]_\Sigma$; and ω_n^+ is proportional to $[H_3O^+]$. Then, equation (5.30) transforms to equation

$$r_{DCE} = \frac{k_2 [CuCl]_\Sigma\, p_{CA}}{\dfrac{k_3}{k_4 [H_3O^+]} + 1 + \dfrac{k_1 [CuCl]_\Sigma}{[H_3O^+]}}.$$ (5.32)

This equation describes the totality of the experimental data.[19] Equations (5.31) and (5.32) give the rate law for r_{TCE}:

$$r_{TCE} = \frac{k_1 k_2 k_4 [CuCl]_\Sigma\, p_{CA}}{k_3 + k_4 [H_3O^+] + k_1 k_4 [CuCl]_\Sigma}.$$

The discussion held in this chapter supports the idea that the topological view of the reaction mechanism is useful for the interpretation of kinetic data and hypothesis discrimination.

References and Notes

1. Yablonskii, G. S.; Bykov, V. I.; Gorban', A. N. *Kinetic Models of Catalytic Reactions*; Nauka: Novosibirsk, 1983 (in Russian).
2. Yablonskii, G. S.; Bykov, V. I. Structured Rate Laws of Complex Catalytic Reactions. *Dokl. Akad. Nauk SSSR* **1978**, *238*, 645–648.
3. Temkin, O. N.; Bruk, L. G.; Bonchev, D. G. Topological Structure of Complex Reaction Mechanisms. *Teor. Eksp. Khim.* **1988**, *23*, 282-291.
4. Temkin, O. N.; Bonchev, D. Classification and Coding of Chemical Reaction Mechanisms. In *Mathematical Chemistry. Chemical Graph Theory. Reactivity and Kinetics*; Bonchev, D.; Rouvray, D. H., Eds.; Gordon and Breach: Chichester, 1992, Vol. 2, Chapter 2.
5. Bonchev, D.; Kamenski, D.; Temkin, O. N. Complexity Index for the Linear Mechanisms of Chemical Reactions. *J. Math. Chem.* **1987**, *1*, 345–388.
6. Emanuel', N. M. Kinetic Features of the Chain Mechanism of Processes of liquid-Phase Oxidation. In *Problems of Chemical Kinetics*; Kondrat'ev, V. N., Ed.; Nauka: Moscow, 1979, pp 118–138 (in Russian).
7. Temkin, O. N.; Bruk, L. G.; Zeigarnik, A. V. Some Aspects of the Methodology of Mechanistic Studies and Kinetic Modeling of Catalytic Reactions. *Kinet. Katal.* **1993**, *34*, 445–462; *Kinet. Catal. Engl. Transl.* **1993**, *34*, 389–405.
8. Bruk, L. G. private communication.
9. Temkin, O. N.; Shestakov G. K.; Treger Yu. A. *Acetylene: Chemistry, Reaction Mechanisms, and Technology*; Khimiya: Moscow, 1991 (in Russian).
10. Brailovskii, S. M.; Temkin, O. N.; Kostyushin, A. S.; Odintsov, K. Y. Kinetics of Catalytic Synthesis of 1,1- and *trans*-1,2-Dichloroethylenes from Acetylene in the CuCl–CuCl$_2$–NH$_4$Cl-H$_2$O System. *Kinet. Katal.* **1990**, *31*, 1371–1376.

11. Brailovskii, S. M.; Huinh Manh Hoan; Temkin, O. N. Kinetics and Mechanism of Additive Oxychlorination of Acetylene in Chloride Solutions of Cu(I) and Cu(II). *Kinet. Katal.* **1994**, *35*, 734–740.

12. Brailovskii, S. M.; Temkin, O. N.; Shestakova, V. S. Kinetics and Mechanism of 1-Chloro-1,3-butadiene Synthesis from Acetylene. *Kinet. Katal.* **1983**, *24*, 848–852.

13. Varfolomeev, S. D., Zaitsev, S.V. *Kinetic Methods in Biochemical Research*; Moscow State Univ.: Moscow, 1976.

14. Berezin, I. V.; Klesov, A. A. *Practical Course of Chemical and Enzymatic Kinetics*; Moscow State Univ.: Moscow, 1976 (in Russian).

15. Parshall, G. W.; Ittel, S. D. *Homogeneous Catalysis*; Wiley: New York, 1992, 2nd ed.

16. Brailovskii, S. M.; Temkin, O. N.; Flid, R. M. Kinetics and Mechanism of the Oxidative Chlorination of Acetylene. *Kinet. Catal. Engl. Transl.* **1971**, *12*, 1025–1029.

17. Temkin, M. I. Kinetics of Complex Steady-State Reactions. In *Mechanism and Kinetics of Complex Catalytic Reactions*; Isagulyants, G. V.; Tret'yakov, I. I., Eds.; Nauka: Moscow, 1970, p. 57–71 (in Russian).

18. Temkin, M. I. The Kinetics of Some Industrial Heterogeneous Catalytic Reactions. In *Advances in Catalysis*, Academic: New York, 1979, Vol. 28, pp 173–291.

19. Shcheltsyn, L. V.; Brailovskii, S. M.; Temkin, O. N. Kinetics and Mechanism of Chloroalkyne Transformations in Solutions of Cu(I) and Cu(II) Chlorides. *Kinet. Katal.* **1990**, *31*, 1361–1370.

Subject Index

—W—

Milton Keynes UK
Ingram Content Group UK Ltd.
UKHW040446071024
449327UK00020B/1038

9 780367 448479